"This volume brings together some of the top researchers in philosophy of neuroscience to address some of the major issues in the field. The novelty and breadth of the contributions are really impressive. Clearly, the authors are not looking to just re-tread old ground but to advance philosophical discussion in a way informed by the most cutting-edge neuroscientific results."

Dan Burnston, *Tulane University, USA*

Neurocognitive Foundations of Mind

This volume provides a cohesive and comprehensive case that cognitive neuroscience is maturing into an integrated, interdisciplinary science that is transforming our understanding of the mind.

The rise of cognitive neuroscience has prompted a rethinking of levels, computation, representation, psychological explanation, and the relation between psychology and neuroscience. Despite these advances, many philosophers and scientists of the mind continue to write as though cognitive neuroscience didn't exist and psychology remains autonomous from neuroscience or, perhaps, they maintain that cognitive neuroscience has not deepened our understanding of the mind. The chapters in this volume showcase important ways in which cognitive neuroscience makes a profound difference to our understanding of the mind. The contributors address a wide range of topics, including explanation, computation, representation, inference, emotion, language, intention, and thought. Together, they demonstrate the ways in which cognitive neuroscience supersedes traditional cognitive science and supports a unified, integrated, multilevel, mechanistic, neurocomputational account of the mind.

Neurocognitive Foundations of Mind is essential reading for scholars and advanced students interested in the foundations of the philosophy of mind and the mind sciences.

Gualtiero Piccinini is Florence G. Kline Professor and Curators' Distinguished Professor of Philosophy at the University of Missouri. He received the Simon Award (2014) and the Barwise Prize (2018). His publications include *Physical Computation* (2015), *Neurocognitive Mechanisms* (2020), and *The Physical Signature of Computation* (with Neal G. Anderson, 2024).

Routledge Studies in Contemporary Philosophy

Respect and Power
The Foundations of Critical Ethics and Politics
Roberto Mordacci

On Frege Puzzles
Why Propositions Are Not the Objects of Our Mental Attitudes
Stefan Rinner

Thought Experiments and Personal Identity
Finding Ourselves
Simon Beck

Against Aristotelian Character Education
Practical Wisdom, Flourishing, and Liberal Democracy
Benjamin Miller

The Ethics of Argumentation
Katharina Stevens

Indirect Freedom
Andrew J. Latham

The Essence of History
Edited by Aaron Turner

Neurocognitive Foundations of Mind
Edited by Gualtiero Piccinini

For more information about this series, please visit: www.routledge.com/Routledge-Studies-in-Contemporary-Philosophy/book-series/SE0720

Neurocognitive Foundations of Mind

Edited by Gualtiero Piccinini

Routledge
Taylor & Francis Group
NEW YORK AND LONDON

First published 2026
by Routledge
605 Third Avenue, New York, NY 10158

and by Routledge
4 Park Square, Milton Park, Abingdon, Oxon, OX14 4RN

Routledge is an imprint of the Taylor & Francis Group, an informa business

ISBN: 978-1-032-60298-1 (hbk)
ISBN: 978-1-032-60297-4 (pbk)
ISBN: 978-1-003-45853-1 (ebk)

DOI: 10.4324/9781003458531

Typeset in Sabon
by Apex CoVantage, LLC

Contents

Acknowledgments *ix*
List of contributors *xi*

1 **New foundations for the philosophy of mind and the mind sciences** 1
GUALTIERO PICCININI

2 **A neuroecological architecture for situated cognizing systems** 18
LUIS H. FAVELA

3 **Confirmation and explanation in neuroscience: Reassessing the relationship between functional and mechanistic approaches** 38
MARCIN MIŁKOWSKI

4 **Cognitive ontology in terms of cognitive homology: The role of brain, behavior, and environment for individuating cognitive categories** 59
BEATE KRICKEL AND MARIEL K. GODDU

5 **Representational vehicles, from regions to cells** 83
ADINA L. ROSKIES

6 **Frames of discovery and the formats of cognitive representation** 98
DIMITRI COELHO MOLLO AND ALFREDO VERNAZZANI

7 **Structural representation as complexity management** 120
MANOLO MARTÍNEZ

8 The mind-brain is a computer, but what is (neural) computation? 144
COREY J. MALEY AND ORON SHAGRIR

9 Inference in (neuro)cognitive systems 164
URTE LAUKAITYTE AND MATTEO COLOMBO

10 Interventionist methods for interpreting deep neural networks 190
RAPHAËL MILLIÈRE AND CAMERON BUCKNER

11 From cognitive semantics to neurosemantics: The neuroscience turn in the empirical study of word meaning 222
FABRIZIO CALZAVARINI

12 Talking to ourselves: Inner speech and natural language as a language of thought 246
WADE MUNROE

13 Working memory and the neural basis of intention 275
WAYNE WU

14 Basic emotion theory meets the brain: Radicals and reformists in the arena of neuroscience 288
MARCO VIOLA AND FAUSTO CARUANA

Index *313*

Acknowledgments

The work that led to this volume was conducted in part on the ancestral land of the Illini Confederacy, Kickapoo, Osage, Otoe-Missouria, and other Indigenous peoples who have lived on and cared for this land for generations.

For many years, I've observed with delight the growing body of philosophical work distilling insights from neuroscience for the philosophy of mind and psychology. Simultaneously, I've watched with mounting frustration as many scholars hold on to the outdated view that the philosophy of mind and psychology can largely ignore neuroscience. This volume emerged from my desire to remedy this disconnect by showcasing some important ways neuroscience enhances our understanding of the mind, featuring contributions from leading voices in this field.

My initial vision was to organize an in-person conference, but lack of funding repeatedly delayed this plan. When the Covid-19 pandemic shifted our collaborative practices toward virtual engagement, I seized the opportunity to organize an online conference without external funding. This approach offered the dual advantages of global accessibility and minimal environmental impact. However, I couldn't do it alone.

Fortunately, since 2018, Fabrizio Calzavarini and Marco Viola have established Neural Mechanisms Online (https://www.neuralmechanisms.org/), an exceptionally successful webinar series in precisely this area. I asked them if they would help me organize and run a conference on *Neurocognitive Foundations of Mind.* Their agreement to help organize and run the conference during October 2022 proved invaluable. The conference's smooth execution and success were largely due to Marco and Fabrizio's exemplary leadership. I'm also grateful to the University of Turin (coincidentally my undergraduate alma mater) for generously providing their facilities and virtual conferencing system.

Thanks to everyone who submitted work, presented research, and participated in the *Neurocognitive Foundations of Mind* conference. Many chapters in this volume evolved from ideas first articulated there. I'm particularly grateful to all contributing authors for their patience and cooperative spirit as we crafted a cohesive volume where each chapter advances our collective mission.

The conference's success catalyzed another significant development: the founding of the International Society for the Philosophy of the Sciences of the Mind in collaboration with Inês Hipólito. Fittingly, Fabrizio Calzavarini and Marco Viola served as program chairs for its inaugural virtual conference.

My work toward this volume was supported in part by a generous grant under the University of Missouri's Provost Great Books Program.

Finally, my sincere appreciation goes to Andrew Weckenmann, Rosaleah Stammler, and the entire Routledge team, whose guidance shepherded both the initial proposal and the subsequent manuscript through the publication process with professionalism and care.

Contributors

Cameron Buckner, *University of Florida, USA*

Fabrizio Calzavarini, *University of Turin, Italy*

Fausto Caruana, *University of Parma, Italy*

Dimitri Coelho Mollo, *Umeå University, Sweden*

Matteo Colombo, *Tilburg University, the Netherlands*

Luis H. Favela, *Indiana University Bloomington, USA*

Mariel K. Goddu, *Stanford University, USA*

Beate Krickel, *Technical University Berlin, Germany*

Urte Laukaityte, *University of California, Berkeley, USA*

Corey J. Maley, *Purdue University, USA*

Manolo Martínez, *University of Barcelona, Spain*

Marcin Miłkowski, *Polish Academy of Sciences, Poland*

Raphaël Millière, *Macquarie University, Australia*

Wade Munroe, *University of Nebraska-Lincoln, USA*

Gualtiero Piccinini, *University of Missouri, USA*

Adina L. Roskies, *University of California, Santa Barbara, USA*

Oron Shagrir, *Hebrew University of Jerusalem, Israel*

Alfredo Vernazzani, *Ruhr-University Bochum, Germany*

Marco Viola, *University of Roma Tre, Italy*

Wayne Wu, *University of Pittsburgh, USA*

1 New foundations for the philosophy of mind and the mind sciences

Gualtiero Piccinini

This volume argues that, since the mind is a function of the brain situated within a body and environment, understanding the mind requires understanding the brain. In other words, evidence about the brain—or more precisely, the nervous system—changes our understanding of the mind. Thus, to improve our understanding of the mind, we must take the brain into account. Different chapters deliver this message on different aspects of the mind, including explanation, representation, computation, inference, language, thought, intention, and emotion. The common theme is that neuroscience makes a profound difference to psychology and the philosophy of mind.

This message might seem obvious to laypeople and many mind scientists. Yet many philosophers, and even some psychologists, have been trained for decades to think otherwise. They've been taught several assumptions that support the opposite conclusion: psychology, or at least the philosophy of mind, is *autonomous* from neuroscience. We can and should investigate the mind while ignoring neuroscience—or paying minimal attention to it.

Here, I cannot examine every assumption and every thread that allegedly supports autonomism, or how autonomism became so prevalent. That would be a book-length project of its own. Instead, I will list some versions of the Autonomist Picture and then review how this volume demonstrates a better way to do philosophy of mind and the mind sciences, contributing to a more empirically adequate understanding of the mind.

1.1 Autonomism

Let's begin with the sort of influential, Classical Autonomist Picture that was defended by Jerry Fodor (e.g., 1997) and his collaborators, among many others:

DOI: 10.4324/9781003458531-1

A Classical Autonomist Picture of the Mind

(1) To perceive, think, and act, minds represent the world (including the organism) and compute over such representations.
(2) Mental representation and computation are theoretical posits within psychological (not neuroscientific) models, which explain perception, thought, and action autonomously from neuroscience.
(3) Mental representation and computation are modeled after representation and computation within (digital) computers.
(4) Representation and computation within (digital) computers are autonomous from physical implementation. Therefore, mental processes are autonomous from neural implementation.

Note that there are at least two places where this Classical Picture alleges autonomy from neuroscience: via supposedly autonomous psychological explanation and via supposedly autonomous representations and computations.

The presumed nemesis of the Classical Picture is Connectionism (e.g., Rumelhart et al. 1986). Perhaps surprisingly, Connectionism can and often has been pursued in a similarly autonomous way (e.g., Weiskopf 2011; cf. several essays in Kaplan 2018). Here is a Connectionist Autonomist Picture that begins with the first two assumptions of the Classical Picture and diverges after that:

A Connectionist Autonomist Picture of the Mind

(1) To perceive, think, and act, minds represent the world (including the organism) and compute over such representations.
(2) Mental representation and computation are theoretical posits within psychological (not neuroscientific) models, which explain perception, thought, and action autonomously from neuroscience.
(5) Mental representation and computation are modeled after representation and computation within artificial neural networks (ANNs).
(6) Representation and computation within ANNs are autonomous from physical implementation. Therefore, mental processes are autonomous from neural implementation.

A fierce opponent of both the Classical and the Connectionist pictures is a cluster of Situationist approaches that, in their radical versions, eschew representation and computation in favor of dynamical coupling between organism and environment. Such approaches emphasize embodiment, embeddedness, and enaction (e.g., Varela et al. 2017). Despite their disagreement with representational and computational approaches, Situationist approaches can and often have been pursued in purported autonomy from neuroscience. Here is a generic formulation of a Situationist Autonomist Picture:

A Situationist Autonomist Picture of the Mind

(7) To perceive, think, and act, minds engage dynamically with the world.
(8) The study of dynamical engagement is a psychological (not neuroscientific) framework, which explains perception, thought, and action autonomously from neuroscience.
(9) Dynamical engagement is autonomous from physical implementation. Therefore, mental processes are autonomous from neural implementation.

Finally, another influential tradition in the philosophy of mind holds that the mind properly so called belongs in the space of reasons, and the space of reasons has to do with norms, not causes (e.g., Sellars 1956; Davidson 1980; McDowell 1994; Brandom 1994). According to this normativist tradition, the philosophy of mind ought to concern itself with norms, not the sort of empirical matters studied by science. Here is a generic formulation of a Normativist Autonomist Picture.

A Normativist Autonomist Picture of the Mind

(1) To perceive, think, and act (rationally), minds ought to have reasons.
(2) The study of how minds ought to have reasons is a normative (as opposed to an empirically descriptive) framework, which explains perception, thought, and action autonomously from *science*.
(3) Norms of rationality are autonomous from physical implementation. Therefore, mental processes are autonomous from neural implementation.

The point of this brief excursus is to show that prominent versions of the most influential frameworks in the philosophy of mind and the mind sciences, despite their many disagreements, agree on autonomy from neuroscience. That's where they go astray.

The purpose of this volume is to show in some detail that, no matter what merits the above frameworks might have, their Autonomism needs to go. Neuroscience makes a profound difference to the philosophy of mind and the mind sciences. Therefore, philosophers of mind and the mind sciences should incorporate neuroscience within their picture of the mind.

Before we get to it, let's take a brief look at the most visible traditional alternative to Autonomism.

1.2 Reductionism

Historically, many philosophers who championed neuroscience defended an explicitly reductionist or eliminativist picture (e.g., P.M. Churchland 1981; P. S. Churchland 1986; Bickle 2003; cf. also Kim 2007). It might be summarized as follows.

A Reductionist Picture of the Mind

(13) To perceive, think, and act, minds undergo brain processes.
(14) Neuroscience observes and manipulates brain processes, which explain perception, thought, and action.
(15) Therefore, any theoretical posits aimed at explaining perception, thought, and action either reduce to brain processes or do not exist.

A corollary of Reductionism is the rejection of Autonomism.

What reduction amounts to, and whether it is warranted, is debatable. What matters most here is that many early champions of neuroscience among philosophers tended to emphasize the empirical observability and manipulability of neurons and neurotransmitters and the putative reduction of higher-level constructs such as propositional attitudes to (or their elimination in favor of) what neurons and neurotransmitters do. This convinced many traditional philosophers of mind and psychology that neuroscience was primarily about low-level stuff—cells and molecules—and that neuroscience-adjacent philosophers were hostile to higher-level theoretical constructs such as *mental* computation and representation, even though at least some of those same neuroscience-adjacent philosophers sometimes talked about *neural* computation and representation (e.g., Churchland and Sejnowski 1992). At any rate, many traditionalists remained convinced that, if the constructs they posited were not easily reducible to lower-level neural structures and functions, they had better be autonomous from neuroscience on pain of elimination.

Reductionism, whatever its merits might be, has its own drawbacks. First and foremost, as I just pointed out, it tends to push traditionalists back into Autonomism. Reductionists may even agree with Autonomists that many higher-level theoretical constructs and the putative explanations they provide are, in principle, autonomous from neuroscience. Reductionists just think that such putatively autonomous higher-level constructs and explanations aren't real. In response, traditionalists recoil into their Autonomist stance.

Crucially, this volume does not advocate either reducing or eliminating higher-level constructs. It does not advocate reduction or elimination at all. What this volume advocates is using what we learn about brains to improve our understanding of the mind, including by revising and refining higher-level theoretical constructs and integrating the explanations they provide with what we learn about brains. This involves rethinking many foundational concepts.

1.3 An integrationist alternative

In recent decades, some of us have developed theoretical frameworks that go beyond traditional Autonomism and Reductionism. I will now briefly sketch an updated version of the sort of framework I have defended previously (e.g., Piccinini 2020). I will do this to show that viable alternatives exist, without expecting that every contributor to this volume endorses every detail of such a framework.

<u>An Integrationist Picture of the Mind</u>

(16) To perceive, think, and act, minds undergo processes at levels of organization ranging from molecular to whole organism-environment dynamics.
(17) Mental processes at different levels are interdependent and explain perception, thought, and action in complex ways that must be studied empirically.
(18) Neuroscience techniques and models are indispensable for studying and understanding multilevel mental processes.

The crucial tenet of the Integrationist Picture is the middle one. It says that processes at different levels of organization explain mental phenomena and are mutually dependent in ways that cannot be determined a priori or by analogy with artifacts such as digital computers. They can only be discovered empirically and, as the third tenet points out, this requires neuroscience techniques and models. There are several ways of reaching this conclusion.

One way is the study of multilevel mechanisms. Roughly, multilevel mechanisms are nested pluralities of organized components whose individual sub-capacities, when organized, constitute the capacities of a whole system. Mechanisms manifest relevant capacities only if they have appropriate components with appropriate sub-capacities organized in appropriate ways. Therefore, the kinds of components and sub-capacities and the way they are organized must be understood in order to understand how the capacities of the whole mechanism bring about phenomena. This requires studying lower-level components and their organization. Importantly, and contra the Classical and Connectionist Autonomist Pictures, this point applies to computing systems too (e.g., Piccinini 2015). Very briefly, the reason is that computing systems contain computing mechanisms, which require computing components with appropriate computational sub-capacities organized in appropriate ways to exhibit the whole mechanism's computational capacities; if the right components, sub-capacities, and organization are different from those needed for a certain computational explanation to apply, that explanation fails. In addition, mechanistic explanation does not require any strong assumptions about localizing *mental* functions in neural structures. Which mental functions are performed by which plurality of neural structures may well be highly variable and context-dependent (cf. Burnston 2021; de Wit and Matheson 2022; Pessoa 2022; McCaffrey 2023; Ritchie and Piccinini 2024). The important point is that, if the mind involves mechanisms, we cannot fully understand a system as complex as a mind without understanding its components, their capacities, and how they are organized. For that matter, we also cannot fully understand what each component contributes to the whole system without understanding the system as a whole. There are compelling reasons to think that minds (neurocognitive systems) are multilevel mechanisms in this sense (e.g., Craver 2007; Bechtel 2008; Milkowski 2013; Piccinini 2020; Cao and Yamins 2024a, 2024b).

Another way to reach the same conclusion is considering the success of multiscale modeling. Multiscale modeling is a series of modeling techniques that are needed to understand how processes at different scales depend on each other and contribute to a phenomenon (e.g., Batterman 2021; Wilson 2022). Roughly, multiscale modeling combines multiple sub-models of a complex phenomenon so that each individual sub-model describes a process occurring at a specific scale, while the composite multiscale model puts the individual sub-models in interaction with each other under appropriate conditions to improve prediction and explanation of the phenomenon. There are several different multiscale modeling techniques (Rice 2024). A special technique involves *scale separation*, namely, the modeling of processes at different scales largely independently of one another. If mental processes were compatible with the right sort of scale separation between higher-level processes and lower-level ones, this might underwrite the modeling of mental processes in terms of higher-level constructs independently of the modeling of lower-level processes. If this were possible, it might vindicate a kind of autonomy of psychology. Yet scale separation is a relatively rare occurrence that depends on how processes occurring at different scales do or do not depend on one another. It cannot be assumed a priori but must be established empirically. Thus, establishing scale separation would require studying mental (neurocognitive) systems empirically at different scales and understanding them well enough to conclude that such scale separation occurs. At any rate, the likelihood that scale separation occurs within nervous systems in a way that underwrites the autonomy of psychology is negligible. The evidence we have indicates that neurocognitive systems are full of interdependencies and constraints between processes at different scales. This very volume demonstrates that what we already know about neurocognitive systems alters our understanding of the mind.

A final way to reach the same conclusion is to consider the rise of cognitive neuroscience. By "cognitive neuroscience," I mean the scientific study of brain processes that underlie mental phenomena at relevant levels of organization. Cognitive neuroscience employs a wide variety of techniques for observing and manipulating brain processes at many scales. It pairs such experimental methods with a variety of modeling techniques, including multiscale modeling. Cognitive neuroscience faces many methodological challenges and includes heated debates about how best to conceptualize the neural basis of cognition. Regardless of how such challenges and debates pan out, this volume shows that much of the evidence coming out of cognitive neuroscience refutes the putative autonomy of psychology and the philosophy of mind.

Notice that the Integrationist Picture I propose rejects Autonomy without endorsing traditional Reductionism. It rejects Autonomy because it recognizes that lower levels of organization constrain higher levels, and so, in general, higher levels cannot be understood solely based on behavioral evidence. How lower-level mechanisms and processes work must be considered. It does not endorse Reductionism, at least as Reductionism is often interpreted, because it recognizes that studying the mind at higher levels of organization may be needed to identify processes that explain mental phenomena.

Another thing to notice is that this Integrationist Picture need not involve the sort of ontological hierarchy that is often presupposed by many traditional pictures of the mind. It does not prioritize molecules or cells (like traditional Reductionism), nor does it prioritize representations and computations modeled after some artifact (like some versions of Autonomism). It does not require that some levels reduce to or be grounded in other levels, or that some levels be more fundamental than others. Fundamentalism about levels, according to which some levels are more fundamental than others, is the sort of empirically unwarranted metaphysical baggage that may be best left behind (cf. Piccinini 2022).

A final thing to notice is that I formulated the Integrationist Picture without mentioning representation and computation, to remain neutral on cognitive ontology. But the Integrationist Picture is compatible with mental representation and computation playing a role. In fact, the Integrationist Picture is compatible with there being many levels of representation and computation, compositionally related to one another. Crucially, if there are mental representations and computations, they don't float free of neural processes. Instead, they are *constituted by* neural representations and computations, and they can only be understood deeply by considering what neuroscience uncovers about them (more on this shortly; cf. Piccinini 2022, 2025; Piccinini and Hetherington 2025).

Before I proceed to an overview of this volume, let me emphasize again that my brief survey of Autonomism is by no means complete. There are other common assumptions that keep philosophers from engaging fruitfully with neuroscience. Here are some:

A. Mental states and processes belong to the whole organism; brain states and processes are not mental (e.g., Wittgenstein 1953, section 281).
B. Attributing mental capacities to a proper part of a person is a (mereological) fallacy (Bennett and Hacker 2022).
C. Intellectualism about knowledge: Knowing-that, and perhaps even knowing-how, requires representing-that in a propositional, language-like format (cf. Fodor 1968; Stanley and Williamson 2001).
D. Intellectualism about inference: Inference requires logical transitions between propositional, language-like representations (cf. Boghossian 2014; Quilty-Dunn and Mandelbaum 2018).

I cannot do justice to these topics. But I can point to recent work showing these assumptions to be flawed. There is room for more good philosophical work in this area.

On (A) and (B), we often ascribe mental states, processes, and capacities to whole organisms, while scientists explain them by attributing them and their parts to our (situated) brains. Far from being a mistake or a fallacy, this is the same strategy that applies to many other states, processes, or capacities of complex systems. Compare how we explain running, digesting, or breathing. These are capacities of whole organisms, which we explain by identifying crucial components (the locomotive, digestive, and respiratory systems, respectively) and figuring out how such components work (cf. Pöyhönen 2014; Figdor 2018; Drayson 2024).

On (C) and (D), how knowledge and inference work is an empirical question. Close examination suggests that knowing-that may be best construed as a kind of knowing-how grounded in neural representation and computation (Piccinini and Hetherington 2025), that insofar as brains process language-like representations, they do it in a nonclassical way that's antithetical to the Classical Autonomist Picture (Piccinini 2025), and that inference is a much broader notion than the notion of logical transitions between language-like representations (Chapter 9 of this volume; more on this below).

Up to this point, I have argued that an Integrationist Picture of the mind that recognizes the indispensable role of neuroscience is preferable to both Autonomism and Reductionism. In the next section, I will flesh out some elements of this Integrationist Picture by summarizing how the chapters in this volume fit together.

1.4 Toward a deeper understanding of the mind

By now, it should be uncontroversial that minds are dynamically situated within a body and an environment and should be studied and understood as such. One approach that emphasizes the dynamical interaction between minds and their environment is ecological psychology (e.g., Gibson 1966). Yet, as Luis H. Favela points out in Chapter 2, traditional ecological psychologists paid relatively scant attention to the brain. That is, ecological psychology is one of those Situationist approaches that has often attempted to proceed in putative autonomy from neuroscience. To make progress, Favela argues that ecological psychology and neuroscience should embrace each other. The approach he articulates and defends emphasizes the interaction between organism and environment (like ecological psychology and other Situationist approaches) but seeks to explain such interaction in terms of relatively low-dimensional neural population dynamics and the synergies between such dynamics and the body of the organism. The important point is that Situationist approaches should incorporate what we know about brains. Meanwhile, we'll see shortly that our understanding of the neurocognitive basis of mental processes is relying more and more on the embodiment and embeddedness of the brain.

Once we look at the brain, one question is *how*, more precisely, brains help explain mental phenomena. As we saw in Sect. 1, however, there are influential traditions to the effect that psychological explanation is autonomous from neuroscience. An especially influential version is that psychological explanation is *functional* as opposed to mechanistic, where functional explanation (or functional analysis) explains in terms of higher-level states (e.g., representations), processes (e.g., computations), or entities (e.g., information processors). Most importantly, such functional explanations are purportedly autonomous from the underlying mechanisms that implement the posited functional states, processes, and entities. In Chapter 3, Marcin Miłkowski argues that, far from being autonomous, functional explanations are elliptical mechanistic explanations that need to be integrated with mechanistic insights and can only be completed by finding their place within mechanistic models of neural processes. This sort of mechanistic integration can and should incorporate the situatedness of the brain (cf. Mougenot and Matheson 2024).

A related and important question is how best to conceptualize (neuro) cognitive processes. This is the question of so-called cognitive ontology. There is a heated debate about how much of a difference neuroscience makes to cognitive ontology and what difference it makes. In Chapter 4, Beate Krickel and Mariel K. Goddu propose to develop our cognitive ontology by investigating cognitive *homologies*—namely, cognitive traits that share structural similarities and descend from common ancestors. To do so, they import from evolutionary-developmental biology a refinement of the notion of homology that relies on causally necessary, nonredundant developmental mechanisms that affect trait identity. Their proposal integrates neuroscience with developmental and evolutionary considerations. This stands in direct opposition to traditional approaches that eschew evolutionary considerations (e.g., Fodor and Piattelli-Palmarini 2010). In contrast, Krickel and Goddu's approach helps integrate the mind sciences with some of the most promising recent developments in biological theory (Pigliucci and Müller 2010).

A core notion within theories of mind is *representation*. Autonomist representational traditions take mental representations to be unobservable entities posited by psychological theories via analogies with public representations. The main analogies are drawn between mental representations and either formal languages (e.g., Fodor 1975) or cartographic maps (e.g., Camp 2007). One apparent advantage of drawing analogies with public representations is that public vehicles (say, marks on paper), formats (language-like vs. iconic), and contents (conceptual vs. nonconceptual) are relatively easy for observers to identify and characterize. One disadvantage is that the extent to which these analogies help in characterizing mental representations remains to be seen. The next three chapters tackle how to go beyond analogies with public representations and reach a deeper understanding of mental representation.

Neuroscientists started positing representations since the 19th century, at least a century earlier than contemporary cognitive psychologists started doing the same thing. Since then, neuroscientists have developed multiple experimental techniques for observing and manipulating representations (e.g., Thomson and Piccinini 2018; Nanay 2022). While the observability and manipulability of neural representations are contested, some objections rely on the sorts of analogies with public representations that are dubious to begin with. For instance, one objection is that there is no evidence of anything in the brain that works in precisely the way that cartographic maps work (cf. Facchin 2024). This is independently plausible because it would be surprising if evolution had stumbled on a representational system that works in exactly the way our artifacts work. It doesn't follow that there are no neural representations.

I would recommend a different approach: check what sorts of representations neuroscientists find within brains and fit our theories of representation to the phenomena. In that spirit, let me sketch a notion of structural representation that fits much of the available evidence. Structural representations in the present sense are internal states with at least the following four properties: (1) they form a system structurally similar to their target, meaning that relations between internal states map onto relations between targets; (2) they maximize mutual information with their target, usually via a causal connection (Hemskeerk 2025); (3) they are decouplable from their target (meaning they can maximize mutual information with their target even in the absence of a causal connection); and (4) their structural similarity to their target allows them to guide behavior with respect to their target. Neural representations have important additional properties: they are learnable and plastic, evolve dynamically as their targets evolve dynamically, and often tend to predict or simulate the evolution of their targets. One recent development in the theory of this sort of structural representations is that their semantic content as well as the causal efficacy of their content are functions of these properties (cf. Piccinini 2022; Ramsey 2023; Nestor 2024).

In Chapter 5, Adina L. Roskies examines some of the methodological challenges involved in identifying the vehicles of neural representations. She points out that, in general, representations of the sort found in brains presumably involve the activity of large populations of neurons, and there is no way to record the activity of all members of such populations. Neuroscientists can record, at best, from *some* neurons within neuronal populations involved in any given representation. In light of these challenges, she recommends reexamining what counts as a neural representational vehicle without relying too heavily on analogies with public representations.

In Chapter 6, Dimitri Coelho Mollo and Alfredo Vernazzani argue that given the evidence we have about mental representations, there are many disanalogies between the formats of public representations—be they formal languages or cartographic maps—and the format of mental representations. Therefore, we need a better way to characterize representational format. They propose to characterize formats in terms of degrees of freedom of the vehicles and ways that different portions of the vehicles constrain one another.

In Chapter 7, Manolo Martínez argues that theoretical frameworks such as computational complexity theory or information theory, which have often been deployed in the context of Autonomist projects, have wider application: they provide universal feasibility limits to which cognitive systems, within any framework including Integrationist ones, are bound. Martínez applies this insight to the way structural representations of the sort found within brains resemble (mirror) their targets and shows that such resemblance can be used to reduce the computational complexity of the processes needed to perform a task. This is both a novel way to show that minds often need representations to solve problems efficiently and an insight into the relation between mental representations and the computations that process them. It leads straight to the question of the type of computation we find in the brain.

When Autonomists posit computations, typically they assume that mental computations are the kind of computations performed by digital computers, which are modeled by classical computability theory. Some Autonomists appeal to ANNs, but they often still assume that the computations such networks perform fit well within classical computability theory. Many traditionalists (my former self included) go so far as to assume that anything that doesn't fit within classical computability theory is not computation at all. In Chapter 8, Corey J. Maley and Oron Shagrir build on recent philosophical work on physical computation to argue that, first, computation is a broader notion than digital computation; second, the empirical evidence we have suggests that neural computation is importantly different from digital computation; and third, even the theory of ANNs often departs from classical computability theory and analyzes networks in terms imported from statistical mechanics and the theory of dynamical systems. To understand the computations minds perform, they argue, we need to rely *not* on classical computability theory but on a broader conception of computation, empirical evidence from neuroscience, and the kinds of models developed by computational neuroscientists.

Some computations perform inferences, and the notion of inference is ubiquitous in the mind sciences. In Chapter 9, Urte Laukaityte and Matteo Colombo argue that accounts of inference that appeal to logical transitions between language-like representations are too narrow and inadequate as a foundation for the mind sciences. Instead, they propose to construe inferences more generally as rationally evaluable transitions from input and current representations to output and subsequent representations. Their notion of inference is broad enough to fit scientific practices, theories, and models in the mind sciences without trivializing the notion of inference.

One challenge in studying complex, multilevel representational and computational systems is to measure relevant observables and show that they make a difference to behavior. To check whether we have found relevant observables and how they contribute to behavior, model building is crucial. In recent years, the deep learning revolution in artificial intelligence has led to a new generation of modeling tools: deep learning neural networks (LeCun et al. 2015). Deep learning involves many layers of processing units that can be trained to perform difficult cognitive tasks in ways that match and may even surpass human performance (e.g., Doerig et al. 2023). Traditional ANNs were notoriously difficult to analyze, which led to the criticism that they were too mechanistically opaque to illuminate mental processes. The new generation of deep learning neural networks involves so many variables interacting in such complex ways that they may be even more difficult to analyze than earlier neural networks.

In Chapter 10, Raphaël Millière and Cameron Buckner dive into recently developed interventionist methods for studying deep neural networks and understanding their structure. Interventionist methods involve manipulating select variables within deep neural networks to observe what difference they make to the behavior of the network. They parallel experimental methods in science more broadly and allow researchers to infer the computational and representational structure of the networks. As Millière and Buckner argue, these newer methods can improve our mechanistic understanding of deep networks, which can help make them more useful as modeling tools within computational cognitive neuroscience. At the same time, such methods vindicate the usefulness of a computational and representational framework for understanding what neural networks do.

The most influential version of the Classical Autonomist Picture relies on the Classical Language of Thought Hypothesis (CLOTH). According to CLOTH, a language of thought analogous to the formal languages processed by digital computers explains some important cognitive processes. For instance, CLOTH holds that we plan by forming computer-like instructions within working memory (Miller et al. 1960; Fodor 1968) and learn natural language and understand its meaning by translating natural language into our innate language of thought (Katz 1966; Fodor et al. 1974; Fodor 1975; cf. Vendler 1972). CLOTH remains influential to this day (e.g., Dehaene et al. 2022; Quilty-Dunn et al. 2023).

CLOTH faces many problems. As we've seen, Chapters 5–8 argue that in general, neural representation and computation are nonclassical. What about higher-level processes such as planning and language processing? Could CLOTH retain a foothold there? In separate work, I defend a negative answer to these questions. I argue that not only are neural computation and representation inherently nonclassical, but there is also every reason to extend this conclusion to all cognitive processes. While some neural representations are language-like at a coarse level of grain, the language in question is natural language; therefore, CLOTH is ruled out on empirical grounds (Piccinini 2025). The next three chapters argue that, in addition, the neuroscience of language processing, discursive thought, and working memory support a more empirically adequate alternative to CLOTH.

In Chapter 11, Fabrizio Calzavarini focuses on how natural-language speakers understand and process the meaning of words. Neuroimaging evidence strongly suggests that speakers represent the meaning of words by combining amodal representations (i.e., representations untied to any particular sensory modality) with vast distributed networks of sensorimotor representations. For example, when speakers represent the meaning of the word "dog," they appear to activate neural populations that are selective for dogs regardless of sensory input along with neural populations that respond to various sensory features of dogs in different modalities (visual, auditory, tactile, olfactory). Here, it is important to realize that while CLOTH *requires* amodal representations, the existence of amodal representations per se does not support CLOTH. In fact, there is plentiful evidence that neural representations may be more or less amodal depending on how many kinds of sensory sources they integrate, and this has nothing specifically to do with CLOTH (e.g., Tamber-Rosenau et al. 2013).

In Chapter 12, Wade Munroe defends a similar conclusion with respect to discursive thought. There is robust empirical evidence that some thoughts are a form of inner speech, and such inner speech modulates and enhances several mental processes. Far from being the sort of amodal formal language posited by CLOTH, however, such inner speech involves natural language encoded via a combination of amodal and sensorimotor representations.

In Chapter 13, Wayne Wu leverages recent empirical findings about working memory to defend a related account of intention. He argues that intentions are more dynamic than traditionally thought and rely on sensory representations that are widely distributed throughout the brain. Once again, this stands in stark contrast to CLOTH.

Finally, in Chapter 14, Marco Viola and Fausto Caruana close the volume by looking at whether neuroscience alters our cognitive ontology of emotions. They focus on what are traditionally considered the six basic emotions: anger, disgust, fear, joy, sadness, and surprise. They point out that the list had a rather unprincipled origin and there is no empirical justification for retaining all and only those six emotions as basic. Regardless of how deeply anyone thinks that neuroscience challenges our cognitive ontology, our understanding of basic emotions will be changed by taking neuroscience into account.

1.5 Conclusion

Perhaps there was a time when psychologists and philosophers of mind could theorize about the mind fruitfully while largely ignoring neuroscience. That time has long passed. The essays in this volume make a powerful case that deepening our understanding of the mind requires integrating psychology and neuroscience.

The picture they paint is that mental (neurocognitive) processes are inextricably tied to our body and environment, so much so that analogies with disembodied artifacts such as digital computers are at best of limited help in understanding mental processes. Instead, mental representation and computation, and mental processes more generally, need to be understood in their own right by looking at how brains work. Neural representation and computation rely heavily on sensorimotor codes that are widely distributed and provide semantic grounding for our language and thought, which also relies on natural language as a powerful thinking tool. Further progress will require an even better understanding of how brain processes at different levels of organization affect each other and produce behavior. The many recently developed experimental and modeling tools that comprise contemporary neuroscience are bound to help.

Note

Thanks to Manolo Martínez and Marco Viola for helpful comments.

References

Batterman, R. W. (2021). *A Middle Way: A Non-Fundamental Approach to Many-Body Physics*. Oxford: Oxford University Press.

Bechtel, W. (2008). *Mental Mechanisms: Philosophical Perspectives on Cognitive Neuroscience*. London: Routledge.

Bennett, M. R., and Hacker, P. M. S. (2022). *Philosophical Foundations of Neuroscience*. 2nd ed. Hoboken: John Wiley & Sons.

Bickle, J. (2003). *Philosophy and Neuroscience: A Ruthlessly Reductive Approach*. Dordrecht: Kluwer.

Boghossian, P. (2014). "What Is Inference?" *Philosophical Studies* 169(1): 1–18.

Brandom, R. (1994). *Making It Explicit: Reasoning, Representing, and Discursive Commitment*. Harvard University Press.

Burnston, D. C. (2021). "Getting Over Atomism: Functional Decomposition in Complex Neural Systems." *The British Journal for the Philosophy of Science* 72(2): 742–772.

Camp, E. (2007). "Thinking with Maps." *Philosophical Perspectives* 21: 145–182.

Cao, R., and Yamins, D. (2024a). "Explanatory Models in Neuroscience, Part 1: Taking Mechanistic Abstraction Seriously." *Cognitive Systems Research* 87: 101244.

Cao, R., and Yamins, D. (2024b). "Explanatory Models in Neuroscience, Part 2: Functional Intelligibility and the Contravariance Principle." *Cognitive Systems Research* 85: 101200.

Churchland, P. M. (1981). "Eliminative Materialism and the Propositional Attitudes." *Journal of Philosophy* 78: 67–90.

Churchland, P. S. (1986). *Neurophilosophy: Toward a Unified Science of the Mind-Brain*. Cambridge, MA: The MIT Press.

Churchland, P. S., and Sejnowski, T. J. (1992). *The Computational Brain*. Cambridge, MA: MIT Press.

Craver, C. F. (2007). *Explaining the Brain*. Oxford: Oxford University Press.

Davidson, D. (1980). *Essays on Actions and Events*. Oxford: Clarendon Press.

De Wit, M. M., and Matheson, H. E. (2022). "Context-Sensitive Computational Mechanistic Explanation in Cognitive Neuroscience." *Frontiers in Psychology* 13: 903960. https://doi.org/10.3389/fpsyg.2022.903960

Dehaene, S., Al Roumi, F., Lakretz, Y., Planton, S., and Sablé-Meyer, M. (2022). "Symbols and Mental Programs: A Hypothesis about Human Singularity." *Trends in Cognitive Sciences* 26(9): 751–766.

Doerig, A., Sommers, R. P., Seeliger, K. et al. (2023). "The Neuroconnectionist Research Programme." *Nature Reviews Neuroscience* 24: 431–450. https://doi.org/10.1038/s41583-023-00705-w

Drayson, Z. (2024). "Personal/Subpersonal Distinction." In Michael C. Frank and Asifa Majid (eds.), *Open Encyclopedia of Cognitive Science*. MIT Press. https://doi.org/10.21428/e2759450.8593aeda

Facchin, M. (2024). "Neural Representations Unobserved—Or: A Dilemma for the Cognitive Neuroscience Revolution." *Synthese* 203(7). https://doi.org/10.1007/s11229-023-04418-6

Figdor, C. (2018). *Pieces of Mind: The Proper Domain of Psychological Predicates*. Oxford University Press.

Fodor, J. (1968). "The Appeal to Tacit Knowledge in Psychological Explanation." *Journal of Philosophy* 65: 627–640.

Fodor, J. (1975). *The Language of Thought*. Cambridge, MA: Harvard University Press.

Fodor, J. (1997). "Special Sciences: Still Autonomous After All These Years." *Philosophical Perspectives, Mind, Causation, and the World* 11: 149–163.

Fodor, J., Bever, T., and Garrett, M. (1974). *The Psychology of Language: An Introduction to Psycholinguistics and Generative Grammar*. New York: McGraw Hill.

Fodor, J., and Piattelli-Palmarini, M. (2010). *What Darwin Got Wrong*. Farrar, Straus and Giroux.

Gibson, J. J. (1966). *The Senses Considered as Perceptual Systems*. Boston, MA: Houghton Mifflin.

Hemskeerk, J. (2025). *How Informational Teleosemantics Works: Towards a Realist Theory of Content*. PhD Dissertation. University of Warwick.

Kaplan, D. M. ed. (2018). *Explanation and Integration in Mind and Brain Science*. Oxford University Press.

Katz, J. J. (1966). *The Philosophy of Language*. Harper & Row.

Kim, J. (2007). *Physicalism, or Something Near Enough*. Princeton University Press.

LeCun, Y., Bengio, Y., and Hinton, G. (2015). "Deep Learning." *Nature* 521(7553): 436–444.

McCaffrey, J. (2023). "Evolving Concepts of Functional Localization." *Philosophy Compass*. http://doi.org/10.1111/phc3.12914

McDowell, J. (1994). *Mind and World*. Cambridge, MA: Harvard University Press.

Milkowski, M. (2013). *Explaining the Computational Mind*. Cambridge, MA: MIT Press.

Miller, G. A., Galanter, E., and Pribram, K. H. (1960). *Plans and the Structure of Behavior*. New York: Holt.

Mougenot, D., and Matheson, H. (2024). "Theoretical Strategies for an Embodied Cognitive Neuroscience: Mechanistic Explanations of Brain-Body-Environment Systems." *Cognitive Neuroscience*. https://doi.org/10.1080/17588928.2024.2349546

Nanay, B. (2022). "Entity Realism About Mental Representations." *Erkenntnis* 87: 75–91.

Nestor, M. J. (2024). *From the Body to the World: Structural Representation and the Mapping Problem*. PhD Dissertation. University of Adelaide.

Pessoa, L. (2022). *The Entangled Brain: How Perception, Cognition, and Emotion are Woven Together*. MIT Press.

Piccinini, G. (2015). *Physical Computation: A Mechanistic Account*. Oxford: Oxford University Press.

Piccinini, G. (2020). *Neurocognitive Mechanisms: Explaining Biological Cognition*. Oxford: Oxford University Press.

Piccinini, G. (2022). "Situated Neural Representations: Solving the Problems of Content." *Frontiers in Neurorobotics* 16: 846979.
Piccinini, G. (2025). "Neural Hardware for the Language of Thought: New Rules for an Old Game." Unpublished manuscript.
Piccinini, G., and Hetherington, S. (2025). "Knowing That as Knowing How: A Neurocognitive Practicalism." *Synthese* 205(5). https://doi.org/10.1007/s11229-024-04843-1
Pigliucci, M., and Müller, G. B. (2010). *Evolution: The Extended Synthesis.* MIT Press.
Pöyhönen, S. (2014). "Intentional Concepts in Cognitive Neuroscience." *Philosophical Explorations* 17: 93–109.
Quilty-Dunn, J., and Mandelbaum, E. (2018). Inferential Transitions. *Australasian Journal of Philosophy* 96(3): 532–547. https://doi.org/10.1080/00048402.2017.1358754
Quilty-Dunn, J., Porot, N., and Mandelbaum, E. (2023). "The Best Game in Town: The Re-emergence of the Language of Thought Hypothesis Across the Cognitive Sciences." *Behavioral and Brain Sciences* 46: e261, 1–75. https://doi.org/10.1017/S0140525X22002849
Ramsey, W. M. (2023). "The Hard Problem of Content is Neither." *Review of Philosophy and Psychology*. https://doi.org/10.1007/s13164-023-00714-9
Rice, C. (2024). "Beyond Reduction and Emergence: Tailoring Multiscale Modeling Techniques to Particular Contexts." *Biology and Philosophy* 39(12). https://doi.org/10.1007/s10539-024-09949-x.
Ritchie, J. B., and Piccinini, G. (2024). "Cognitive Computational Neuroscience." In Nora Heinzelmann (ed.), *Advances in Neurophilosophy*. Bloomsbury, pp. 151–178.
Rumelhart, D. E., McClelland, J. M., and the PDP Research Group. (1986). *Parallel Distributed Processing: Explorations in the Microstructure of Cognition.* Cambridge, MA: MIT Press.
Sellars, W. (1956). "Empiricism and the Philosophy of Mind." In Herbert Feigl and Michael Scriven (eds.), *Minnesota Studies in The Philosophy of Science, Vol. I: The Foundations of Science and the Concepts of Psychology and Psychoanalysis.* Minneapolis: University of Minnesota Press, pp. 253–329.
Stanley, J., and Williamson, T. (2001). "Knowing How." *The Journal of Philosophy* 98: 411–444.
Tamber-Rosenau, B. J., Dux, P. E., Tombu, M. N., Asplund, C. L., and Marois, R. (2013). "Amodal Processing in Human Prefrontal Cortex." *Journal of Neuroscience* 33(28): 11573–11587. https://doi.org/10.1523/JNEUROSCI.4601-12.2013
Thomson, E. E., and Piccinini, G. (2018). "Neural Representation Observed." *Minds and Machines* 28(6): 1–45.
Varela, F. J., Thompson, E., and Rosch, E. (2017). *The Embodied Mind.* Revised ed. Cambridge, MA: MIT Press.
Vendler, Z. (1972). *Res Cogitans: An Essay in Rational Psychology*. Cornell University Press.
Weiskopf, D. (2011). "Models and Mechanisms in Psychological Explanation." *Synthese* 183: 313–338.
Wilson, M. (2022). *Imitation of Rigor: An Alternative History of Analytic Philosophy*. Oxford: Oxford University Press.
Wittgenstein, L. (1953). *Philosophical Investigations*. Blackwell.

2 A neuroecological architecture for situated cognizing systems

Luis H. Favela

2.1 A little bit of history and a challenge

Situated approaches to cognition generally adhere to the view that intelligent systems are always embodied and embedded (e.g., Robbins & Aydede, 2009; Roth & Jornet, 2013; Walter, 2014). Though there is variety and controversy in how the term is used, as a starting point, "embodied" refers to the idea that cognition requires a body to occur (e.g., Varela et al., 1991). In a more specific sense, this means that there tends not to be a sharp distinction between cognition and noncognition in the body and, relatedly, that sensorimotor processes are important for cognition (Favela & Chemero, 2016). There is a spectrum in how weak or strong these positions are. At the weaker end of the spectrum, cognition is primarily a brain event, but the brain areas associated with the body's sensorimotor processes play important roles in causing and enabling even "higher" forms of cognition, such as abstract reasoning and language (e.g., Barsalou, 1999; Lakoff & Nunez, 2000). At the stronger end of the spectrum, cognition is not primarily a brain event, with the body's (i.e., nonneural) structures and processes causing and constituting even those previously mentioned "higher" forms of cognition (e.g., Di Paolo et al., 2017). In the current context, the term "embedded" is also fraught with variety and controversy, but as a starting point it refers to the idea that cognition always occurs in a situation that is deeply tied to nonbodily features, such as the environment (e.g., light and temperature; e.g., Richardson et al., 2008) and social communities (e.g., classroom; e.g., Heft, 2013). So too does embeddedness fall along a spectrum. At the weaker end, cognition remains primarily a brain event, but nonbodily features, such as those in the environment, can play significant causal roles in cognition. At the stronger end, cognition both cannot occur without nonbodily features like a social situation and is constituted by those elements, such as thinking with nonneural tools (e.g., only being able to solve math problems with the aid of pen and paper).

DOI: 10.4324/9781003458531-2

The broad class of situated approaches to cognition have exerted an immense influence on the cognitive and psychological sciences. Since the earliest days of *scientific* psychology, people like John Dewey and William James promoted various forms of *functionalism*, or the idea that the aim of psychology is to study mind and behavior from the perspective of their roles in an organism's abilities to adapt to their environment (Hergenhahn & Henley, 2014).[1] This form of functionalism stressed the need for explanations of psychological phenomena to include what is happening in the body and environment (e.g., James, 1885).[2] Soon after, *behaviorism* became the dominant framework for scientific psychology. Commonly associated with the work of people like B. F. Skinner and John Watson, behaviorism had three main commitments: psychology is the science of behavior, the sources of behavior are not internal (i.e., mind) but are external (i.e., environment), and any use of mental concepts should be replaced by behavioral concepts (e.g., Graham, 2023). Granted that behaviorism had its share of problems, the point in mentioning it here is that, again, in the history of scientific psychology, the dominant approach emphasized the roles of the body and environment.

The emphasis on the body and environment would come to be relegated to the background in favor of the "autonomous" treatments of cognition (cf. Fodor, 1997), with developments in the mid-1900s that contributed to the *cognitive revolution* that dethroned behaviorism (Boden, 2006). Taken together, developments in computer science (McCarthy et al., 2006), information theory (Shannon & Weaver, 1949/1964), and linguistics (Chomsky, 1965/2015) motivated psychology—and the new *cognitive science*—to shift focus to the internal states that contribute to behavior and thought. The *cognitivist* framework that grew from those foundations treated scientific psychology's target of investigation as a kind of *information processing* (e.g., Thagard, 2023). In this context, "information processing" refers to the claim that cognition is fundamentally computational and representational in nature. The basic idea is that cognition is defined as procedural operations (i.e., computations) over meaningful structures (i.e., representations) to constitute or produce decision-making, perception, problem-solving, and the like (e.g., Neisser, 1967/2014). Notwithstanding the fact that most cognitive science and psychology textbooks trace a single historical trajectory along the lines just described—that is, functionalism to behaviorism to cognitivism—it is a fact that other fruitful approaches continued to carry the torch of situatedness. One of the most influential situated approaches was ecological psychology.

Ecological psychology is a research program and theoretical approach to perception and action originating with James J. Gibson in the mid-1900s (e.g., Gibson, 1966). While initially an investigative approach to perceptual psychology, Gibsonian ecological psychology has been a major influence on a wide range of fields, including, but not limited to, botany (Calvo, 2016), developmental psychology (Thelen, 2000), human-machine interaction (Favela, 2019), and virtual reality (Regia-Corte et al., 2013). Ecological psychology can be understood via the four primary principles at its core (Favela, 2024; cf. Chemero, 2013; Gibson, 1986/2015):

First, *perception is direct*. An organism's perceptual capacities can make unmediated contact with its environment in order to detect ecological information. Within the context of ecological psychology, "information" does not mean what it commonly does in other cognitive and psychological sciences, such as a measure of uncertainty (Shannon & Weaver, 1949/1964) or data (i.e., representations) processed in a formal way that can be communicated (Henderson, 2009). Most importantly, it does not involve any sense of "computation" or "representation" as being the defining nature of cognition or perception. In ecological psychology, "information" is the basis of meaningful interactions between an organism and its environment. Ecological information is "those patterns that uniquely specify properties of the world" (Richardson et al., 2008, p. 177; Gibson, 1986/2015, p. 67). It is not mere stimulus but is the distribution of energy surrounding an organism. Information patterns are higher-order properties in that they are necessarily spatiotemporal in nature, such that an organism perceives them in their surrounding space over time (Gibson, 1966), for example, optic flow. The sense of "unmediated" at the heart of directness refers to the claim that perception does not require inferential processes to both perceive the environment and its meaningful elements for the organism (i.e., affordances).

Given that representational understandings of perception are the norm in the cognitive and psychological sciences (e.g., Edelman, 1999; Marr, 1982/2010), it is worth elaborating more upon the meaning of "direct perception" in ecological psychology. In the context of perception, "representations" are understood as mediating stimulation from the world that impinges upon sensory apparatuses like eyes. Such mediation is postulated as necessary because the stimulation is meaningless, that is to say, it is unstructured, which has the consequence of being unable to participate in guiding successful action. As a result, the perceiving system must transform the meaningless stimulation into meaningful representations. This line of thought is analogous to the poverty of stimulus (PoS) argument in linguistics (e.g., Berwick et al., 2011). In short, the PoS argues that there is not enough structure in the stimulation—e.g., sounds from parents speaking—to account for the speed and sophistication with which children learn and develop language capabilities. Thus, there must be inferences made on the stimulation within the system to make them meaningful. In the context of perception, the PoS argument claims that there is not enough structure in the stimulation—e.g., light reflecting from a surface impinging upon retinal cells in eyes—to account for the ability of organisms to successfully act in the world (e.g., grasp a branch, walk through an aperture). Thus, there must be inferences (i.e., transformations) made on the unstructured stimulation within the system to make them meaningful. Gibsonian ecological psychologists understand perception as *direct* in that they reject the notion that environmental energies—that is, *information* as defined above—cannot guide meaningful action because they are unstructured. On the contrary, they claim that environmental (i.e., ecological) information is *rich* with structure—rich enough to guide action without needing to infer meaning in the form of representations.[3]

Second, *perception and action are continuous*. An organism's perceptual capabilities were selected (i.e., evolutionarily) to guide action, and so were action capabilities selected to enable perception. For the Gibsonian approach, it does not make sense to attempt to understand either perception in isolation from actions or actions in isolation from perception.

Third is *affordances*. As a consequence of perception being direct and the continuity of perception and action, it follows that detected ecological information can specify meaningful opportunities for action, namely, affordances. In short, affordances are perceivable opportunities for behavior (Chemero, 2013; Gibson, 1986/2015; Heras-Escribano, 2019). To properly utilize affordances as part of ecological psychology research, one must understand the significance of both the organism and the environment in acts of perceiving and acting. Affordances are not just features of the external world (i.e., consistent with behaviorism), nor are they just features of the internal organism (i.e., consistent with cognitivism). As Gibson put it:

> [A]n affordance is neither an objective property nor a subjective property; or it is both if you like. An affordance cuts across the dichotomy of subjective-objective. . . . It is equally a fact of the environment and a fact of behavior.
>
> (Gibson, 1986/2015, p. 121)

While that statement may sound cryptic, it makes a clear point: explanations of affordances cannot prioritize or neglect the contributions of the organism or the environment, which leads to the final primary principle.

Fourth, psychology is about the *organism-environment system*. The spatiotemporal scale of organism and environment interactions is the proper point of inquiry for investigating, explaining, and understanding the previous three principles. Treating the organism-environment system as the proper scale at which affordances must be explained has facilitated expansive knowledge in perceptual psychology and has facilitated the development of various sophisticated quantitative techniques. Consequently, ecological psychology provides a scientifically compelling research program for investigating and understanding perception-action events.

Two points should be clear at this point: one is that from its origins in the mid-1900s to today—and like the earlier approaches to scientific psychology (e.g., Jamesian functionalism)—ecological psychology has maintained that the body and environment are necessary elements when investigating and understanding intelligence in its various forms. Taken together with its offering of a rich set of concepts and theories and sophisticated quantitative methods, it is clear why ecological psychology is a major influence on situated approaches in the cognitive and psychological sciences (e.g., see discussion in Varela et al., 1991). The other point is that there has been no mention of the brain's role in the phenomena investigated by ecological psychology. Notwithstanding the fact that brains are parts of many organism-environment systems, and while Gibson himself spent some time discussing the brain's role in perception (e.g., Gibson, 1966, 1986/2015), the vast majority of experimental and theoretical work by ecological psychologists has not focused on brains. For that reason, among others, it becomes clearer why situated approaches have not had as much impact on the neurosciences as they have on the cognitive and psychological sciences. Specifically, insofar as situated approaches are explicitly or implicitly adhering to the core principles of Gibsonian ecological psychology, their investigative reach is limited more to the body and environmental scales and less to the spatial and temporal scales relevant to neurobiological structure and function.

Proponents of situated approaches, especially those at the "stronger" end of the spectrum, have defended ecological psychology and other explicitly non-brain-centric approaches to cognition by arguing that brains and nervous systems have always been part of their theories of perception. For example, it is regularly pointed out that ecological psychologists have always treated perception as occurring in *perceptual systems*. As Gibson put it:

> The eye is considered to be an instrument of the mind, or an organ of the brain. But the truth is that each eye is positioned in a head that is in turn positioned on a trunk that is positioned on legs that maintain the posture of the trunk, head, and eyes relative to the surface of support. Vision is a whole perceptual system, not a channel of sense. . . . One sees the environment not with the eyes but with the eyes-in-the-head-on-the-body-resting-on-the-ground. Vision does not have a *seat* in the body in the way that the mind has been thought to be seated in the brain. The perceptual capacities of the organism do not lie in discrete anatomical parts of the body but lie in systems with nested functions.
>
> (Gibson, 1986/2015, p. 195; italics in original)

Accordingly, perceptual systems like mammals have a brain and nervous system that partially constitute the organism part of organism-environment systems. Despite that point, critics have noted that Gibson's engagement with the neurosciences was unjustifiably minimal, being limited mostly to criticisms (e.g., "Neurophysiologists . . . are still under the influence of dualism, however much they deny philosophizing"; Gibson, 1986/2015, p. 215) and vague metaphors (e.g., "the brain resonates"; Gibson, 1966, p. 260). This has led to scathing criticisms of Gibsonian ecological psychology, which include caricaturing the Gibsonian view of organisms as being "stuffed with foam rubber" (Pribram, 1982, p. 370) and treating the brain with awe as if it were a "hunk of wonder tissue" (Dennett, 1984, pp. 149–150).[4]

Some critics may be right that situated approaches to cognition—especially of the strong variety like ecological psychology—have unjustifiably ignored the brain's contributions or made too little effort to engage with other research areas that do. While there has been some work by self-identified ecological psychologists to connect their research to the neural scale, those are few and far between.[5] Moreover, that work is typically not experimental in nature.[6] The majority of the experimental work in neuroscience that is about "affordances" does not adhere to the core principles of ecological psychology, such as perception being direct (i.e., nonrepresentational; e.g., Cisek, 2007). This situation raises an interesting challenge: to offer a framework that provides the neural scale part of the story for investigating and understanding phenomena that ecological psychologists study without compromising any of its core principles. The next section offers just such a framework: the NeuroEcological Nexus Theory (NExT; Favela, 2024). NExT provides an explicit account of the brain's contributions to an affordance event and does so while maintaining all four core principles of ecological psychology.[7] In doing so, NExT offers a cognitive architecture for even the strongest versions of situated approaches in the cognitive, psychological, and neural sciences. If compelling, NExT will provide a step toward a more complete understanding of mind, one that spans the cognitive and psychological sciences, as well as the neurosciences. Moreover, such progress should not be viewed as occurring within the neurosciences alone and ought to be welcomed by ecological psychologists and proponents of situated approaches as well.

2.2 A NeuroEcological cognitive architecture

The NExT (Favela, 2024) is a theory about cognitive systems and a theoretical program for the scientific investigation of such phenomena. NExT is appropriate for the investigation of cognizing organism-environment systems that have nervous systems. Since cognitive systems are always situated, NExT is committed to the position that such systems necessarily include organisms (i.e., embodied) that are tightly coupled with or constituted by their ecologies (i.e., embedded). The *nexus* part refers both to the framework as a point of convergence for ecological psychology and neuroscience and as a "nexus" that constitutes the primary causal and/or constitutive links or junctions that connect the various elements of a cognizing organism-environment system. NExT is defined here via five hypotheses that provide a way to integrate ecological psychology (especially its four core principles) and neuroscience (i.e., neural spatiotemporal scales), while not treating one as being epistemically or metaphysically more fundamental than the others during investigative work.[8]

Hypothesis 1 is that the organism-environment system is the privileged spatiotemporal scale of description to understand cognizing. NExT provides a scientific framework that aims to explain overlapping phenomena studied by ecological psychology and neuroscience while not privileging the spatiotemporal scales of either. Consequently, a comprehensive investigation of cognition is not achieved by either neuroscience (e.g., neuronal activity) or ecological psychology (e.g., ambient arrays) alone. This does not mean rejecting what has been successful about each, such as computational modeling in neuroscience or relative measurements (e.g., body scaled) in ecological psychology. Additionally, treating the organism-environment system as a single spatial and/or temporal scale of investigation would be incorrect, as cognitive phenomena are *multiscale* in nature.[9]

Hypothesis 2 is that neural population dynamics generate the relevant states for cognizing organism-environment systems. While the single neuron continues to be a significant aspect of understanding neural systems, much work in the neurosciences has come to investigate "populations" as the primary unit for explaining various activities. It has been argued that the neuron doctrine ought to be replaced by the "neural population doctrine" (Ebitz & Hayden, 2021; Saxena & Cunningham, 2019). Here, neural populations are defined in terms of "well-specified function" (Kohn et al., 2016, p. 237), which is further illuminated via Gerald Edelman's theory of neuronal group selection, or Neural Darwinism (Edelman, 1987). Neural Darwinism explains brain structure and function in terms of population thinking, or selectionism, which says that variance within biological populations is necessary for the process of evolution, such that those individuals who can best cope with their environment will reproduce more successfully (Edelman, 1988). Neural Darwinism offers a rich justification for the hypothesis that neural populations generate the most relevant dynamics for organism-environment system activity and does so in a way that is linked to well-specified function (cf. Kohn et al., 2016). Accordingly, a *neural population* is a neuronal group that has been selected for over species development timescales, individual organism's lifetime, and moment-to-moment experience timescales, which have been selected for due to their contributions to synchronized and integrated brain-body-environment structures that have facilitated evolutionary advantages.

Hypothesis 3 is that cognizing is based on low-dimensional neural dynamics. The relevant dynamics occur at the neural population scale and are low-dimensional in terms of manifold theory, especially topological manifolds. Topology is the mathematical study of the properties of objects that are maintained despite changing their shape (e.g., twisting), without compromising their integrity (e.g., ripping; Weisstein, 2022). The "objects" of topology are called *topological spaces*. When two or more topological spaces have the same properties (e.g., shape), then they are called *homeomorphic*. These concepts underlie the idea of manifolds. In topology, a *manifold* is a mathematical object that looks locally like Euclidean space but globally may have a more complicated structure. This means that if a small part of a manifold looks flat and Euclidean when zoomed in upon, zooming out and seeing the entire space of the manifold may reveal more complicated features (e.g., holes). The classic example of a homeomorphic topological space is the coffee mug and donut, or torus. *Manifold theory* is the study of such mathematical objects, with a focus on features like curvature and differentiability (Edgar, 2008; Ma & Fu, 2012).

The concepts and tools from manifold theory, and topology more broadly, have come to be increasingly utilized in data science. In particular, this is evident in widening applications of the *manifold hypothesis*, which is the claim that very high-dimensional datasets have much lower-dimensional manifolds that capture their principal structure (Fefferman et al., 2016). Here, "very high-dimensional datasets" can be understood as referring to any large amount of recorded data from sources like neural activity (Cunningham & Yu, 2014). The *neural manifold hypothesis* is the claim that very high-dimensional datasets—specifically, neural population dynamics—have much lower-dimensional manifolds that capture their principal structure (i.e., "neural modes") that generate specific behaviors (Gallego et al., 2017). The neural manifold hypothesis claims that the spatiotemporal scope of neural activity causally related to and/or constitutive of a range of phenomena (e.g., motor control) may seem incredibly large but is in fact confined to a much smaller scale. In line with Hypothesis 2, NExT hypothesizes that those neural modes are found at the mesoscale of population activity. This commitment claims that relevant neural activity is not found in individual neurons or in brain regions. Thus, statistical methods that average across activity of individual neurons or across larger brain regions will not suffice. As Gallego and colleagues put it:

> Here we argue that the underlying network connectivity constrains these possible patterns of population activity . . . and that the possible

> patterns are confined to a low-dimensional manifold . . . spanned by a few independent patterns that we call "neural modes." These neural modes capture a significant fraction of population covariance. It is the activation of these neural modes, rather than the activity of single neurons, that provides the basic building blocks of neural dynamics and function.
>
> (Gallego et al., 2017, p. 978)

Identifying these modes requires extracting ω variables from a recorded set of α variables. Because ω is not directly observed, they are called *latent variables*. In neuroscience, the α variables typically take the form of neuronal time series. Thus, the neural manifold hypothesis expects that because α neurons are part of an interconnected network, only ω will be needed to explain their activity. These neural population dynamics are processed with various dimensionality reduction techniques to produce a *neural manifold*, which is a set of activity plotted in an *n*-dimensional state space (e.g., Chaudhuri et al., 2019). After, the latent dynamics across the plotted states are weighted to assess if there are patterns in the activity or persistent homologies. If the neural manifold hypothesis is supported by the data, then this analysis should reveal a topological manifold in the noisy cloud of data points.

Consider the example of a mouse's head movement (Figure 2.1). The aim is to identify the neural populations that play causal and constitutive roles in head movement. The idea is that particular neurons will be active depending on where the mouse is orientated (Figure 2.1a). The neural manifold hypothesis claims that if the population dynamics embody a

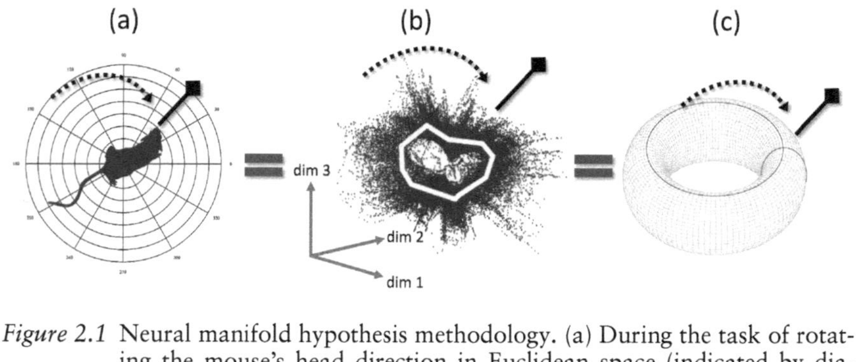

Figure 2.1 Neural manifold hypothesis methodology. (a) During the task of rotating the mouse's head direction in Euclidean space (indicated by diamond shape-tipped arrow), activity from the relevant neural population is recorded. (b) Dimensionality reduction techniques are applied to isolate the neural activity and then further analyzed to identify the latent variables via a generated spline. (c) Movement in Euclidean space and along the spline is equivalent to the location on the abstracted torus.

variable of particular dimension and topology, then those dynamics would be focused on a subspace of corresponding manifold dimension and topology (Figure 2.1b). If head direction is a single variable (i.e., topologically one-dimensional), then the expectation is that among the high-dimensional data there will be a low-dimensional spline that is also a single variable (i.e., topologically one-dimensional). In the current context, a "spline" is a mathematical function applied to interpolate or smooth data and can be helpful in identifying trajectories and regions of data that are of particular interest when explaining a phenomenon. This is exactly what is seen in this research. The mouse's head direction (indicated by diamond shape-tipped arrow; Figure 2.1a–c) occurs across a one-dimensional topology within Euclidean space (Figure 2.1a). So too, do the latent variables track along a one-dimensional topology (Figure 2.1b), which, if straightened out, conforms to a torus (Figure 2.1c). This point cannot be stressed enough: the topology of the actual mouse's head direction and its causal and constitutive neural population activity are homeomorphic, not just in shape (i.e., torus) but in direction (i.e., the neural population activity along the spline tracks identically with the direction of the mouse's head). In short, the mouse's head direction in real space, neural population activity in the brain, neural population activity in the state space along the spline, and the location along an abstract torus are equivalent. Examples like this motivate the reasonable assumption that such successes will generalize the neural manifold hypothesis to other neural-involving cognizing events as well.[10]

Hypothesis 4 is that the body organizes into low-dimensional synergies to generate relevant states. In the context of bodily action, *synergies* are functional assemblies of parts (e.g., neurons, muscles) that are temporally constrained to act as a single unit (Kelso, 2009). The body can be understood as a synergy when its various parts are organized into a functional unit, where "functional" is understood in terms of tasks conducted by the unit (Kugler et al., 1980). For instance, when somebody with a visual impairment is wielding their cane, the particular assemblage of their skeleton and muscles, along with the tool, are a synergy in that they are a functional unit aimed at accurately navigating the world (cf. Favela et al., 2021).

The uncontrolled manifold (UCM; Scholz & Schöner, 1999) is a methodology for assessing the variability of movements with regard to tasks, or the aim of functional units. Here, a distinct configuration space is defined that is populated by variables hypothesized to capture a particular movement. UCM treats motor control as being fundamentally about the stabilization of performance variable values. Those values are quantified in terms of their being compensatory and uncompensatory with regard to the task. The methodology generates a state space where a manifold depicts the variables contributing to the task and quantifies the amount of constancy among those variables. The result is a "synergy index" (Latash et al., 2010), where variance along the manifold are compensatory variables when they maintain task performance (i.e., "good variability") and variance perpendicular to the manifold are uncompensatory variables when they result in loss of performance (i.e., "bad variability"; Latash, 2012).

NExT claims that a body can self-organize into synergies. As a synergy, the body is partially caused and constituted by neural population dynamics across a flexible range of variance that is constrained only by their ability to facilitate successful task completion. This is consistent with the selectionist view of evolution (see above discussion of Neural Darwinism). So too does the environment contribute to the body's performance in flexible ways (e.g., adapting to new forms of ecological information to guide action). UCM methodology allows investigators to empirically assess the presence of and degree to which a system is a synergy. This allows for defining synergies without adhering to a priori definitions about where the boundaries of an intelligent system must be drawn. Variables that contribute to synergies can be located in the brain, arm, or environment, as long as they facilitate successful task completion. Accordingly, bodies are properly understood as being part of cognizing organism-environment systems.

Hypothesis 5 is that cognizing fundamentally emerges at low-dimensional states of organism (neural, body)-environment activity. NExT claims that cognizing is a multiscale phenomenon that emerges from systematic relationships among brain, body, and world. Each spatiotemporal scale is high dimensional, but the principal structure of each is low dimensional (Figure 2.2b). This is motivated by the neural manifold hypothesis regarding neural populations and uncontrolled manifold regarding body synergies. From those methods, targets of investigative interest at each spatiotemporal scale can be modeled and depicted via phase space plots that present the state space of empirically verified activity. As neural populations and body synergies are dynamical systems, these data can take the form of dynamical landscapes (Figure 2.2c). A *dynamical landscape* is a qualitative description of system's states based on quantitative data from characteristics like directionality and strength. As brain and body spatiotemporal scale activity has low-dimensional structure within high-dimensional data sets, so too does the environment (Figure 2.2c, bottom). Here, the "high dimensionality" of the environment refers to the flood of potential stimulation constituting the world—e.g., light, odors, trees—and "low dimensional" refers to ecological information that specifies affordances.

With the hypotheses explained, all the pieces can be put together and demonstrate how NExT is applied to an affordance event. Consider the case of a mouse deciding if a hole affords pass-through-ability (Figure 2.2a). The mouse detects ecological information about the hole

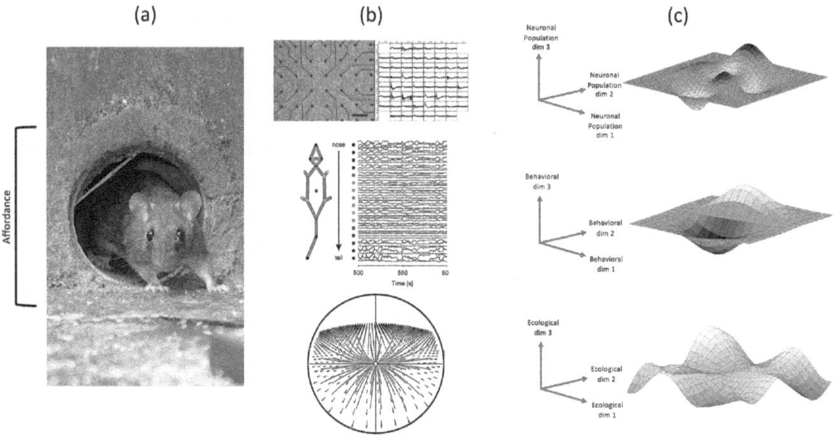

Figure 2.2 The NExT applied to the affordance *pass-through-able*. (a) The affordance *pass-through-able* is exhibited when a mouse walks through a hole and involves features of the environment (e.g., hole in wall), organism (e.g., shoulder width), and brain (e.g., sensorimotor neural populations). (b) Each row depicts a potential data source that contributes to a NExT-based explanation. Top: Time series from multielectrode recording of relevant neural populations. Middle: Motion tracking markers on mouse body and trace data. Bottom: Optic flow as an example of relevant ecological information. (c) Each row depicts a dynamical landscape generated from low-dimensional principal structure identified from high-dimensional data recorded from sources such as those depicted in column (b).

(e.g., optic flow; Figure 2.2b, bottom), which informs its actions (i.e., head movement; Figure 2.2b, middle), which in turn alters the structure of the ecological information it is detecting, and so on (ecological psychology core principles 1 and 2). Contributions by neural spatiotemporal scales are accounted for by the neural manifold hypothesis (Figure 2.1). Given the research aim to identify the neural populations most relevant to the affordance *pass-through-able*, the neural population that causes and constitutes head movement will be one of the focuses because it plays a significant role in the mouse's engagement with ecological information that guides the most applicable action (Figure 2.1 and Figure 2.2b, top). The activity of these particular neurons will be dependent on the mouse's orientation in Euclidean space (Figure 2.1a-b). The neural manifold hypothesis claims that the principal structure of these population dynamics will be expressed by particular topological dimensions, such that head direction is one variable (i.e., one-dimensional topology) that fits on a low-dimensional spline within the high-dimensional cloud of data (Figure 2.1b). Prior research (e.g., Chaudhuri et al., 2019; Gallego et al., 2017) has demonstrated location equivalence in Euclidean space of the mouse's body (Figure 2.1a), location of activity along the low-dimensional spline (Figure 2.1b), and the abstracted torus (Figure 2.1c).

Although the current discussion thus far seems to emphasize neural population activity, the affordance event is multiscale across the organism-environment system (principle 4), and the mutuality of perception-action (principle 2) is maintained. A crucial reason for this is that the activity of neural population dynamics does not hold a unidirectional causal relationship with the body or environment. The state of the body (i.e., head location; Figure 2.1a) and the environment (i.e., aperture edge; Figure 2.2a) inform and constrain the state of the neural activity. If the head cannot move any further to the left because it hits up against the hole's edge, then the neural population will also not continue activity in that direction, as exhibited by its location on the low-dimensional manifold (Figure 2.1b and Figure 2.2c, top). So too does the state of the neural activity inform and constrain the state of the body and environment, such that where the direction of the head points will provide the perspective from which the body (e.g., eyes) will detect ecological information (e.g., light reflecting from the surface edges of the hole). In that way, direct perception (principle 1) is also maintained because neural activity is informed and constrained by the direct engagement with ecological information. In this way, NExT offers an approach to investigating and explaining an affordance event (i.e., *pass-through-able*), while maintaining the four core principles of ecological psychology and offering an account of the contributions at the neural spatiotemporal scale via the neural manifold hypothesis.

2.3 Conclusion

Situated approaches to cognizing are highly influential in the cognitive and psychological sciences. I argued that the influence of ecological psychology on such approaches has undermined their influence in the neurosciences. One reason is that ecological psychology motivates a strong version of situatedness, where cognizing is not primarily a brain event and cannot occur without nonneural elements such as those in the body and environment. A major criticism of this approach is that even in organism-environment systems constituted by organisms with brains, those neural elements play crucial roles in cognizing events, such as those involving affordances. Thus, the challenge taken up by the current work was to offer a framework for investigating and understanding cognizing systems, one that does not compromise any of the four core principles of ecological psychology and suggests a plausible account of contributions from the neural spatiotemporal scale.[11] The NExT is such a framework. By way of the example of the affordance *pass-through-able*, NExT provides a plausible account that honors both ecological psychology and neuroscience. Accordingly, NExT bolsters even the stronger versions of situated approaches to cognizing by offering the neural part of the story of strongly embodied and embedded systems.

In closing, to say that cognizing fundamentally emerges at the low-dimensional scales of organism (neural, body)-environment systems is to adhere to a set of metaphysical and epistemological claims. Metaphysically, it is to maintain that cognizing is caused and constituted by an organism's body—including its nervous system—and spatiotemporal structures of the environment. Epistemologically, it is to maintain that more complete explanations of cognizing systems necessitate concepts, methods, and theories that do not privilege "lower" (e.g., neuronal activity) or "higher" scales (e.g., ambient arrays) but integrate understanding across multiple scales. Taken together, the metaphysical and epistemological aspects of NExT contribute to its ability to offer an architecture for situated cognizing systems, one that provides a more complete understanding of minds and, thus, benefits both the cognitive and psychological sciences (including ecological psychology) and the neurosciences.

Notes

1 For the sake of brevity, I am focusing on the strand of scientific psychology predominantly associated with the United States. Of course, other regions of the world developed scientific psychologies and, sometimes, concurrently with those mentioned in the present work, such as the structuralism of Edward Titchener and Wilhelm Wundt in Europe (Hergenhahn & Henley, 2014).

2 It is important not to confuse the functionalism of Dewey and James with the "functionalism" of 20th-century philosophy of mind, namely, a theory about the nature of mental states, defined by what they do (e.g., pain is defined along the following lines: if bodily damage is detected and unwanted, then remove that body part from the source of damage) rather than by what they are made of (e.g., neurons; Putnam, 1975).

3 Claire Michaels and Claudia Carello go to great pains to explain what "direct perception" means to ecological psychologists and attempt to dispel a naïve interpretation. As they state:

> We certainly do not mean "direct" as in the "unmediated" phone-to-phone example cited; that is, we do not go along with Plato's notion that objects can be directly apprehended without reference to the senses. Rather we mean that the structure of environmental objects and events as they relate to a behaving animal is preserved in the energy patterns stimulating perceptual systems. The converse of this view, which we have attributed to more traditional theories, is that the structure of environmental objects and events is lost (in the

> light or at the receptor surface or in the stimulation) and must be restored by the brain.
>
> (Michaels & Carello, 1981, p. 184)

4 The full quotes are worth reading in full:

> Gibson tends to leave the organism, if not empty, apparently stuffed with foam rubber.
>
> (Pribram, 1982, p. 370)

> J. J. Gibson's theory of perception, for instance, seems to treat the whole visual system as a hunk of wonder tissue, for instance, resonating with marvelous sensitivity to a host of sophisticated "affordances."
>
> (Dennett, 1984, pp. 149–150)

5 For recent exceptions to this, see de Wit and Withagen (2019), Favela (2024), and Raja (2018).
6 For a rare recent exception, see van der Weel et al. (2022).
7 A reasonable concern can already be made about the current project: Why preserve the principles of Gibsonian ecological psychology, especially the more controversial principle that perception is direct (i.e., nonrepresentational)? This chapter aims to maintain these principles for five reasons. First, the NeuroEcological Nexus Theory (NExT) is intended to be part of ecological psychology. As such, it attempts to preserve that which makes an investigative approach "Gibsonian," namely, to abide by the four principles. Second, and conversely, if NExT did not attempt to maintain the four principles, then it would not be "ecological" in the sense highlighted in the current work, namely, Gibsonian *ecological* psychology. Third, if, as claimed above, ecological psychology was a major influence on the broader category of situated approaches, then maintaining the principles of ecological psychology via NExT ought to preserve what was appealing about the Gibsonian approach and may thus make NExT appealing to other situated approaches as well. Fourth, if it turns out that NExT is not compelling, then it could be due to any or all of the four principles. Conversely, if NExT is compelling, then that would be further support for said principles. Finally, and regarding the specific principle that perception is direct and its related commitment to antirepresentationalism, there are reasons to view the concept of representation as unnecessary to both experimental work and theorizing—especially concerning perception-action—even in the neurosciences. One reason comes from experimental work suggesting that the concept of representation plays unclear and confused roles in the neurosciences (cf. Favela & Machery, 2023; see also Favela & Machery, 2025a, 2025b, and related commentaries by Rosa Cao, Frances Egan, and John Krakauer). Another reason is that the concept of representation may not be necessary to explain at least some brain activity (e.g., motor control; e.g., Shenoy et al., 2013). Consequently, approaching perception-action as direct (i.e., nonrepresentational) may not be viewed as quite so controversial after all.
8 A fuller explanation of NExT is provided in Favela (2024). There, NExT is defined by six hypotheses. The sixth hypothesis is not crucial for the current work.
9 Admittedly, the claim that "the organism-environment system is the privileged spatiotemporal scale of description to understand cognizing" lends itself to misinterpretation. Moreover, it seems, at first glance, contradictory to the last sentence

in the same paragraph: "cognitive phenomena are *multiscale* in nature." How can a phenomenon have a privileged scale of description *and be* multiscale? The two claims are not contradictory if one is viewed as primarily an epistemic claim and the other metaphysical. The first statement is intended to be epistemic, that is, from the perspective of NExT, cognizing (i.e., intelligence, mind) is fruitfully approached from the perspective that organisms act in environments for ends such as fighting and reproduction. As such, while understanding such phenomena can involve investigations at "smaller" (e.g., immune system) or "larger" (e.g., city) spatiotemporal scales, such work must be in the service of illuminating how those scales contribute to the cognizing event (e.g., affordances). The second statement is intended to be metaphysical; that is, from the perspective of NExT, cognizing organism-environment systems are caused and constituted by features at multiple scales (e.g., ambient temperature and cells).

10 It could be argued that the neural manifold approach looks a lot like typical work by computational cognitive neuroscientists (CCNs). To the extent that CCN is the use of computational modeling in neuroscience, CCN and NExT overlap. However, a more substantial difference is clear when the theoretical commitments of CCN are highlighted, especially if CCNs understand the brain as *being* a computer. Within NExT, brains are not computers, at least not in any substantial sense (e.g., Turing machine). If CCNs merely mean that the brain processes with regularities, then NExT accepts that the brain is a "computer." One way to think about how CCN and NExT differ is that they have radically distinct *motifs*, or guiding ideas that characterize their work (Raja, 2024).

11 It is worth drawing attention to the fact that others have attempted to integrate considerations of situatedness in neuroscience without adhering to the principles of ecological psychology—especially the first, that perception is direct—and do so without eliminating concepts like computation and representation (for an illustrative example, see Piccinini, 2018, 2022, 2024).

References

Barsalou, L. W. (1999). Perceptual symbol systems. *Behavioral and Brain Sciences*, *22*(4), 577–660. https://doi.org/10.1017/S0140525X99002149

Berwick, R. C., Pietroski, P., Yancama, B., & Chomsky, N. (2011). Poverty of the stimulus revisited. *Cognitive Science*, *35*(7), 1207–1242.

Boden, M. A. (2006). *Mind as machine: A history of cognitive science* (Vols. 1–2). New York, NY: Oxford University Press.

Calvo, P. (2016). The philosophy of plant neurobiology: A manifesto. *Synthese*, *193*(5), 1323–1343. https://doi.org/10.1007/s11229-016-1040-1

Chaudhuri, R., Gerçek, B., Pandey, B., Peyrache, A., & Fiete, I. (2019). The intrinsic attractor manifold and population dynamics of a canonical cognitive circuit across waking and sleep. *Nature Neuroscience*, *22*(9), 1512–1520. https://doi.org/10.1038/s41593-019-0460-x

Chemero, A. (2013). Radical embodied cognitive science. *Review of General Psychology*, *17*(2), 145–150. https://doi.org/10.1037/a0032923

Chomsky, N. (1965/2015). *Aspects of a theory of syntax*. Cambridge, MA: The MIT Press.

Cisek, P. (2007). Cortical mechanisms of action selection: The affordance competition hypothesis. *Philosophical Transactions of the Royal Society B*, *362*(1485), 1585–1599. https://doi.org/10.1098/rstb.2007.2054

Cunningham, J. P., & Yu, B. M. (2014). Dimensionality reduction for large-scale neural recordings. *Nature Neuroscience*, *17*(11), 1500–1509. https://doi.org/10.1038/nn.3776

de Wit, M. M., & Withagen, R. (2019). What should a "Gibsonian neuroscience" look like? Introduction to the special issue. *Ecological Psychology*, *31*, 147–151. https://doi.org/10.1080/10407413.2019.1615203

Dennett, D. C. (1984). Cognitive wheels: The frame problem of AI. In C. Hookway (Ed.), *Minds, machines and evolution: Philosophical studies* (pp. 129–151). Cambridge, MA: Cambridge University Press.

Di Paolo, E. A., Buhrmann, T., & Barandiaran, X. E. (2017). *Sensorimotor life: An enactive proposal*. Oxford, UK: Oxford University Press.

Ebitz, R. B., & Hayden, B. Y. (2021). The population doctrine in cognitive neuroscience. *Neuron*, *109*(19), 3055–3068. https://doi.org/10.1016/j.neuron.2021.07.011

Edelman, G. M. (1987). *Neural Darwinism: The theory of neuronal group selection*. New York, NY: Basic Books.

Edelman, G. M. (1988). *Topobiology: An introduction to molecular embryology*. New York, NY: Basic Books.

Edelman, S. (1999). *Representation and recognition in vision*. Cambridge, MA: The MIT Press.

Edgar, G. (2008). *Measure, topology, and fractal geometry* (2nd ed.). New York, NY: Springer.

Favela, L. H. (2019). Soft-assembled human-machine perceptual systems. *Adaptive Behavior*, *27*(6), 423–437. https://doi.org/10.1177/1059712319847129

Favela, L. H. (2024). *The ecological brain: Unifying the sciences of brain, body, and environment*. New York, NY: Routledge. https://doi.org/10.4324/9781003009955

Favela, L. H., Amon, M. J., Lobo, L., & Chemero, A. (2021). Empirical evidence for extended cognitive systems. *Cognitive Science: A Multidisciplinary Journal*, *45*(11), e13060, 1–27. https://doi.org/10.1111/cogs.13060

Favela, L. H., & Chemero, A. (2016). The animal-environment system. In Y. Coelllo & M. H. Fischer (Eds.), *Foundations of embodied cognition: Volume 1: Perceptual and emotional embodiment* (pp. 59–74). New York, NY: Routledge. https://doi.org/10.4324/9781315751979

Favela, L. H., & Machery, E. (2023). Investigating the concept of representation in the neural and psychological sciences. *Frontiers in Psychology: Cognition*, *14*(1165622), 1–13. https://doi.org/10.3389/fpsyg.2023.1165622

Favela, L. H., & Machery, E. (2025a). The concept of representation in the brain sciences: The current status and ways forward. *Mind & Language*. https://doi.org/10.1111/mila.12531

Favela, L. H., & Machery, E. (2025b). Contextualizing, eliminating, or glossing: What to do with unclear scientific concepts like representation. *Mind & Language*. https://doi.org/10.1111/mila.12533

Fefferman, C., Mitter, S., & Narayanan, H. (2016). Testing the manifold hypothesis. *Journal of the American Mathematical Society*, *29*(4), 983–1049. https://doi.org/10.1090/jams/852

Fodor, J. (1997). Special sciences: Still autonomous after all these years. *Philosophical Perspectives*, *11*, 149–163. https://doi.org/10.1111/0029-4624.31.s11.7

Gallego, J. A., Perich, M. G., Miller, L. E., & Solla, S. A. (2017). Neural manifolds for the control of movement. *Neuron*, *94*(5), 978–984. https://doi.org/10.1016/j.neuron.2017.05.025

Gibson, J. J. (1966). *The senses considered as perceptual systems*. Boston, MA: Houghton Mifflin.
Gibson, J. J. (1986/2015). *The ecological approach to visual perception* (classic ed.). New York, NY: Psychology Press.
Graham, G. (2023). Behaviorism. In E. N. Zalta & U. Nodelman (Eds.), *The Stanford encyclopedia of philosophy* (spring 2023 ed.). Stanford, CA: Stanford University. Retrieved April 6, 2024, from https://plato.stanford.edu/archives/spr2023/entries/behaviorism/
Heft, H. (2013). An ecological approach to psychology. *Review of General Psychology*, *17*(2), 162–167. https://doi.org/10.1037/a0032928
Henderson, H. (2009). *Encyclopedia of computer science and technology* (revised ed.). New York, NY: Infobase Publishing.
Heras-Escribano, M. (2019). *The philosophy of affordances*. Cham, Switzerland: Palgrave Macmillan.
Hergenhahn, B. R., & Henley, T. B. (2014). *An introduction to the history of psychology* (7th ed.). Belmont, CA: Wadsworth.
James, W. (1885). On the function of cognition. *Mind*, *10*(37), 27–44. https://doi.org/10.1093/mind/os-X.37.27
Kelso, J. A. S. (2009). Synergies: Atoms of brain and behavior. In D. Sternad (Ed.), *Progress in motor control: A multidisciplinary perspective* (pp. 83–91). New York, NY: Springer. https://doi.org/10.1007/978-0-387-77064-2_5
Kohn, A., Coen-Cagli, R., Kanitscheider, I., & Pouget, A. (2016). Correlations and neuronal population information. *Annual Review of Neuroscience*, *39*, 237–256. https://doi.org/10.1146/annurev-neuro-070815–013851
Kugler, P. N., Kelso, J. A. S., & Turvey, M. T. (1980). On the concept of coordinative structures as dissipative structures. I. Theoretical lines of convergence. In G. E. Stelmach & J. Requin (Eds.), *Tutorials in motor behavior*. New York, NY: North: Holland Publishing Co.
Lakoff, G., & Nunez, R. E. (2000). *Where mathematics comes from: How the embodied mind brings mathematics into being*. New York, NY: Basic Books.
Latash, M. L. (2012). The bliss (not the problem) of motor abundance (not redundancy). *Experimental Brain Research*, *217*, 1–5. https://doi.org/10.1007/s00221-012-3000-4
Latash, M. L., Levin, M. F., Scholz, J. P., & Schöner, G. (2010). Motor control theories and their applications. *Medicina*, *46*(6), 382–392. https://doi.org/10.3390/medicina46060054
Ma, Y., & Fu, Y. (Eds.). (2012). *Manifold learning theory and applications*. Boca Raton, FL: CRC Press.
Marr, D. (1982/2010). *Vision: A computational investigation into the human representation and processing of visual information*. Cambridge, MA: The MIT Press.
McCarthy, J., Minsky, M. L., Rochester, N., & Shannon, C. E. (2006). A proposal for the Dartmouth summer research project on artificial intelligence, August 31, 1955. *AI Magazine*, *27*(4), 12–14. https://doi.org/10.1609/aimag.v27i4.1904
Michaels, C. F., & Carello, C. (1981). *Direct perception*. Englewood Cliffs, NJ: Prentice-Hall.
Neisser, U. (1967/2014). *Cognitive psychology* (classic ed.). New York, NY: Psychology Press.
Piccinini, G. (2018). Computation and representation in cognitive neuroscience. *Minds & Machines*, *28*, 1–6. https://doi.org/10.1007/s11023-018-9461-x

Piccinini, G. (2022). Situated neural representations: Solving the problems of content. *Frontiers in Neurorobotics*, *16*(846979), 1–13. https://doi.org/10.3389/fnbot.2022.846979

Piccinini, G. (2024). Embodied (4EA) cognitive computational neuroscience. *Cognitive Neuroscience*, *15*(3–4), 119–121. https://doi.org/10.1080/17588928.2024.2405192

Pribram, K. H. (1982). Reflections on the place of brain in ecology of mind. In W. B. Weimer & J. S. Palermo (Eds.), *Cognition and the symbolic processes* (Vol. 2, pp. 361–381). Hillsdale, NJ: Erlbaum.

Putnam, H. (1975). The nature of mental states. In *Mind, language, and reality: Philosophical papers* (Vol. 2, pp. 429–440). Cambridge, UK: Cambridge University Press.

Raja, V. (2018). A theory of resonance: Towards an ecological cognitive architecture. *Minds and Machines*, *28*(1), 29-51. https://doi.org/10.1007/s11023-017-9431-8

Raja, V. (2024). The motifs of radical embodied neuroscience. *European Journal of Neuroscience*, *60*(5), 4738–4755. https://doi.org/10.1111/ejn.16434

Regia-Corte, T., Marchal, M., Cirio, G., & Lécuyer, A. (2013). Perceiving affordances in virtual reality: Influence of person and environmental properties in perception of standing on virtual grounds. *Virtual Reality*, *17*(1), 17–28. https://doi.org/10.1007/s10055-012-0216-3

Richardson, M. J., Shockley, K., Fajen, B. R., Riley, M. R., & Turvey, M. T. (2008). Ecological psychology: Six principles for an embodied-embedded approach to behavior. In R. Calvo & T. Gomila (Eds.), *Handbook of cognitive science: An embodied approach* (pp. 161–187). Amsterdam: Elsevier Science.

Robbins, P., & Aydede, M. (Eds.). (2009). *The Cambridge handbook of situated cognition*. New York, NY: Cambridge University Press.

Roth, W. M., & Jornet, A. (2013). Situated cognition. *Wiley Interdisciplinary Reviews: Cognitive Science*, *4*(5), 463–478. https://doi.org/10.1002/wcs.1242

Saxena, S., & Cunningham, J. P. (2019). Towards the neural population doctrine. *Current Opinion in Neurobiology*, *55*, 103–111. https://doi.org/10.1016/j.conb.2019.02.002

Scholz, J. P., & Schöner, G. (1999). The uncontrolled manifold concept: Identifying control variables for a functional task. *Experimental Brain Research*, *126*, 289–306. https://doi.org/10.1007/s002210050738

Shannon, C. E., & Weaver, W. (1949/1964). *The mathematical theory of communication*. Urbana, IL: The University of Illinois Press.

Shenoy, K. V., Sahani, M., & Churchland, M. M. (2013). Cortical control of arm movements: A dynamical systems perspective. *Annual Review of Neuroscience*, *36*, 337–359. https://doi.org/10.1146/annurev-neuro-062111–150509

Thagard, P. (2023). Cognitive science. In E. N. Zalta & U. Nodelman (Eds.), *The Stanford encyclopedia of philosophy* (winter 2023 ed.). Stanford, CA: Stanford University. Retrieved April 6, 2024, from https://plato.stanford.edu/archives/win2023/entries/cognitive-science/

Thelen, E. (2000). Motor development as foundation and future of developmental psychology. *International Journal of Behavioral Development*, *24*(4), 385–397. https://doi.org/10.1080/016502500750037937

van der Weel, F. R. R., Sokolovskis, I., Raja, V., & van der Meer, A. L. H. (2022). Neural aspects of prospective control through resonating taus in an interceptive timing task. *Brain Sciences*, *12*(1737), 1–16. https://doi.org/10.3390/brainsci12121737

Varela, F. J., Thompson, E., & Rosch, E. (1991). *The embodied mind: Cognitive science and human experience*. Cambridge, MA: The MIT Press.
Walter, S. (2014). Situated cognition: A field guide to some open conceptual and ontological issues. *Review of Philosophy and Psychology*, *5*(2), 241–263. https://doi.org/10.1007/s13164-013-0167-y
Weisstein, E. W. (2022). Topology. *MathWorld—A Wolfram Web Resource*. Retrieved July 1, 2022 from https://mathworld.wolfram.com/Topology.html

3 Confirmation and explanation in neuroscience

Reassessing the relationship between functional and mechanistic approaches

Marcin Miłkowski

3.1 Introduction

Functional explanation, as used in the philosophy of cognitive science, is a method for explaining the roles and contributions of components within a system. This type of explanation focuses not on the historical reasons for a structure's presence in a system but on how the structure contributes to the system's current capacities and dispositions. Functional explanations are common in both everyday language and scientific discourse, particularly in the life sciences. Robert Cummins (1975) introduced a specific approach to functional explanation centered on the analysis of a system's capacities. According to Cummins, the goal of functional analysis is to explain complex dispositions and capacities of a system in terms of the simpler capacities and dispositions of its parts. This approach is distinct from etiological considerations, which Cummins argues are irrelevant for determining function.

In contrast, the new mechanistic approach explains phenomena by recourse to mechanisms, which may involve historical or etiological dependencies or organized temporal processes comprising entities and their activities and interactions, which are jointly responsible for the phenomena (Craver, 2007; Illari & Williamson, 2012; Machamer et al., 2000).

Advocates of the functional approach contend that functional explanations are autonomous, independent of what they term "mere mechanistic detail." They claim that discovering the nitty-gritty of how cognitive phenomena are realized does not change the level of abstraction at which functional explanations operate.

The most radical defense of functional explanation dismisses the necessity of studying mechanisms, implying that functionality is all we need (Shapiro, 2017, 2019). However, this stance faces a challenge: practitioners of cognitive science or psychology no longer consider neuroscientific mechanisms irrelevant to their enterprise (Boone & Piccinini, 2016), even if psychology does not become more neuroscientific over time (Haslam et al.,

DOI: 10.4324/9781003458531-3

2021). A less radical and intriguing functional response is that mechanistic details play merely a confirmatory, rather than explanatory, role for functional explanations, and this will be the focus of this chapter.

In what follows, we will present a dilemma for the latter functional approach: confirmation requires "mechanistic details" to be described in the explanation, but functional explanations are supposed to be devoid of them. This implies that the relation between the explanation and "mechanistic detail" is mere logical consistency, committing what Max Coltheart dubbed the "consistency fallacy" (Coltheart, 2013).

We will explore possible escape routes for defenders of the functional approach, such as arguing that mechanistic explanations are kinds of functional explanations, that realizers confirm functional explanations, that functional explanations are elliptical mechanistic explanations, or that mechanistic detail is irrelevant to confirmation. However, we will show that these responses are either unattractive, as they undermine the autonomy of functional explanations, or go against established scientific practice.

We will then argue for a proper mechanistic approach, holding that mechanisms are explanatory if confirmed in a way that warrants structural validity. The crucial point is that what is considered irrelevant detail is relative to one's theoretical perspective. Thus, it is hardly possible to state in advance which facts must be included in a given answer to an explanatory request. In cognitive science, despite the lack of a common theoretical background in this interdisciplinary endeavor, there is a minimal standard provided by the Minimal Cognitive Grid (MCG) (Lieto, 2021). By examining it, we will argue that the functional approach, with its emphasis on the explanatory autonomy of functional analysis, is both descriptively and normatively inadequate vis-à-vis current scientific practice.

We will conclude that confirmation cannot be easily dissociated from explanation and that the prospects for preserving the autonomy of functional explanations are dim. Mechanistic evidence can confirm functional explanations, making them mechanistic. This is why functional explanations are incomplete (Piccinini & Craver, 2011). In this view, which will also be defended here, there is no genuine opposition between functional and mechanistic explanations because, strictly speaking, the former are a kind of the latter, only elliptical ones (Piccinini, 2020). While in the past, defenders of the autonomy of functional explanation concentrated on distinguishing between explanations in psychology and those in neuroscience, our emphasis is on neuroscience, which allows us to address the core of the debate.

3.1.1 Overview of subsequent sections

Section 3.2 presents examples from cognitive neuroscientist John Krakauer, who argues that "mechanistic details" serve only to confirm functional explanations. Section 3.3 introduces a dilemma for the functional approach, showing that it either conflates logical consistency with confirmation or commits the "consistency fallacy." In Section 3.4, we explore possible ways out for defenders of the functional approach, while Section 3.5 argues for a proper mechanistic approach based on structural validity and the MCG. Finally, Section 3.6 concludes that confirmation cannot be easily dissociated from explanation, casting doubt on the prospects for preserving the autonomy of functional explanations.

3.2 Krakauer's defense of the autonomy of functional explanations

In this section, we will examine the arguments put forth by cognitive neuroscientist John Krakauer in defense of the autonomy of functional explanations. Many objections to Piccinini and Craver's (2011) argument focus on comparing explanatory practices in psychology and neuroscience (e.g., Bantegnie, 2022, 2024; Barrett, 2014; Shapiro, 2017; Stinson, 2016; Weiskopf, 2016), aiming to refute the claim that psychology's autonomy should be discarded. Regarding the descriptive accuracy of the lack of distinctness, there is little textual evidence that psychology's research practices have become more neuroscientific over time (Haslam et al., 2021), despite the adoption of neuroimaging in psychology labs. However, to address the normative issue, it is more useful to focus on the purported explanatory power of functional explanations rather than on disciplinary practices, which may be entrenched for various reasons (some of which may contribute to questionable research practices, such as small sample sizes in developmental psychology). These functional explanations need not originate from psychology. Krakauer (personal communication) argues that in neuroscience, mechanistic details can confirm functional explanations.

Krakauer provides two examples to illustrate his point. The first example is from the work of Hodgkin and Huxley on the action potential. As Krakauer states:

> [W]hen Hodgkin and Huxley wrote their famous paper in 1952 they had no idea about the existence of voltage-sensitive ion channels. Their circuit diagram called for voltage-dependent changes in membrane permeability. The subsequent discovery confirmed the mechanism of their differential equations . . . if the ion channel data changes the parameter values or even adds a term to the differential equation for the action potential—the form of the equation—its level of functional abstraction—will NOT change.

The second example Krakauer provides is from the field of reinforcement learning. He notes, "Temporal difference learning is a mathematical theory of model-free reinforcement learning that posits a reward prediction error. Later, Wolfram Schultz discovered a phasic dopamine signal that acts like such an error—confirmation again."

For Krakauer, these examples demonstrate that mechanistic evidence serves to confirm functional explanations in neuroscience by showing that the proposed functional components and their interactions are realized in the brain. However, he argues that this confirmation does not require the functional explanation to incorporate the mechanistic details. Instead, the functional explanation can remain abstract, focusing on the computational level rather than the implementation level, to use Marr's (1982) distinction.

The Hodgkin-Huxley (H&H) model, developed by Alan Hodgkin and Andrew Huxley in the early 1950s, is a mathematical model that describes how action potentials are generated and propagated in neurons. The model is based on the idea that the membrane of a neuron contains voltage-gated ion channels that open and close in response to changes in the membrane potential. Hodgkin and Huxley used a series of elegant experiments on the giant axon of the squid to measure the ionic currents flowing across the membrane during an action potential. They then used these data to construct a set of differential equations that describe the dynamics of the membrane potential and the gating of sodium and potassium channels. The H&H model provided a quantitative understanding of the biophysical mechanisms underlying neuronal excitability. It remains a cornerstone of modern neuroscience and has inspired countless studies on the function and dysfunction of neurons in health and disease.[1]

Temporal difference (TD) learning is a class of reinforcement learning algorithms that learn to predict future rewards based on past experiences. The key idea behind TD learning is to update the predicted value of a state based on the difference between the predicted value and the actual reward received in the next state. This "temporal difference" error is used to adjust the predictions so that they become more accurate over time. In neuroscience, TD learning has been proposed as a model of how the brain learns to predict rewards and guide behavior. In particular, the activity of dopamine neurons in the midbrain has been shown to resemble the TD error signal, suggesting that these neurons may play a role in reinforcement learning. This was first proposed by Schultz and colleagues in a groundbreaking paper published in *Science* (Schultz et al., 1997) and further elaborated in a subsequent study published in the *Journal of Neurophysiology* (Schultz, 1998). TD learning has also been used to explain various aspects of decision-making, such as the role of uncertainty and risk in choice behavior. Overall, TD learning provides a powerful framework for understanding how the brain learns from experience to adapt and optimize behavior in a changing environment.

According to Krakauer, in the case of Hodgkin and Huxley (1952), the subsequent discovery of voltage-sensitive ion channels confirmed the mechanism underlying their differential equations, but it did not alter the form of the equations themselves or the level of abstraction at which they operate. Similarly, the discovery of a phasic dopamine signal that acts like a reward prediction error confirmed the functional explanation provided by temporal difference learning, but it did not change the mathematical theory itself.

Krakauer's examples suggest that functional explanations have a degree of autonomy from mechanistic explanations. While mechanistic evidence can support functional explanations, it does not necessarily constrain them, which seems at least surprising. If this is right, functional explanations can thus be developed and tested independently of mechanistic details, at least to some extent. This is a different claim from the one defended by Barrett, for example, who argues that "genuine confirmation of functional analyses does not inevitably require neuroscientific information" (Barrett, 2014, pp. 2710–2711).

If Krakauer is correct, his view has important implications for cognitive science. It suggests that researchers can fruitfully investigate cognitive phenomena in terms of their functionality without understanding the underlying neural mechanisms. Functional explanations can guide research and provide insights into cognitive processes, even if the mechanistic details remain unknown.

However, Krakauer's view also raises questions about the relationship between functional and mechanistic explanations. If mechanistic evidence merely confirms functional explanations without influencing their content, how do functional explanations relate to mechanistic explanations? Are they truly autonomous, or are they ultimately dependent on mechanistic details in some way? These questions will be explored in the following sections.

3.3 The dilemma for the functional approach

Krakauer's defense of the autonomy of functional explanations presents a dilemma for his functional approach. On the one hand, if mechanistic evidence is to confirm a functional explanation, it seems that the explanation must incorporate or at least reference the relevant empirical facts. On the other hand, functional explanations are supposed to be abstract and independent of mechanistic details. This section will explore this dilemma and its implications for the relationship between functional and mechanistic explanations. Notably, the dilemma will not appear for functional approaches that deny that mechanistic evidence can support functional explanations, arguing that functional explanations are confirmed by observing behavioral patterns.

3.3.1 The requirement of empirical content

Even if philosophical accounts of confirmation are fraught with difficulties, we will assume the following: for mechanistic evidence to confirm a functional explanation, the explanation must have some empirical content related to mechanisms that can be confirmed (or disconfirmed). In other words, the explanation must make predictions or have implications that can be tested against empirical data regarding particular spatiotemporal entities and activities. If the explanation is entirely abstract and divorced from such empirical facts, it is unclear how it could be confirmed or disconfirmed by mechanistic evidence.

This requirement poses a challenge for the functional approach, which aims to develop explanations that are independent of mechanistic details. If functional explanations must incorporate or refer to empirical facts to be confirmable, they may not be as autonomous as Krakauer and others suggest.

Consequently, this leads to the following dilemma. Consider the argument below:

1. Confirmation cannot be obtained between an explanation E and empirical facts F unless F is described in E (explicitly or in terms of logical consequences or statistically implied claims).
2. But functional explanations are supposed to be devoid of mechanistic detail.

Thus, the relation between E and F, if F is merely "mechanistic detail," is that of mere logical consistency.

If F is merely "mechanistic detail," then F is merely logically consistent with E. But if it is merely logically consistent, it cannot confirm E. Conversely, if F confirms E, then E cannot be true or accurate independently of F, undermining the notion of explanatory autonomy in functional explanations.

In summary, the functionalist must concede that mechanistic explanations are either merely logically consistent and thus not genuinely confirmatory, or they are confirmatory, which implies a lack of autonomy. If the functionalist takes the first horn, they must assume that attempts to provide mechanistic explanations for neuroscientific and psychological phenomena are explanatorily irrelevant, which seems to go against widespread scientific practice. In general, the discovery of mechanisms, such as the place cells as implementing the cognitive map (Tolman, 1948; for a mechanistic story, see Bechtel, 2016), is considered epistemically desirable.

Thus, it seems prima facie unattractive, even if some functionalists seem to embrace this position (Shapiro, 2017, 2019; Weiskopf, 2016). The second horn, however, involves undermining the autonomy of functional explanations because of the confirmatory significance of "mechanistic detail," as we will soon see. This is equally unattractive for functionalists.

The crux of the problem is indeed the notion of autonomy. It is incompatible with various accounts of confirmation, whether we take confirmation to be a logical relation of entailment, as the received view in the philosophy of science might suggest, or a probabilistic relationship. In both cases, the problem remains similar—logical and statistical independence make confirmation impossible.

If we were to adopt a simple covering-law model (Hempel & Oppenheim, 1953), then confirmation would require inferring the description of the explanandum from the equation provided by Hodgkin and Huxley or from a TD model. This description cannot contain any terms that are not introduced in the equation or the TD model (or additional definitions). Successful predictions, which have exactly the same logical form in this account of explanation, could be viewed as providing confirmation of this model. Thus, no confirmation is possible. In addition, if we introduce additional terms regarding mechanistic features of the ion channels in our premises describing the explanatory factors F, then we will be able to infer the mechanistic explanandum, but the explanation will speak of mechanisms. However, this is merely a toy example, not only because this account of explanation is problematic in itself (Salmon, 1984), but also because it conflates prediction with confirmation (even if novel predictions are treated as confirmatory).

The prospects for functional explanation might look less dim from a mainstream perspective on confirmation, which is Bayesian. From the Bayesian point of view, we can think of the functional explanation as a hypothesis (H) and the mechanistic evidence as data (D). According to Bayes' theorem, the posterior probability of the hypothesis given the data, P(H|D), is proportional to the product of the prior probability of the hypothesis, P(H), and the likelihood of the data given the hypothesis, P(D|H):

$$P(H|D) \propto P(H) \times P(D|H)$$

In the case of the H&H model, the discovery of ion channels can be seen as data that are consistent with the model's assumptions about voltage-dependent changes in membrane permeability. While the model did not explicitly predict the existence of ion channels, it is plausible that the likelihood of observing ion channels is higher under the model than under alternative hypotheses that do not involve voltage-dependent changes in permeability.

Formally, we can say that $P(D|H) > P(D|{\sim}H)$, where ~H represents the negation of the hypothesis (i.e., alternative hypotheses that do not involve voltage-dependent changes in permeability). This means that the discovery of ion channels provides evidence in favor of the H&H model, even if the model did not directly imply their existence.

Moreover, if we assign a nonzero prior probability to Hodgkin and Huxley's model, $P(H) > 0$, then the discovery of ion channels should increase the posterior probability of the model, $P(H|D)$, according to Bayes' theorem. This is because the likelihood ratio, $P(D|H) / P(D|{\sim}H)$, is greater than 1, which means that the data is more likely under the model than under alternative hypotheses.

In this sense, the discovery of ion channels can be seen as confirming the H&H model, even if there is no direct logical entailment from the model to the existence of ion channels. The mechanistic evidence is not independent of the functional explanation but instead changes the conditional probability of the explanation being true.

The problem is that functionalism assumes autonomy. If the H&H model is autonomous from details of how ion channels work, then we should assume it is statistically independent of mechanistic evidence. But then no confirmation is possible again.

We suggest, therefore, that the purported functional autonomy should be simply rejected: yes, mechanistic evidence does confirm functional explanations because they are elliptical mechanistic explanations (Piccinini, 2020; Piccinini & Craver, 2011). But before we can show this fully, we must first demonstrate why consistency is not sufficient for confirmation.

3.3.2 *The consistency fallacy*

One way to resolve the tension between autonomy and confirmation is to argue that functional explanations need not incorporate empirical facts directly but must merely be *consistent* with them. On this view, mechanistic evidence confirms a functional explanation by showing that it is consistent with the underlying mechanisms, without requiring the explanation to refer to those mechanisms explicitly.

However, this view risks committing what Max Coltheart has dubbed the "consistency fallacy." The consistency fallacy occurs when mere consistency between an explanation and empirical data is taken as evidence for the explanation, even though the explanation does not entail the data. In other words, the fallacy conflates logical consistency with empirical confirmation. Consistency is insufficient for confirmation.

Consistent data might come from observations from an unrelated set (e.g., the observation of black ravens is consistent with the theory that all bears are brown), or their consistency may rely on the theory's degree of adaptability. Thus, a body of data cannot provide evidence for a theory only in virtue of its consistency. It must also weigh against competing hypotheses. The data *d* support a hypothesis only if the conditional probability of hypothesis H, given that data *d* were accurately observed, exceeds the conditional probability of an alternative (i.e., $P(H|d) > P(\sim H|d)$). If competing hypotheses are equally capable of accommodating the data, the study is nondiagnostic.

If functional explanations are merely required to be consistent with mechanistic evidence, they may not be genuinely confirmed by that evidence. Many functional explanations could be consistent with the same mechanistic data, but this does not mean that all those explanations are equally supported or confirmed by the data.

3.3.3 *Implications for the autonomy of functional explanations*

The dilemma posed by the requirement of empirical content and the risk of the consistency fallacy has important implications for the autonomy of functional explanations. If functional explanations must incorporate empirical facts to be confirmable, they may not be as independent of mechanistic details as the functional approach suggests. Conversely, if functional explanations are merely required to be consistent with mechanistic evidence, they may not be genuinely confirmed by that evidence.

In summary, the functional approach faces a dilemma in its attempt to maintain the autonomy of functional explanations while allowing for their confirmation by mechanistic evidence. If functional explanations must have empirical content to be confirmable, they may not be as independent of mechanistic details as claimed. If they are merely required to be consistent with mechanistic evidence, they risk committing the consistency fallacy and may not be genuinely confirmed by that evidence.

This suggests that the relationship between functional and mechanistic explanations may be more complex than Krakauer's view implies. Functional explanations may not be *entirely* autonomous from mechanistic details but may instead depend on them in subtle ways. The exact nature of this dependence will be explored in the following sections.

3.4 Responses to the dilemma

The dilemma outlined in the previous section challenges the autonomy of functional explanations and their relationship to mechanistic evidence. In this section, we will consider possible responses to this dilemma and their implications for the functional approach.

3.4.1 *The abstractness response*

One possible response is to argue that despite their abstraction, functional explanations can still possess empirical content related to mechanisms.

On this view, functional explanations need not refer to specific mechanistic details but can still make predictions or have implications that can be tested against empirical data.

For example, a functional explanation of memory might posit the existence of a short-term store and a long-term store, without specifying the neural mechanisms that implement these stores. This explanation can still make predictions about the capacity and duration of short-term and long-term memory, which can be tested empirically.

This response suggests that functional explanations can be autonomous from mechanistic details while still being empirically confirmable. However, it raises questions about the degree of abstraction that is appropriate for functional explanations. If an explanation is too abstract, it may not make sufficiently specific predictions to be testable. Conversely, if an explanation incorporates too many mechanistic details, it may lose its autonomy and become indistinguishable from a mechanistic explanation.

According to Craver (2007), textual evidence suggests that Hodgkin and Huxley thought their model was merely phenomenological, with some predictive power: "the success of the equations is no evidence in favor of the mechanism of permeability change that we tentatively had in mind when formulating them" (Hodgkin & Huxley, 1952, p. 541). Indeed, the predictive capacity of a model is not indicative of its explanatory power. Some models have provably higher predictive power *because* they ignore explanatorily relevant factors (Shmueli, 2010). In addition, the H&H model was produced "by fitting solutions to the experimental curves relating sodium or potassium conductance to time at various membrane potentials" (Hodgkin & Huxley, 1952, p. 543). The debate on possible mechanisms would continue for several years, with a number of physiologists skeptical of the very idea of channels.

Craver thus considers the H&H model, at best, a "how possibly" sketch of the action potential, stressing that Huxley, in his Nobel Lecture, was instrumentalist toward the model (Craver, 2007, p. 57). There is a slight tension here with his description of the model as phenomenological and treating it as explanatory, but given that the same scientific artifact might serve various purposes, this is not necessarily an inconsistency.

In contrast to Craver, Levy (2014) argues that the model is explanatory but lumps entities using "aggregative abstraction." He points out that Hodgkin and Huxley did not see their results as merely descriptive:

> The point that we do consider to be established is that fairly simple permeability changes, in response to alterations in membrane potential, of the kind deduced from the voltage clamp results, are a sufficient explanation of the wide range of phenomena that have been fitted by solutions of the equations.
>
> (Hodgkin & Huxley, 1952, p. 541)

So, instead of thinking of the H&H model as a gappy description of the mechanism, Levy considers it a strategic abstraction that provides the model with explanatory power. But he considers the model mechanistic.

This is not to say that abstractions in the H&H model can be understood only mechanistically (Carrillo & Knuuttila, 2023)—but it is difficult to see how it could be a functional explanation that remains autonomous from further research on ion channels, in which Hodgkin and Huxley were obviously interested. In fact, Boone and Piccinini (2016) argue that the H&H model is a mechanism sketch that evolved into a mechanism schema.

Levy's interpretation elucidates why equations in the H&H model were supposed to track components but lump them together to screen off detail that was unknown at the time. One way to resolve this difference is to see again that we should not consider a scientific representation as capable of playing just one role in research. Depending on one's theoretical interest, the same representation could be used for descriptive purposes as a phenomenological model, as a predictive model, or as an abstract explanatory model. Only given a purpose can we assess whether the explanation is complete, that is, it refers to all and only the relevant factors of the explanandum phenomenon (Craver & Kaplan, 2020).

The H&H model is limited when explaining the properties of various ion channels, including those beyond sodium and potassium channels. It is also partially outdated, as it does not account for key features of cortical neuron action potentials, such as their rapid initiation and variable onset potential, which are outside the behaviors described by the model (Naundorf et al., 2006). Consequently, it fails to explain action potential initiation observed in cortical neurons in vivo or in vitro (Colwell & Brenner, 2009). Additionally, all classes of ion channels exhibit "modal gating behavior," meaning they spontaneously change their activity level without external stimulus. These phenomena and their underlying mechanisms are essential for understanding pharmacological manipulations. However, the H&H model remains largely successful in explaining other phenomena explicitly mentioned by its authors (Hodgkin & Huxley, 1952, pp. 543–544).

All in all, the subsequent debate on this example shows that mechanistic explanations should not necessarily "bottom out" at the molecular or cellular detail. In fact, the very notion of "mechanistic detail" should always be used in scare quotes, as this is not a technical term in the new mechanistic account of explanation. Mechanisms are, after all, temporal processes, whose causal organization explains phenomena. Causal relevance is the sole determining factor for where we "bottom out": the level of grain depends on how we frame the phenomenon to be explained, and that is driven by our theoretical interest (Craver, 2009; Pöyhönen, 2013). Of course, we may be wrong when framing phenomena theoretically one way rather than another, but it is a misunderstanding to think that there is some specific "mechanistic level of detail" in neuroscience or psychology, irrespective of our explanatory interests.

As a consequence, the abstractness response misses the mark: it is indeed an abstract explanation, but clearly causal and mechanistic. The good news is that depending on what we want to explain, some facts about mechanisms may be both confirmatory and explanatory.

Moreover, mechanisms posited in the life sciences are typically functional (Garson, 2013). It is no coincidence that critical intuitions behind functionalism can be elucidated in mechanistic terms, leading some to proclaim that mechanism provides a better foundation for functionalism (Maley & Piccinini, 2013). In essence, abstract functional mechanisms are already part and parcel of the standard mechanistic framework.

3.4.2 *The response that realizers merely confirm*

One possible response to the dilemma is that a functional explanation H can be confirmed by evidence er about its physical realizers, if the conditional probability $P(H|er) > P(H)$, all without providing a mechanistic explanation. In other words, learning about how H is physically implemented provides evidence in favor of H being true.

However, this type of confirmation is only meaningful for functional hypotheses whose physical realization is uncertain or in doubt. For example, a novel AI system for machine translation could have its functional architecture confirmed by evidence of how it is implemented in code and hardware. But this is more of a proof-of-concept scenario, demonstrating that the proposed functionality is physically and computationally possible.

In contrast, for most functional explanations in neuroscience, the physical realizability is not in serious doubt. Few would question whether phenomena like neuronal signaling or reinforcement learning can be physically implemented. The uncertainty lies more in the *specific* functional architecture and algorithms.

Moreover, given the thesis of multiple realizability, embraced by proponents of functionalism,[2] a functional hypothesis H is likely to have many possible physical realizers. These different realizers may correspond to quite different mechanistic explanations, some of which may turn out to be false for the actual phenomenon. Returning to the machine translation example, even if the AI system confirms the functional hypothesis, it may be implemented very differently than the neural mechanisms underlying human language translation.

If proof-of-concept demonstrations, or "how possibly" explanations, are acceptable for the functional approach, the approach faces a serious challenge. For instance, explaining John Kennedy's assassination (his capacity to be killed) by suggesting it was possible to poison him or bomb Dallas with nuclear weapons indicates possible causes of his death. However, these scenarios, while featuring physically possible realizers, do not actually explain his death. They simply describe potential causal realizers. Functional analysis thus risks suggesting phenomena that have no real effect or influence (Miłkowski, 2013, Chapter 3).

As a result, for functionalism, evidence about specific realizers may have limited relevance for adjudicating between competing functional hypotheses. At best, realizer evidence could rule out functional hypotheses that are impossible to implement physically. However, the space of possible (and even plausible) realizers is still vast, and it can be pruned only by assuming a less liberal approach to explanatory norms, which would avoid the problem of epiphenomenal realizers (such as nuclear weapons in Dallas). But this would invalidate the purported autonomy of functional explanations. In other words, if realizers indeed confirm, they also provide explanatory power, providing at least elliptical mechanistic explanations.

What seems necessary for substantive confirmation is the ability to comparatively evaluate functional hypotheses based on their empirical adequacy and other theoretical virtues. This may involve testing their differing predictions, assessing their consistency with established mechanistic constraints, and judging their simplicity, unification, and fruitfulness. Only through this broader consideration can the evidential impact of realizer-based confirmation be properly situated.

3.4.3 *The response that mechanistic details are irrelevant*

A final possible response from defenders of functional explanation is to simply deny that mechanistic evidence plays any confirmatory role at all (Barrett, 2014; Shapiro, 2017, 2019). On this view, functional explanations are confirmed solely by their ability to capture the abstract computational principles underlying a phenomenon, without any need to ground these principles in neural mechanisms.

However, this response seems to go against established scientific practice (Boone & Piccinini, 2016). If mechanistic details were truly irrelevant to confirmation, why would neuroscientists like Wolfram Schultz bother to investigate the neural basis of computational theories like TD learning? The discovery of phasic dopamine signals that resemble reward prediction errors was taken as important evidence in favor of TD models, not merely as an implementation detail.

More generally, the practice of comparing computational models to neural data is widespread in cognitive neuroscience. Researchers routinely test whether the internal variables and dynamics of their models correspond to observed patterns of neural activity, using techniques like representational similarity analysis and model-based fMRI. The clear assumption behind this work is that mechanistic evidence can constrain and inform computational theories, not just implement them.

Denying the confirmatory relevance of mechanistic evidence would thus require rejecting a central methodological tenet of modern neuroscience. It would imply that the extensive research program aimed at grounding computational models in the brain is fundamentally misguided. While some defenders of functional explanation might be willing to embrace this iconoclastic view, it is a heavy cost to bear.

Moreover, even if one accepts that mechanistic evidence is irrelevant to confirmation, it is unclear how functional explanations could be tested and evaluated. Without any contact with neural data, functional theories risk becoming unconstrained and unfalsifiable, able to accommodate any behavioral phenomenon by positing ad hoc computational mechanisms. Some grounding in "mechanistic details" seems necessary to rein in the degrees of freedom of functional explanations and enable them to make substantive empirical claims.

3.5 Implications for the autonomy of functional explanations and the role of the Minimal Cognitive Grid

The responses to the dilemma discussed in the previous section have important implications for the autonomy of functional explanations and their relationship to mechanistic details. In this section, we will explore these implications and consider their consequences for the functional approach, as well as discuss the role of the MCG in providing a proper mechanistic way to confirm models.

3.5.1 The Minimal Cognitive Grid

The paradigmatic way to analyze explanations in cognitive science is to rely on Marr's (1982) three-level account of computational explanation. His account is problematic for our purposes, as it is grounded in functionalism and does not provide a clear answer to the question of how to connect questions about cognitive algorithms and representations with questions about their physical implementation (McClamrock, 1991; Miłkowski, 2013). Instead, we need a more modern and general framework, in particular for confirmation purposes. While generic norms of mechanistic explanation, particularly the completeness norm (Baetu, 2015; Craver & Kaplan, 2020), are useful, they seem too abstract and theory-neutral to provide substantial guidance for (cognitive) neuroscience. We need a more substantive approach, which is compatible with a diverse variety of theoretical approaches but provides advice regarding minimal standards of explanatory and confirmatory acceptability. To this end, we propose to rely on the MCG.

Lieto (2021) introduces the MCG as consisting of three key dimensions that characterize the relationship between a model and its biological or cognitive target system:

- **Functional/Structural Ratio:** This dimension concerns the balance between functional and structural components in the model. It evaluates the extent to which the model relies on abstract functional descriptions versus detailed structural mechanisms. A lower ratio indicates a more mechanistic model, while a higher ratio suggests a more functional approach.
- **Generality:** This dimension assesses the breadth of phenomena that the model can represent. A highly general model can be applied to a wide range of cognitive functions or biological systems, while a narrow model is tailored to a specific task or domain.
- **Performance Match:** This dimension involves a direct comparison between the model's performance and that of the target system. It considers not only the overall accuracy of the model but also the similarity of its errors and execution times to those of the biological or cognitive system. A close performance match suggests a higher degree of psychological or biological plausibility.

In addition to these core dimensions, the MCG also considers secondary aspects, such as the physical medium used to implement the model and the level of abstraction employed.

The MCG provides a way to evaluate the structural accuracy of models. The structural accuracy used in the MCG resembles the notion of structural validity discussed by Zeigler (1976). A model is structurally valid if it not only reproduces the observed behavior of the real system but also truly reflects the way in which the real system operates to produce this behavior. This is closely related to Fodor's (1968) concept of strong equivalence, which requires a model not only to mimic the input–output behavior of a system but also to mirror its internal structure and processes.

By requiring structural validity or strong equivalence, the MCG ensures that functional explanations are grounded in mechanistic details to some extent (lumping variables, required by aggregative abstraction, is admissible, but it can lower the accuracy, depending on the aim of the model). It provides a minimal standard for the confirmation of models, ensuring that they are supported by evidence about the internal workings of the system. This provides a framework for understanding the complex relationship between functional and mechanistic explanations in cognitive science, suggesting that functional explanations must be grounded in mechanistic details.

This framework may supplant, or at least complement, Marr's influential but vague concept of "levels." The MCG suggests that models in neuroscience must ultimately be supported by mechanistic evidence in one way or another.

3.5.2 *Analyzing Krakauer's examples using the Minimal Cognitive Grid*

Let us consider the two examples provided by Krakauer and analyze them using the dimensions of the grid. When using the MCG, we must remember that it should be applied to some particular phenomenon; the H&H model and the TD model can address various phenomena with varying degrees of success.

Example 1: The Hodgkin-Huxley Model of the Action Potential

Functional/Structural Ratio: The H&H model has a moderate functional/structural ratio. While it relies on abstract voltage-dependent permeability changes, it also specifies the ionic currents and conductances that underlie these changes. The subsequent discovery of ion channels provided a more detailed structural grounding for the model's functional components.

Generality: The H&H model has a high degree of generality, as it can represent action potentials across a wide range of neurons and species. However, if we assume that the target phenomenon of the model are features of cortical neuron action potentials specifically, it fails to capture some of them.

Performance Match: The H&H model achieves a strong performance match with the observed behavior of the action potential, accurately reproducing its shape, threshold, and refractory period. However, it does not account for the variability and rapid initiation seen in cortical neurons.

Example 2: Temporal Difference Learning and the Phasic Dopamine Signal

Functional/Structural Ratio: TD learning has a high functional/structural ratio, as it focuses on the abstract computational principles of reinforcement learning without specifying the neural mechanisms that implement them. The discovery of the phasic dopamine signal provided a potential structural substrate for the reward prediction error but did not alter the model's functional form.

Generality: TD learning is highly general, as it can be applied to a wide range of reward-based learning tasks across both biological and artificial agents.

Performance Match: TD learning achieves a good performance match with the observed behavior of reward-based learning, accurately

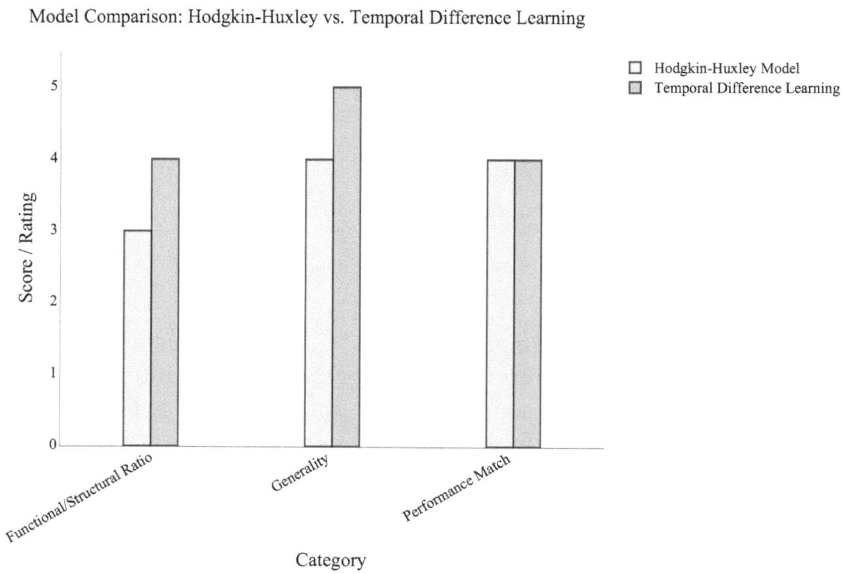

Figure 3.1 A grouped bar chart comparing the Hodgkin-Huxley (H&H) model and temporal difference (TD) learning across the three core dimensions of the MCG: Functional/Structural Ratio, Generality, and Performance Match. The H&H model has a lower functional/structural ratio and higher specificity, while TD learning has a higher functional/structural ratio and greater generality.

Source: Diagram © 2024 Marcin Miłkowski

predicting the patterns of choices and response times. The similarity of the phasic dopamine signal to the reward prediction error further supports this match.

This analysis using the MCG highlights the complementary strengths and weaknesses of the two models. The H&H model has a stronger structural grounding, but when taken to target cortical neurons, it is somewhat limited in its generality, failing to account for specific features of cortical neuron action potentials, while TD learning is more abstract and general but lacks a detailed mechanistic implementation (see Figure 3.1). The analysis suggests that Krakauer's examples, while successful, are not entirely satisfactory.

3.5.3 *Implications for the autonomy of functional explanations*

The application of the MCG to Krakauer's examples suggests that functional explanations in neuroscience are not fully autonomous from mechanistic details but instead exist on a continuum of abstraction. While some models, like TD learning, may rely more heavily on functional principles, they are still constrained by the need to match the performance of the biological system and to find plausible structural implementations.

Moreover, the MCG's emphasis on structural validity implies that functional explanations should aim to capture the internal workings of the system to some extent, even if they do so at a high level of abstraction. This challenges the notion that functional explanations can be entirely independent of mechanistic details.

At the same time, the MCG allows for a degree of functional autonomy, as it recognizes that models can vary in their functional/structural ratio and level of abstraction. Abstract functional models can still be useful and informative, as long as they satisfy the other dimensions of the grid, such as generality and performance match.

The MCG thus provides a nuanced perspective on the relationship between functional and mechanistic explanations in neuroscience. It suggests that these two approaches are not fundamentally distinct but rather lie on a spectrum of abstraction and integration. The most powerful explanations are those that combine functional principles with mechanistic grounding, as exemplified by the progressive elaboration of the H&H model with the discovery of ion channels. Functional models can maintain a degree of independence from specific mechanistic details, but they are still accountable to the broader constraints of structural validity and empirical confirmation.

By situating functional explanations within this broader landscape, the MCG suggests that the most complete and powerful explanations in neuroscience will be those that bridge the gap between function and mechanism, combining the generality of computational models with the specificity of neural implementations.

3.6 Concluding remarks

In this chapter, we have challenged the view that mechanistic evidence merely confirms, without explaining, functional explanations in neuroscience. We have argued that functional explanations are fundamentally elliptical mechanistic explanations whose explanatory power ultimately depends on their integration with mechanistic evidence.

We have addressed a critical dilemma: functional explanations, which are supposed to lack mechanistic detail, require empirical content for confirmation. This has led us to the view that the relationship between functional explanations and mechanistic evidence is one of mere logical consistency, which risks committing the "consistency fallacy"—where confirmation is mistakenly equated with explanation.

To address this, we have adopted the MCG to analyze examples from neuroscience and demonstrate that abstract mechanistic explanations can be developed independently of detailed mechanisms, yet their validity is limited by the need for biological relevance and empirical confirmation.

Our analysis has shown that mechanistic evidence not only confirms but also mechanizes functional explanations. The MCG has provided a framework for understanding this complex relationship, offering a multidimensional approach for evaluating the confirmation and explanatory power of functional models in neuroscience.

Furthermore, we have demonstrated that the mechanistic approach to explanation is not confined to cellular or molecular details but can remain abstract, similar to functional explanations, as long as it accounts for all relevant factors influencing the phenomena. This blurs the lines between functional and mechanistic explanations, with some proponents of the new mechanistic explanation aligning themselves with functionalism.

Some earlier functional explanations, particularly from artificial intelligence and cognitive science in the 1970s, served only as conceptual demonstrations and lacked significant empirical grounding, with limited relevance to actual phenomena. These were, at best, preliminary, how-possibly mechanistic explanations that did not fully capture the causal mechanisms involved. But functional explanation can do better by following the explanatory norms of the new mechanistic framework. Some of the work in cognitive neuroscience already does.

Acknowledgments

The author is grateful to John Krakauer for allowing him to cite his personal correspondence and discuss it in this chapter. His criticisms were instrumental in revising it as well. Detailed remarks from Gualtiero Piccinini and Antonio Lieto were also very useful.

Notes

1 Interestingly, the H&H model was analyzed in some detail by Carl Craver, who argued against interpreting it as an instance of a covering-law explanation (Craver, 2007, pp. 49–61), which was proposed by Weber (2008). We will return to his discussion in Section 3.4.

2 Advocates of mechanism need not support this thesis (Haimovici, 2013; Miłkowski, 2016; Polger & Shapiro, 2023).

References

Baetu, T. M. (2015). The Completeness of Mechanistic Explanations. *Philosophy of Science*, *82*(5), 775–786. https://doi.org/10.1086/683279

Bantegnie, B. (2022). Psychology and Neuroscience: The Distinctness Question. *Erkenntnis*, *87*(4), 1753–1772. https://doi.org/10.1007/s10670-020-00272-x

Bantegnie, B. (2024). Psychology and Neuroscience. The (New) Autonomy Question. *Erkenntnis*. https://doi.org/10.1007/s10670-024-00833-4
Barrett, D. (2014). Functional Analysis and Mechanistic Explanation. *Synthese*, *191*(12), 2695–2714. https://doi.org/10.1007/s11229-014-0410-9
Bechtel, W. (2016). Investigating Neural Representations: The Tale of Place Cells. *Synthese*, *193*(5), 1287–1321. https://doi.org/10.1007/s11229-014-0480-8
Boone, W., & Piccinini, G. (2016). The Cognitive Neuroscience Revolution. *Synthese*, *193*(3), 1509–1534. https://doi.org/10.1007/s11229-015-0783-4
Carrillo, N., & Knuuttila, T. (2023). Mechanisms and the Problem of Abstract Models. *European Journal for Philosophy of Science*, *13*(3), 27. https://doi.org/10.1007/s13194-023-00530-z
Coltheart, M. (2013). How Can Functional Neuroimaging Inform Cognitive Theories? *Perspectives on Psychological Science*, *8*(1), 98–103. https://doi.org/10.1177/1745691612469208
Colwell, L. J., & Brenner, M. P. (2009). Action Potential Initiation in the Hodgkin-Huxley Model. *PLoS Computational Biology*, *5*(1), e1000265. https://doi.org/10.1371/journal.pcbi.1000265
Craver, C. F. (2007). *Explaining the Brain: Mechanisms and the Mosaic Unity of Neuroscience*. Oxford University Press.
Craver, C. F. (2009). Mechanisms and Natural Kinds. *Philosophical Psychology*, *22*(5), 575–594. https://doi.org/10.1080/09515080903238930
Craver, C. F., & Kaplan, D. M. (2020). Are More Details Better? On the Norms of Completeness for Mechanistic Explanations. *The British Journal for the Philosophy of Science*, *71*(1), 287–319. https://doi.org/10.1093/bjps/axy015
Cummins, R. (1975). Functional Analysis. *The Journal of Philosophy*, 72(20), 741. https://doi.org/10.2307/2024640
Fodor, J. A. (1968). *Psychological Explanation*. Random House.
Garson, J. (2013). The Functional Sense of Mechanism. *Philosophy of Science*, *80*(3), 317–333. https://doi.org/10.1086/671173
Haimovici, S. (2013). A Problem for the Mechanistic Account of Computation. *Journal of Cognitive Science*, *14*(2), 151–181.
Haslam, N., Vylomova, E., Murphy, S. C., & Wilson, S. J. (2021). The Neuroscientification of Psychology: The Rising Prevalence of Neuroscientific Concepts in Psychology from 1965 to 2016. *Perspectives on Psychological Science*, *17*(2), 519–529. https://doi.org/10.1177/1745691621991864
Hempel, C. G., & Oppenheim, P. (1953). The Logic of Explanation. In H. Feigl & M. Brodbeck (Eds.), *Readings in the Philosophy of Science* (pp. 319–352). Appleton-Century-Crofts, Inc.
Hodgkin, A. L., & Huxley, A. F. (1952). A Quantitative Description of Membrane Current and Its Application to Conduction and Excitation in Nerve. *Bulletin of Mathematical Biology*, *117*, 500–544. https://doi.org/10.1016/S0092–8240(05)80004–7
Illari, P. M., & Williamson, J. (2012). What is a Mechanism? Thinking about Mechanisms Across the Sciences. *European Journal for Philosophy of Science*, *2*(1), 119–135. https://doi.org/10.1007/s13194-011-0038-2
Levy, A. (2014). What was Hodgkin and Huxley's Achievement? *The British Journal for the Philosophy of Science*, *65*(3), 469–492. https://doi.org/10.1093/bjps/axs043
Lieto, A. (2021). *Cognitive Design for Artificial Minds*. Routledge/Taylor & Francis Group.
Machamer, P., Darden, L., & Craver, C. F. (2000). Thinking about Mechanisms. *Philosophy of Science*, 67(1), 1–25. https://doi.org/10.1086/392759

Maley, C., & Piccinini, G. (2013). Get the Latest Upgrade: Functionalism 6.3.1. *Philosophia Scientae*, *17*(2), 135–149. https://doi.org/10.4000/philosophiascientiae.861
Marr, D. (1982). *Vision*. W. H. Freeman and Company.
McClamrock, R. (1991). Marr's Three Levels: A Re-Evaluation. *Minds and Machines*, *1*(2), 185–196. https://doi.org/10.1007/BF00361036
Miłkowski, M. (2013). *Explaining the Computational Mind*. MIT Press.
Miłkowski, M. (2016). Computation and Multiple Realizability. In V. C. Müller (Ed.), *Fundamental Issues of Artificial Intelligence* (Vol. 376, pp. 29–41). Springer International Publishing. https://doi.org/10.1007/978-3-319-26485-1_3
Naundorf, B., Wolf, F., & Volgushev, M. (2006). Unique Features of Action Potential Initiation in Cortical Neurons. *Nature*, *440*(7087), 1060–1063. https://doi.org/10.1038/nature04610
Piccinini, G. (2020). *Neurocognitive Mechanisms: Explaining Biological Cognition*. Oxford University Press.
Piccinini, G., & Craver, C. (2011). Integrating Psychology and Neuroscience: Functional Analyses as Mechanism Sketches. *Synthese*, *183*(3), 283–311. https://doi.org/10.1007/s11229-011-9898-4
Polger, T. W., & Shapiro, L. A. (2023). The Puzzling Resilience of Multiple Realization. *Minds and Machines*. https://doi.org/10.1007/s11023-023-09635-z
Pöyhönen, S. (2013). Carving the Mind by Its Joints: Culture-Bound Psychiatric Disorders as Natural Kinds. In M. Miłkowski & K. Talmont-Kaminski (Eds.), *Regarding the Mind, Naturally: Naturalist Approaches to the Sciences of the Mental* (pp. 30–48). Cambridge Scholars Publishing.
Salmon, W. C. (1984). *Scientific Explanation and the Causal Structure of the World*. Princeton University Press.
Schultz, W. (1998). Predictive Reward Signal of Dopamine Neurons. *Journal of Neurophysiology*, *80*(1), 1–27. https://doi.org/10.1152/jn.1998.80.1.1
Schultz, W., Dayan, P., & Montague, P. R. (1997). A Neural Substrate of Prediction and Reward. *Science*, *275*(5306), 1593–1599. https://doi.org/10.1126/science.275.5306.1593
Shapiro, L. A. (2017). Mechanism or Bust? Explanation in Psychology. *The British Journal for the Philosophy of Science*, *68*(4), 1037–1059. https://doi.org/10.1093/bjps/axv062
Shapiro, L. A. (2019). A Tale of Two Explanatory Styles in Cognitive Psychology. *Theory & Psychology*, *29*(5), 719–735. https://doi.org/10.1177/0959354319866921
Shmueli, G. (2010). To Explain or to Predict? *Statistical Science*, *25*(3), 289–310. https://doi.org/10.1214/10-STS330
Stinson, C. (2016). Mechanisms in Psychology: Ripping Nature at Its Seams. *Synthese*, *193*(5), 1585–1614. https://doi.org/10.1007/s11229-015-0871-5
Tolman, E. C. (1948). Cognitive Maps in Rats and Men. *Psychological Review*, *55*(4), 189–208.
Weber, M. (2008). Causes without Mechanisms: Experimental Regularities, Physical Laws, and Neuroscientific Explanation. *Philosophy of Science*, *75*(5), 995–1007. https://doi.org/10.1086/594541
Weiskopf, D. A. (2016). Integrative Modeling and the Role of Neural Constraints. *Philosophy of Science*, *83*(5), 674–685. https://doi.org/10.1086/687854
Zeigler, B. P. (1976). *Theory of Modelling and Simulation*. Wiley.

4 Cognitive ontology in terms of cognitive homology

The role of brain, behavior, and environment for individuating cognitive categories

Beate Krickel and Mariel K. Goddu

4.1 Introduction

The *cognitive ontology question* is how to carve up cognition so that the resulting carving (i.e., ontology, taxonomy, categorization) enables successful scientific predictions, explanations, and generalizations.[1] The issue with contemporary cognitive science, however, is that its categories are ill-defined and ambiguous (Poldrack et al., 2011; Zagaria et al., 2020) and not derived from rigorous scientific reasoning but from intuition, speculation, or "dreamed-up vocabulary" (Buzsáki, 2019, p. 6) (see also Vanderwolf, 2007). According to Buzsáki and Vanderwolf, William James introduced this vocabulary in 1890, and, indeed, it still broadly corresponds to the categories that guide cognitive scientific investigations today: "self-consciousness," "attention," "association," "perception," "memory," "imagination," "reasoning," "production of movement," and "emotion" (James, 1890). Many theorists and researchers agree that these categories should be revised (e.g., Anderson, 2014; Cisek, 2019; De Brigard, 2017; Dewhurst, 2021; Figdor, 2018; Poldrack, 2010; Poldrack et al., 2011; Price & Friston, 2005). The challenge, then, is to find a new way (or ways) of categorizing cognitive capacities. This is what scientists and philosophers who engage with the cognitive ontology question aim to do.

There are two broad strategies for addressing this challenge (for a discussion of accounts corresponding to these two strategies, see Francken et al., 2022; McCaffrey & Wright, 2022). *Incremental approaches* take existing cognitive categories and revise them. These approaches are hypothesis-driven and iterative: roughly, they hypothesize that some cognitive category X corresponds to a cognitive capacity, outline specific predictions that follow from this hypothesis, and conduct empirical research to test them. Then, they keep, revise, or reject X in light of the findings. The challenge for incremental approaches is that they are prone to inheriting

DOI: 10.4324/9781003458531-4

the problems of the existing categories. *Radical approaches,* by contrast, reject the current categories altogether. They start from scratch, constructing new cognitive categories that might depart dramatically from the ones posited by contemporary cognitive science. Radical approaches face two main challenges: first, they must provide reasonable bases for deriving the novel categories, and second, they must justify why the resulting categories are indeed *cognitive* categories (rather than, e.g., purely mathematical entities; see, e.g., Yeo et al. [2015]—and for a critical discussion along these lines, see McCaffrey and Wright [2022, p. 446]).

A key question in the cognitive ontology literature concerns the relationship between cognitive categories and neural categories (such as brain regions). One central issue is whether one should anticipate finding one-to-one correspondences between cognitive functions and neural structures, or whether more complex mappings, such as one-to-many, many-to-one, or many-to-many, might be all one can get (Viola, 2017). We will not address this issue here, as it is less relevant to the objectives of this chapter. Another significant issue is whether or not the development of a new cognitive ontology should begin with the brain. *Bottom-up approaches* to cognitive ontology seek to derive cognitive categories from neural categories, usually anatomically individuated brain regions. Radical approaches to cognitive ontology typically adopt this bottom-up perspective. In contrast, *top-down approaches* start with hypotheses about cognitive categories and then—if at all—examine how these cognitive categories are realized in the brain. Incremental approaches to cognitive ontology are usually top-down.

One feature that most approaches to cognitive ontology in the current literature share is that they tend to focus solely on *human* cognition (Figdor, 2022). This is a severe shortcoming. To the best of our scientific knowledge, cognition is a product of evolution (Cisek, 2019). An approach to cognitive ontology informed by basic considerations from evolutionary biology should assume that (some form of) cognition is found in many species.

In this chapter, we propose a new answer to the cognitive ontology question that takes the evolution of cognition into account. We suggest that cognitive carvings should be based on *cognitive homologies*, cognitive capacities that are "the same across species" by the standards of evolutionary biology. Still, our approach is incremental. We suggest investigating existing cognitive capacities to discover whether they are cognitive homologies. If presently distinct cognitive capacities are found to be homologous, revision is necessary: they collapse into one and the same capacity. If cognitive capacities that are currently considered to be the same turn out *not* to be homologous, revision is also necessary: they are different cognitive capacities. In that sense, our approach is top-down. Still, as we show, investigating the brain and neural mechanisms is crucial for determining cognitive homologies and thus for cognitive ontology.

More specifically, we aim to develop a novel account of cognitive homology by adopting a recent account of "trait identity" from evolutionary-developmental biology (DiFrisco, 2023; DiFrisco et al., 2020). This account supplements standard conceptions of homology by introducing "character identity mechanisms" (ChIMs): developmental mechanisms that guarantee the stability of trait identity across phylogeny. Similarly, our account proposes to individuate cognitive capacities based on the developmental mechanisms that bring them about—what we will call *cognitive ChIMs*. As we will show, cognitive ChIMs are developmental mechanisms that crucially involve brain mechanisms as well as the organism's interactions with its environment.

Here, the goal is primarily to provide a proof of concept. We argue that a plausible notion of cognitive homology can be developed based on the notion of a *cognitive ChIM,* a concept which can then be used in answering the cognitive ontology question. Turning the idea into a full-fledged approach to cognitive ontology constitutes a novel interdisciplinary research program—of which this chapter can only be a starting point.

The chapter proceeds as follows. Section 4.2 introduces the biological concept of homology and presents two conceptions of it: the *phylogenetic conception* and the *ontogenetic conception* that will be crucial for our account of cognitive homology. In Section 4.3, we review two previous accounts of cognitive homology. We argue that they are unsuccessful mainly because they attempt to define a notion of structural similarity for entities that are characterized functionally. In Section 4.4, we argue that an account of cognitive homology that incorporates the ontogenetic conception of homology avoids the problems afflicting existing accounts. Section 4.5 concludes this chapter.

4.2 Homology in biology

> Countless philosophy students have been taught that the definition of the heart as "the organ whose purpose is to pump the blood" is a typical example of biological classification. . . . [T]his picture "is utterly false . . ." . . . [I]n comparative anatomy the parts of organisms are primarily defined as *homologues*.
>
> (Griffiths, 2007, p. 644)

Contrary to a commonly held view among philosophers, biologists tend not to individuate biological entities—such as organs, tissues, and cells—in terms of what they do. Rather, they individuate them in terms of *homology*.

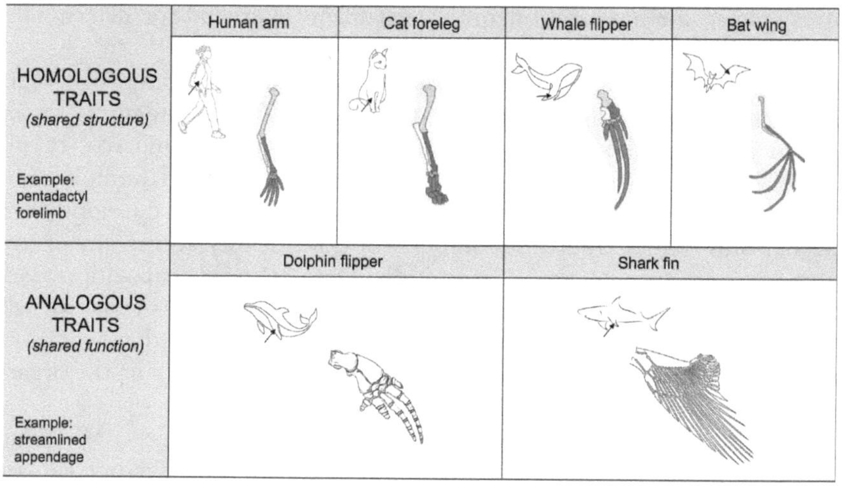

Figure 4.1 Examples of homologous versus analogous traits.

Source: Own illustration.

Homologous traits (also called "homologues" or "homologies") share a kind of identity: "[w]hen two characters[2] or traits are homologous, they are fundamentally the same kind of character" (DiFrisco, 2023, p. 1; see also Fusco, 2022).

Homology is typically defined for morphological features (body parts) but need not be restricted to it (see Section 4.3). One example of a morphological homology is the "pentadactyl forelimb," the general forelimb form of species that have five "fingers." Instances of this trait include the human arm, the cat's foreleg, the whale's flipper, and the bat's wing (see Figure 4.1).

These forms look different: they are different *phenotypes* or *phenotypical expressions* of the "pentadactyl forelimb" trait. They have different shapes and fulfill rather different functions—reaching, walking, swimming, and flying. However, the sameness relevant for homology of traits concerns their *structural* similarity rather than their outward appearance or function. Structural similarity of morphological entities involves sameness of internal organization, positions relative to other structures, degree of complexity, and position within a series of trait transformations in evolution (García, 2010; Bergeron, 2021; DiFrisco, 2023).

The conceptual foil to homologous traits is *analogous* traits (also called "analogues" or "analogies"). In contrast to homologues, where the similarities are structural, analogues have merely functional similarities. Some species develop superficially similar traits because they adapted to similar environments. For example, the ancestors of dolphins and sharks faced adaptive pressures such that those individuals with streamlined appendages had a survival benefit. However, despite their common function (swimming) and apparent similarity of form (streamlined shape), the dolphin's flipper and the shark's fin are structurally *dis*similar (see Figure 4.1). These body parts are not instances of the same trait.

Biological sameness or trait identity thus concerns *structural sameness* rather than *functional sameness*. Since the late 19th century (and already present in Darwin's work), a further crucial aspect of homology is *shared ancestry*. Species with homologous traits are descended from a common ancestor that had the trait in question and are linked to that ancestor by a continuous lineage of ancestors that all had that trait. Common descent and the continuous line of ancestors with the trait are assumed to be the cause and explanation for the structural similarities between them (Wagner, 1989; Griffiths, 2007). Call this concept of homology the *phylogenetic conception* (DiFrisco, 2023). The phylogenetic conception of homology is the standard view of homology.

Recently, the phylogenetic conception has become the target of various criticisms (for an overview, see DiFrisco, 2023). One problem is that there is simply no continuity in the presence of a trait across generations: almost all traits need to develop anew in each individual. Another criticism is that the phylogenetic conception cannot address *serial homology*, the sameness of traits within the same individual organism. For example, the fore and hind legs of tetrapods are iterated versions of the same body part. Thus, they are homologous but do not have an obvious phylogenetic connection. A further problem for the phylogenetic conception of homology is that it does not provide a satisfactory explanation of what it means for traits to "maintain their identity" despite vast changes across evolution (Wagner, 1989; for details, see DiFrisco, 2023). Finally, the phylogenetic conception is problematic for epistemic reasons. Shared ancestry and continuous inheritance are not directly accessible; rather, they must be inferred from fossil records or comparative anatomical or genetic analyses (DiFrisco et al., 2023).

A conception of homology that aims to address these problems may be called the *ontogenetic* or *developmental conception* of homology. This is seen as a supplement rather than an alternative to the phylogenetic conception (see DiFrisco, 2023).[3] It is based on adopting the "developmental homology principle":

> *Developmental homology principle (DHP)*: The identity and classification of morphological homologies is determined, at least in part, by the specific developmental factors that cause them in ontogeny.
>
> (DiFrisco, 2023, p. 4)

DHP provides a developmental basis for individuating and identifying homologous traits. Different accounts of ontogenetic homology make use of DHP in different ways. Here, we focus on an account due to DiFrisco et al. (2020) that centers on ChIMs. The concept of a ChIM follows from two hypotheses:

(1) There is a general, recognizable mechanistic architecture in development that explains the traceability of characters [traits].
(2) These mechanisms are themselves cohesive units that can be traced as homologues in evolution (DiFrisco et al., 2020, p. 7).

Traceability of traits is a challenge for researchers: as outlined above, the expression of the same trait (homologue) in different species can look superficially different (compare a human arm and a bat wing). This gives rise to two questions. First, how can homologues be detected and identified? And second, in what sense can a homologue be said to maintain its identity (i.e., remain the "same trait"), even if it undergoes drastic changes?

According to DiFrisco et al., ChIMs provide answers to both questions because they are themselves traceable. They are highly evolutionarily conserved and resistant to change. This is because ChIMs have special characteristic features, as we will now explain.

ChIMs are a special type of biological mechanism. According to the so-called new mechanists (whom DiFrisco et al. follow here), *mechanisms are entities and activities organized such that they bring about a phenomenon* (see, e.g., Craver, 2007). Mechanisms are individuated based on their entities, activities, organizational features, and the phenomena they bring about.[4] For example, the mechanisms for neurotransmitter release consist of ions, ion channels, and vesicles (the entities) that open, close, diffuse, and move (the activities) at specific locations and times and in specific orders (their organization) such that neurotransmitters are released into the synaptic cleft (the phenomenon). DiFrisco et al. (2020) provide a detailed discussion of ChIM individuation that is based on this general framework. For lack of space, we won't repeat these considerations here but refer the reader to their article.

ChIMs are *developmental* mechanisms. Many developmental mechanisms—such as those that govern the formation of particular feathers in birds or wing patterns in butterflies—"micro-manage" the expression of species-specific traits. That is, they are responsible for the development of a particular *phenotypical expression of a trait* in a species. Changes to such mechanisms might cause slightly different feathers or wing patterns to develop. By contrast, ChIMs regulate *trait identity*. Changes to a ChIM might result in the total absence of feathers or wing patterns at all. One example of a ChIM is the mechanism involving *Ultrabithorax*, a gene that regulates hindwing development in insects. Evidence from "knockdown"

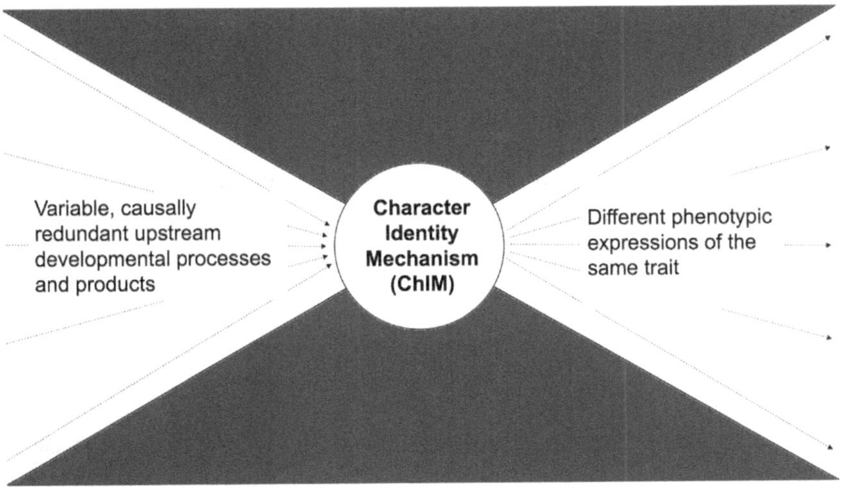

Figure 4.2 A ChIM is a developmental mechanism with a specific position in a causal process: it is the knot of a "bowtie" causal topology (in white on gray background). ChIMs are complex mechanisms that take a variety of variable, causally redundant upstream processes as inputs and transform them into a single output: the specific character that the ChIM produces. This output then typically undergoes subsequent developments. Due to their causal necessity and non-redundancy, ChIMs are highly conserved across evolution.

Source: Own illustration

experiments (in which *Ultrabithorax* is "switched off") shows that this mechanism controls character identity rather than phenotype: the knockdown causes the development of a forewing in the hindwing location on the body without altering the size, shape, or any other structural features of the developing structure (Wagner, 2007, p. 475; DiFrisco et al., 2020, p. 3). The gene *Ultrabithorax* thus determines whether a wing *is* a hindwing (trait identity) rather than, for example, the specific color, shape, or structure of the hindwing (trait expression).

ChIMs have the power to determine trait identity because of their "mechanistic architecture" (DiFrisco et al., 2020, p. 2). They are characterized by their location within a causal topology (the "shape" of a causal process). Specifically, ChIMs are the knot in a bowtie-like causal structure (Figure 4.2). They occupy this position because they are causally necessary and nonredundant. ChIMs are causally necessary because there is no other way to turn the "inputs" from the left bowtie wing (the "upstream" developmental processes) into the "outputs" on the right (the "downstream" developmental processes) other than via the ChIM. They are causally nonredundant because they are not replaceable by any other causal mechanism: there is no back-up mechanism that can fill the position of the ChIM in case the ChIM fails. This is due to the complexity of ChIMs that is necessary to transform the ChIM's inputs into its outputs. It is evolutionarily very unlikely that an organism develops a "better-safe-than-sorry" system that instantiates such complexity twice.

Due to these characteristic features, ChIMs can be seen as causing and explaining trait identity throughout the phylogenetic tree. Because ChIMs' necessity and complexity render them less replaceable than other mechanisms, their

> "developmental outcomes . . . are subject to strong stabilizing selection." The result is that ChIMs "are more likely to be evolutionarily conserved than other developmental mechanisms."
>
> (DiFrisco et al., 2020, pp. 8–9)

Individuals with a defective ChIM do not develop whatever trait the ChIM is causally responsible for and are thus likely to die early and not reproduce. All living individuals who developed the trait also had the ChIM, which makes it very likely that they pass on the ChIM to their offspring. Figure 4.3 presents the ChIM for the vertebrate paired limbs trait as an example.

The vertebrate limb ChIM has the hallmark features described by DiFrisco et al.: it is a developmental mechanism that is causally necessary and nonredundant for the transformation of "upstream" developmental inputs (here, pluripotent mesenchymal cells in early embryos) into "downstream" developmental outputs (here, various possible phenotypical expressions of a limb in different species). The ChIM's central role in generating the limb trait explains the structural similarities across species. By contrast, analogous traits—such as the streamlined appendages of sharks and dolphins mentioned above—are generated by different mechanisms and are, thus, not of the same type.

To summarize, biological traits are carved up in terms of homology. The *phylogenetic conception of homology* holds that homologies share structural similarities that can be traced back to a common ancestor. The *ontogenetic conception of homology* supplements this approach to solve several problems afflicting a purely phylogenetic conception of homology. A recent ontogenetic account introduces the concept of ChIMs: causally necessary, nonredundant developmental mechanisms that are responsible for trait identity. ChIMs are conserved across evolution. On this account, two traits are the same in different species if only if they are produced by the same ChIM.

Now that we have a better grasp of homology and its role in trait identification in biology, we are prepared to ask: Can this concept be applied to cognitive capacities?

4.3 Cognitive homology—existing accounts

The idea of cognitive homology might seem like a non-starter. Cognitive capacities are characterized by their functional profiles (e.g., "memory is

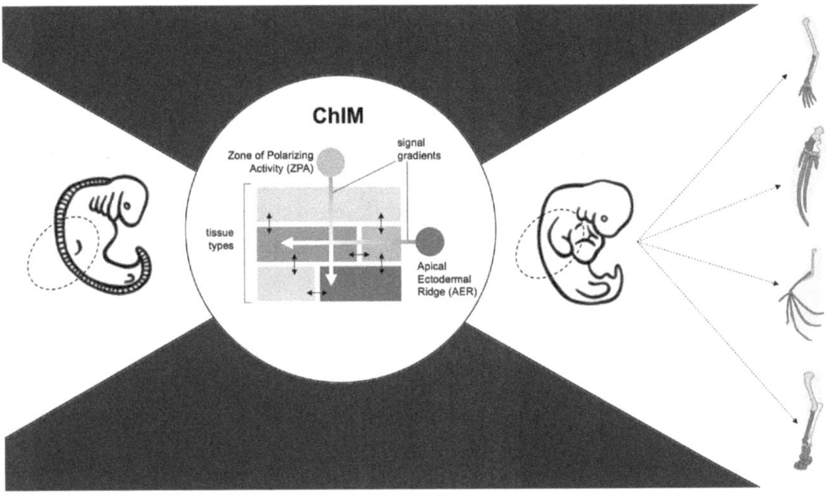

Figure 4.3 Schematic illustration of the "paired gnathostome appendage ChIM." The development of limb buds in vertebrate embryos precedes the development of paired appendages (e.g., fins and limbs). Bi-directional arrows schematically depict local, hormone-driven interactions between tissue types; limb development is orchestrated by four signaling centers throughout the limb bud. The two most well-known signaling centers are included here as circles with arrows indicating signal gradients. The Zone of Polarizing Activity (ZPA) establishes anterior-posterior organization of the limb, and the Apical Ectodermal Ridge (AER) establishes proximal-distal organization. These signaling centers form a complex, causally nonredundant, and necessary mechanism that jointly determines limb identity. Their activities are interdependent and mutually reinforce each other. Upstream of the ChIM (its causal "input") is relatively undifferentiated, loosely organized mesodermal embryonic tissue; downstream of the ChIM (its causal "output") is a structure with the identifiable tissue types and organization of a limb. This basic limb structure is then modified by further processes to determine a broad spectrum of possible limb phenotypes, such as the range of pentadactyl forelimb phenotypes discussed above.

Source: Own illustration adapted from DiFrisco et al. (2020)

the capacity to recall information that was acquired in the past") rather than their structural features. Given that homology concerns the sameness of structure rather than function (see Section 4.2), the idea that functionally characterized entities might be homologous seems to be a contradiction in terms (García, 2010; Love, 2007). How are we to make sense of the idea of cognitive homologies?

In this section, we present two existing accounts of cognitive homology—that is, Claudia García's (2010) and Vincent Bergeron's (2021)—that try to make sense of cognitive homology by explaining how functionally defined entities can still have structural features. The discussion of these accounts illustrates the challenges that arise for accounts of cognitive homology that seek to base the relevant notion of "homology" on structural features of functional entities.

Both Bergeron and García begin their analyses with the already-established notion of "behavioral homology." Homologous behaviors are the same across species due to common ancestry. Tail-wagging behaviors in different species of birds and web-weaving behaviors in different species of spiders are examples of behavioral homologies. The notion of sameness relevant for behavioral homology is the same as for morphological homology: it is *structural* sameness that matters (García, 2010, p. 126). Behaviors can have similar internal organizations, relative positions in behavioral sequences, and degrees of complexity. Thus, behavioral homology makes plausible notions of "structure" and "structural similarity" that are not restricted to material structure. This makes it an apparently promising starting point for cognitive homology.

Despite their common starting point, García and Bergeron wind up with different accounts of cognitive homology. The crucial difference lies in the structural features they identify relative to which structural similarity, and thus homology, can be spelled out.

Garcia's account characterizes cognitive homology as a particular type of what she calls "functional homology." Functional homology, according to her account, can hold between *cognitive systems* such as neural networks. Two cognitive systems S1 and S2 are functionally homologous if and only if two conditions hold:

(i) "S1 and S2 perform the same function, albeit possibly in different ways, that is, whether differently implemented . . . or subsystemically differently performed . . . ."[5]
(ii) "S1 and S2 have a common ancestor."

(García, 2010, p. 128)

On García's account, the evidence that two cognitive systems are indeed functionally homologous and the criteria for structural similarity between functions is the same as in morphological homology (see Section 4.2):

- S1 and S2 have the same position in a causal topology of two larger systems T1 and T2 that perform the same function.
- The subsystemic operations of S1 and S2 have the same complexity, and the functions of S1 and S2 have the same specificity.
- The functions have the same place in an evolutionary series of traits.

In other words, García's account consists in a direct translation of the phylogenetic concept of homology to cognitive systems—but without referring to material structure (such as brain structures). Her account allows that the material implementation of the functions can differ between homologous functions.

García's account presents several challenges. One key issue is the ambiguity surrounding condition (ii), which is that "S1 and S2 have a common ancestor." Since S1 and S2 are cognitive systems, this must mean that the *organisms implementing* these systems share a common ancestor. However, this alone does not establish homology. Tracing back far enough in evolutionary history, every organism shares a common ancestor with every other organism. Therefore, an important addition to García's condition is that the common ancestor must have already possessed a precursor version of S1 and S2 (call it "S0"), which must have performed the same function as S1 and S2. This requirement poses a significant epistemic challenge. Unlike morphological structures, cognitive functions do not leave fossil records. Even if a common ancestor with S0 can be identified, determining whether S0 performed the same cognitive function as S1 and S2 remains elusive. One potential counterargument might suggest that, if we have already established that S1 and S2 perform the same function and that a common ancestor had S0, it is highly improbable that S0 did not perform a similar function to S1 and S2.

A second issue arises with condition (i) and the criteria for structural sameness of function. If we require that S1 and S2 share the same relative position within their systems, we risk falling into an infinite regress. According to García, the relative positions of S1 and S2 are determined relative to larger systems, T1 and T2, of which S1 and S2 are subsystems, where T1 and T2 *perform the same function*. Thus, to establish structural sameness between the functions of S1 and S2, we need to assess the functions of T1 and T2 and determine whether they are the same. However, under the current proposal, this is the case only if T1 and T2 occupy the same relative position within yet further larger systems. This leads to a regress that can only stop when we reach a system that is not a subsystem of any larger system or when T1 and T2 are subsystems of the same larger system (e.g., sharing the same ecological niche or ecosystem). In such cases, the present account cannot determine whether the functions are structurally the same. Eliminating the requirement for relative position does not resolve the problem, as the remaining criteria are insufficient to fully define structural sameness of functions. Surely, a more detailed analysis of these criteria is needed, but that is beyond the scope of this chapter.

Bergeron's account avoids the problems of García's account because it takes the underlying neural structures to be essential for determining cognitive homology and does not rely on an account of structural similarities of functions. On Bergeron's account, structural sameness is defined in terms of sameness of material, that is, brain structure. Because there is already plenty of evidence for homologous brain structures in extant species, it is in principle possible to use comparative anatomical or genetic analyses to determine whether two species that share a brain structure shared a common ancestor that already had a version of it. Cognitive homologies, then, according to Bergeron, are the "cognitive workings" of homologous brain structures.

> [T]he notion of cognitive homology can be understood as the cognitive workings of a homologous (brain) structure, regardless of how these cognitive workings are being used in different species—the same cognitive workings in different species under every variety of cognitive uses, where "the same cognitive workings" means "the cognitive workings of a homologous brain structure".
>
> (Bergeron, 2021, p. 36)

Bergeron adopts a distinction between "cognitive use" and "cognitive working." To illustrate, consider the difference between the use and the working of a forelimb. The *use* of a forelimb can be different across species—for example, in cats they are for running, in whales they are for swimming, and in bats they are for flying. Still, the forelimbs' *workings* are the same: there are similar bones, muscles, and ligaments in similar relative positions that move and interact in similar ways. "Cognitive workings" of brain structures, for Bergeron, are "the basic cognitive contribution [the brain structure] makes to the various cognitive functions it participates in" (Bergeron, 2021, p. 35), or the basic contributions a brain structure makes to its different uses. According to Bergeron, the basic cognitive contributions of a brain structure can be described in different ways—such as in psychological, computational, or neural terms—as well as in a "domain-specific" or "domain-neutral" way. The preferred description depends on the experimental, theoretical, or clinical contexts. So, for Bergeron, the recipe for identifying cognitive homologies seems to (1) identify a brain structure that is homologous in different species and then (2) determine its "basic cognitive contribution"—that is, what the brain structure contributes to its various cognitive uses in different species.

Bergeron's account can only be successful if we assume that homologous brain regions always make the same basic cognitive contribution. It is unclear why this should be the case, that is, whether it should be impossible that homologous brain regions make different basic contributions to the various cognitive uses that they are involved in. Indeed, there seems to be empirical evidence that homologous brain structures can change their basic cognitive contribution. For example, small structural differences in the internal organization of *nucleus laminaris* (a homologous brainstem auditory region) between barn owls and chickens led to a change in the computational contribution of that brain region, allowing owls to detect interaural time differences with greater accuracy (Striedter, 2002, p. 241). This computational contribution is just the basic cognitive contribution of *nucleus laminaris* which, thus, seems to have changed in owls compared to chickens. Now, Bergeron might argue that the basic cognitive contribution of *nucleus laminaris* indeed did *not* change in owls—if individuated in a sufficiently coarse-grained way—the basic cognitive contribution of *nucleus laminaris* is "detecting interaural time differences," not the specific computational contribution that it makes—and that is still the same in owls and in chickens. This, however, hints at the second problem for Bergeron's account.

The second issue with Bergeron's account is that he does not provide a generalized account of what "basic contributions" might be and how they are to be identified. He provides two examples: the basic cognitive contributions of the amygdala and Broca's area. He suggests that the amygdala's fundamental role is (roughly) "relevance detection," while Broca's area is responsible for (roughly) "translating information into complex motor actions." He argues that there is substantial evidence indicating that structures homologous to the human amygdala and Broca's area exist in non-human animals and that these structures fulfill the same basic cognitive functions. However, a critical question remains: How can we determine that "relevance detection" and "translating information into complex motor action" are indeed the basic cognitive contributions of the amygdala and Broca's area?

Bergeron's approach seems to be similar to Price and Friston's (2005) strategy for determining the function of a brain structure. According to Price and Friston, scientists should ascribe functions to brain regions that explain "all patterns of activation" of that brain region (Price & Friston, 2005, p. 268). To do so, they claim, scientists should look at all tasks that elicit activity in the brain region of interest. Then, scientists should identify what these tasks have in common when that brain region is most active. As an example, Price and Friston set out to analyze the function of the *left posterior lateral fusiform*. They conclude it is not—as many researchers have argued—primarily a "visual word form area," an area "sensitive to the visual attributes of animals," or a "tactile-visual region." Rather, they argue, it is only on the assumption that the left posterior lateral fusiform "integrates sensory cues with motor output" that one can explain all its activations during the different tasks.

In the context of Bergeron's account, one could apply a similar strategy. One could compare the activity of the same (i.e., homologous) brain region in members of different species across similar sets of tasks and ask which basic cognitive contribution should be ascribed to the homologous brain region such that the region's activity is best explained.

The problem that arises for Bergeron's account of cognitive homology, then, is the same problem that has been raised in the context of Price and Friston's account of brain function. Recall that our overall goal is to answer the cognitive ontology question: how to carve up the cognitive domain in ways that promote robust scientific prediction, explanation, and generalization. Even if it were adequate to ascribe the basic contribution of "sensory-motor integration" to a brain region, this ascription is not "cognitively interesting" (Klein, 2012, p. 955). First, it is rather difficult to find a brain region that does *not* perform sensory-motor integration (Klein, 2012). Second, categories like "sensory-motor integration" are too broad to increase the predictive and explanatory power of a scientific model or theory of cognition. Such broad function ascriptions are useless for research and theory building—yet this is exactly what we aim to improve by rethinking cognitive ontology.

Take Bergeron's example of the amygdala. Bergeron writes that "the amygdala can be described as a relevance detector, a structure that contributes to increasing vigilance and attention based on stimulus saliency, ambiguity and unpredictability" (Bergeron, 2021, p. 9). However, the amygdala is also involved in tasks that do not obviously fall under that description. For example, it is also involved in memory consolidation. Ferry et al. (1999) conclude that the amygdala (or rather one of its nuclei) "is part of a neuromodulatory system that regulates the strength of memories in relation to their emotional significance." Here, the amygdala increases neither vigilance nor attention, and it does not react to external stimuli at all. Surely, one could now try to find a more general description that covers all scenarios in which the amygdala is activated. For example, one could argue that the amygdala's basic cognitive contribution is that it "orchestrates the allocation of processing resources (sensorimotor, cognitive, emotional, endocrine) to a given stimulus based on its relevance, or value, to the organism in a particular context" (Bergeron, 2021, p. 10). To capture the role of the amygdala in memory consolidation, one would have to add something like "to an external stimulus *or internal representation.*"

This function description, however, again seems too vague and general to be relevant for a cognitive ontology, similar to "sensory-motor integration." Many brain structures orchestrate the allocation of processing resources to a given stimulus or internal representation based on their relevance or value to the organism in a particular context. For example, the prefrontal cortex (PFC) plays a crucial role in evaluating the relevance of stimuli and making decisions based on the agent's goals. It is heavily involved in orchestrating processing resources (Miller & Cohen, 2001). The basal ganglia are essential for decision-making processes and for selecting actions based on reward and value. The basal ganglia work closely with the PFC to help prioritize which actions or thoughts should be pursued (O'Reilly & Frank, 2006). While the primary sensory cortices are focused on processing specific types of sensory information, they are also involved in allocating processing resources to stimuli that are deemed relevant or valuable (Moore & Zirnsak, 2017). This is often modulated by top-down influences from regions like the PFC. The thalamus acts as a relay station, filtering and prioritizing sensory information before it reaches the cortex, thus playing a role in the allocation of processing resources (Saalmann & Kastner, 2011). The anterior cingulate cortex (ACC) is involved in error detection, conflict monitoring, and assessing the value of different actions, which contributes to the allocation of cognitive and emotional resources (Bush et al., 2000). The hypothalamus is involved in orchestrating the body's response to internal states (e.g., hunger, stress) and ensuring that resources are allocated to maintain homeostasis, which includes both endocrine and autonomic functions (Saper & Lowell, 2014). Thus, the overall process of allocating resources based on relevance and value is a general, distributed function of the brain rather than being isolated to specific regions.

A broader issue with Bergeron's account reflects a common concern about all similarly radical, bottom-up approaches to cognitive ontology: it begins with predefined brain delineations (in this case, based on the homology of brain regions) and then attempts to identify the functions of these regions. Various authors have expressed different critiques of such approaches. Beyond the challenge of ensuring that the resulting categories remain cognitively meaningful, some have argued that the idea of finding one-to-one mappings between cognitive entities and neural entities is illusory (Viola, 2017). If at all, such one-to-one mappings can only be established when accepting that the functions of brain regions are context-dependent (Klein, 2012). Additionally, it has been suggested that brain regions may not be the appropriate neural entities for deriving a cognitive ontology. Instead, cognitive functions might be better understood in terms of neural networks (Klein, 2012), neural mechanisms (Francken et al., 2022; Krickel, 2024), or neural processes (Viola & Zanin, 2017). Notably, neural mechanisms and processes may not be easily decomposable into brain regions as defined by homology.

Both García's and Bergeron's accounts follow the same general strategy to make sense of the notion of cognitive homology. The core of these accounts are different ways of cashing out the notion of "structural sameness" for cognitive entities. García attempts to locate structural sameness in the functional organization of cognitive systems and runs into a regress. Bergeron reduces the structural sameness of "cognitive workings" to structural sameness of brain regions and thereby runs the risk of missing the cognitive.

In the next section, we introduce an account of cognitive homology that is not based on structural sameness of cognitive entities and thus avoids the problems facing these accounts.

4.4 Cognitive ontology in terms of cognitive ChIMs

In this section, we apply DiFrisco, Wagner, and Love's notion of a ChIM to cognitive capacities to develop a novel account of cognitive homology. The core idea is this: If we find the same ChIM producing a cognitive capacity in different species, we can infer that the cognitive capacities are the same. But this raises an important question: what are ChIMs in the cognitive domain?

To introduce the idea of *cognitive ChIMs*, let us first consider an intuitive example. Suppose that you hypothesize that humans' cognitive capacity to form grammatically correct sentences in language is the same as songbirds' cognitive capacity to form correct song patterns (Berwick et al., 2011; Figdor, 2022, p. 11). The capacities seem to share some similarities: for example, both speech and birdsongs have specific, recurring patterns, and both are used in communicative interactions between conspecifics. However, human syntactic capacities and songbird syntactic capacities are also quite different. Human syntax is famously generative: it allows for productivity (an infinite set of grammatically correct sentences can be formed) and compositionality (the elements of a sentence can be recombined to form novel sentences that get their meanings from the meaning of their elements). By comparison, birdsong syntax is far more limited. Furthermore, these cognitive capacities manifest in a different range of situations for the different species: whereas human language (and thus syntax) occurs in a wide variety of social interactions, birdsong is primarily more narrowly involved in specific behavioral contexts, such as mating, deterring rivals, or defending territory (Berwick et al., 2011). So, are the cognitive capacities underlying human syntax and birdsong syntax different phenotypic expressions of the same cognitive trait (e.g., a capacity to produce regularly patterned stimuli for the purpose of communication), or are they different cognitive traits?

According to our account, this question is to be addressed by investigating the mechanisms by which the capacities *develop* in each species. If and only if they are produced by the same trait-determining, causally necessary, and nonredundant developmental mechanism—that is, the same cognitive ChIM—then they are instances of the same cognitive capacity.

In DiFrisco et al.'s original account (2020; see Section 4.2), ChIMs underlie the development of morphological traits and thus largely make their contributions during embryogenesis. In contrast, the mechanisms underlying the development of *cognitive* traits are largely manifest during the organisms' postnatal life and often involve learning. Learning involves robust changes in the neural and behavioral setup of an organism. Thus, cognitive ChIMs will involve neural mechanisms as well as the organisms' patterns of interactions with their respective ecological niches (Quartz & Sejnowski, 1997; Sirois & Karmiloff-Smith, 2009; Anderson, 2014). For example, the typical development of cognitive capacities underlying human syntax and birdsong syntax both depend on what is called *experience-expectant* input from the environment during early life (Greenough et al., 1987), such as exposure to adult speech or birdsong (Louder et al., 2024). They also depend on infants' and birds' own attempted production and practice of language and songs (Johnson & Whitney, 2005; Tchernichovski & Marcus, 2014). Furthermore, it has been found that specific neurons in the auditory forebrain, the caudomedial nidopallium (NCM), are crucial for song learning (Louder et al., 2024). It could be demonstrated that there are compelling computational and physiological similarities between corresponding neuronal cell types in the mammalian cortex (Spool et al., 2021). Thus, cognitive ChIMs involve not only organism-internal neural entities and activities but also species-typical interactions between organisms and their environments.

Determining whether the ChIMs that give rise to the cognitive capacities for linguistic and birdsong syntax are the same would thus involve two steps. First, you must identify the developmental mechanisms in each species on the neural and the behavioral level. Here, this means that you must identify the set of causally necessary and nonredundant biological endowments and learning experiences that cause humans and birds, respectively, to develop their species-specific capacity to produce regularly patterned stimuli for the purpose of communication (talking/singing). Second, you must compare these mechanisms to determine whether they are instances of the same ChIM. That is, you must determine whether the developmental mechanisms that control the development of the capacity (trait identity) in both species have a common mechanistic architecture.

These considerations, along with the general characterizations of ChIMs outlined in Section 4.2, suggest three desiderata for an account of cognitive ChIMs:

- (i) The notion of a cognitive ChIM must take into account that cognitive development rests on neural change.
- (ii) The notion of a cognitive ChIM must account for the fact that many cognitive capacities develop through experience-expectant inputs from early, species-typical interactions with the environment that crucially involve the organism's own *behavior*.
- (iii) Cognitive ChIMs must be *specific enough* to provide a convincing and useful categorization of cognitive capacities, while at the same time

> being *general enough* to be possibly present in different species, to abstract away from the phenotypic specificities of the different expressions of that capacity, and to account for neural plasticity.[6]

Having sketched this preliminary outline of our ontogenetic approach to cognitive homology, we now turn to a more specific possible candidate for a cognitive ChIM as a proof of concept. Here, we consider the developmental mechanism underlying *episodic memory*. Contemporary cognitive science characterizes episodic memory as the capacity to recall specific experiences along with details about the time and place of their occurrence (Tulving, 1993).

Cognitive psychologists Arthur M. Glenberg and Justin Hayes (2016) propose an account for a common mechanism by which episodic memory develops in multiple species. The immediate goal of Glenberg and Hayes' account is to provide an explanation for *infantile amnesia*, the phenomenon that humans cannot remember the first few years of their lives. Interestingly, there is evidence that infantile amnesia occurs in other species, such as monkeys and rodents, as well. Glenberg and Hayes' explanation of infantile amnesia essentially proposes a common developmental mechanism for episodic memory in different species. Our main purpose here is to show how their account could (in principle) be construed as an account of an "episodic memory ChIM."

The starting point for Glenberg and Hayes' account is that species with infantile amnesia have another feature in common: young juveniles do not self-locomote but rather are carried around by their caregivers. Based on this observation, Glenberg and Hayes propose that infantile amnesia results from the fact that "the pre-locomotor infant does not have the opportunity to tune his hippocampal place and grid cells [brain regions associated with spatial representation] to the environment" (Glenberg & Hayes, 2016, p. 3): in order for this to happen, self-locomotion is necessary. Because episodic memory involves memory for places, an organism with an "untuned" hippocampus will lack the capacity to form such memories. Once a juvenile begins to locomote, "[r]eciprocal connections between the hippocampus and sensorimotor and emotional cortices" develop (Glenberg & Hayes, 2016, p. 2).

The explanation of infantile amnesia that Hayes and Glenberg propose relies on the assumption that the cognitive capacity "episodic memory," which is shared by multiple species, is acquired via a specific developmental mechanism that seems to have the role of a ChIM. Assuming species-typical neonate brain anatomy and physiology (i.e., being born with a species-typical hippocampus and sensorimotor cortex), the development of episodic memory is driven by the development of self-locomotion. This proposed mechanism is the same in different species: the development of self-locomotion tunes their hippocampal place and grid cells to regularities in the environment. This mechanism also essentially involves the organism's brain, behavior, and the environment.

A caveat to this "proof of concept" example is that it is incomplete: the "tuning" of hippocampal place and grid cells to regularities in the environment via self-locomotion is likely only necessary rather than sufficient for episodic memory to develop. In addition, the "development of self-locomotion" and "tuning" would also need to be spelled out in far more detail. Finally, it is also very likely that the "tuning" mechanism described is not specific to episodic memory but rather necessary for the development of other cognitive capacities as well. As mentioned above, at least further reciprocal connections between the hippocampus and sensorimotor and emotional cortices must be established to be able to form episodic memories.

However, despite these caveats, the example suggests that there are indeed plausible candidates for cognitive ChIMs. That is, there are developmental mechanisms that underlie the production of cognitive capacities and are shared across species. The example also illustrates the way that a cognitive ontology based on a notion of cognitive homology grounded in ChIMs can be put to good scientific work: the common developmental mechanism both accounts for the common cognitive capacity (episodic memory), and it also explains another phenomenon of interest (infantile amnesia) in a way that generalizes to all altricial species (not just humans).

Providing full details for a general account that satisfies the two desiderata outlined above must be left for future work. However, this example suggests that an ontogenetic account of cognitive homology based on cognitive ChIMs is indeed plausible.

4.5 Conclusion

This chapter developed a proof of concept for a novel account of cognitive homology based on an ontogenetic account of homology from evolutionary-developmental biology. While several previous authors have attempted to shed light on cognitive homology, our account avoids their pitfalls by incorporating insights about *cognitive development*. We propose that cognitive capacities should be individuated by the causally necessary, nonredundant developmental mechanisms that give rise to them—cognitive ChIMs. These mechanisms crucially involve not only brain structures but also behavior and environmental interactions.

We began with the *cognitive ontology question*: How should cognitive scientists "carve up" cognition to generate robust predictions, explanations, and generalizations? Following a growing literature, we suggested that a human-centered cognitive ontology based on outdated folk psychological categories is untenable. We assume that a solution to the cognitive ontology question must incorporate the fact that cognition is the product of evolution. As such, we argue that cognitive categories should, like biological kinds, be individuated in terms of homology: two traits *x* and *y* are the same if and only if *x* and *y* are homologous.

After showing that two existing accounts of cognitive homology run into trouble (due to efforts to characterize *structural similarities* in entities that are functionally individuated), we argued that an account of cognitive homology grounded in ontogenetic mechanisms has the potential to avoid these problems. According to our account, cognitive capacities should be grounded in, and individuated by, cognitive developmental mechanisms: in particular, two cognitive capacities are the same if they are produced by the same cognitive ChIM.

Our account of cognitive ChIMs suggests that, unlike the account of (morphological) ChIMs on which our account is based (DiFrisco et al., 2020), cognitive ChIMs consist in interactions between organisms' brains, behaviors, and environments. Consequently, while investigating the brain is crucial for cognitive ontology, it cannot be the only consideration. On our account, cognitive homologies (and thus cognitive ontology) cannot be constructed by mapping brain *regions* to cognitive capacities. Rather, we must identify developmental mechanisms for cognitive capacities that situate the brain in the context of behavior and environment.

Notes

1 The cognitive ontology question has a metaphysical and an instrumental (epistemological, or philosophy of science) reading. Understood metaphysically, the cognitive ontology question asks how to "carve cognition at its joints," identifying natural kinds that exist independently of our interests. On an epistemic or instrumental reading, the cognitive ontology question asks how to develop categories that best serve our interests, such as increasing the explanatory and predictive power of scientific models and theories. These readings might differ more in emphasis than in content: for example, the explanatory and predictive power of our models and theories will likely increase depending on how well they capture the actual causal structure of the world (Khalidi, 2023). Here, our primary aim is to contribute to the epistemic or instrumental project: we approach the cognitive ontology question as a means for improving the explanations, predictions, models, and theories of the cognitive sciences while being indeterminate on whether we thereby "cut nature at its joints."

2 The notion of a "character" in evolutionary biology is notoriously difficult to define. Roughly, a character is a trait, structure, property, or process that can be "seen" by evolution. The intuition behind this concept can be illustrated with help of an example. Scientists disagree on whether the chin—a forward pointing part of the bone structure below the lower lip—is a character. If chins have been selected for due to an evolutionary benefit, such as resistance against mechanical pressures to the mouth region or a sexual selection advantage, chins are a

character. Alternatively, the chin might be simply a side effect of other evolutionarily selected features, such as a shortening of the breadth of the dental arch (Pampush & Daegling, 2016)—in which case it is not a character.

3 We do not aim to contribute to the debate between proponents of the phylogenetic conception and proponents of the ontogenetic conception. Instead, we follow DiFrisco's (2023) convincing argument that a satisfying account of homology will likely incorporate both conceptions. For present purposes, what is important is that the ontogenetic conception provides an explanation of trait identity that is objective, rooted in causal mechanisms, and allows for detection and tracing of traits in practice.

4 When individuating mechanisms, some degree of vagueness cannot be avoided. For example, two mechanism tokens that produce neurotransmitter release might still be considered the "same type" of mechanism if they involve different amounts of sodium and potassium ions, and they might be considered different mechanisms if one involves ion channels and the other does not (since this would entail different "entities," "activities," and "organization," even if the "phenomenon" of neurotransmitter release were shared; see Section 4.2). In between the two extremes might lie various mechanism tokens for which questions of "sameness" are less clear. In practice, such vagueness is often resolved based on researchers' explanatory interests.

5 García also refers to a distinction between "structured" and "unstructured" systems. A system is structured, according to García, if it is "constituted by functional subsystems," whereas an unstructured system is not so constituted (García, 2010, p. 128). We will ignore this distinction here as cognitive systems are plausibly always or even necessarily "structured" (see, e.g., Weiskopf & Adams, 2015, p. 39).

6 We thank Gualtiero Piccinini's students for highlighting that the account needs to deal with neural plasticity.

References

Anderson, M. L. (2014). *After Phrenology—Neural Reuse and the Interactive Brain*. MIT Press.

Bergeron, V. (2021). Carving the mind at its homologous joints. *Biology & Philosophy*, *36*(4), 36. https://doi.org/10.1007/s10539-021-09812-3

Berwick, R. C., Okanoya, K., Beckers, G. J. L., & Bolhuis, J. J. (2011). Songs to syntax: The linguistics of birdsong. *Trends in Cognitive Sciences*, *15*(3), 113–121. https://doi.org/10.1016/j.tics.2011.01.002

Bush, G., Luu, P., & Posner, M. I. (2000). Cognitive and emotional influences in anterior cingulate cortex. *Trends in Cognitive Sciences*, *4*(6), 215–222. https://doi.org/10.1016/S1364–6613(00)01483–2

Buzsáki, G. (2019). *The Brain from Inside Out*. Oxford University Press. https://doi.org/10.1093/oso/9780190905385.001.0001

Cisek, P. (2019). Resynthesizing behavior through phylogenetic refinement. *Attention, Perception, and Psychophysics*, *81*(7), 2265–2287. https://doi.org/10.3758/s13414-019-01760-1

Craver, C. F. (2007). *Explaining the Brain: Mechanisms and the Mosaic Unity of Neuroscience*. Oxford University Press.

De Brigard, F. (2017). Cognitive systems and the changing brain. *Philosophical Explorations*, *20*(2), 224–241. https://doi.org/10.1080/13869795.2017.1312503

Dewhurst, J. (2021). Folk psychological and neurocognitive ontologies. In *Neural Mechanisms* (pp. 311–334). https://doi.org/10.1007/978-3-030-54092-0_14

DiFrisco, J. (2023, December). Toward a theory of homology: Development and the de-coupling of morphological and molecular evolution. *The British Journal for the Philosophy of Science*, 000–000. https://doi.org/10.1086/714959

DiFrisco, J., Love, A. C., & Wagner, G. P. (2020). Character identity mechanisms: A conceptual model for comparative-mechanistic biology. *Biology & Philosophy*, *35*(4), 44. https://doi.org/10.1007/s10539-020-09762-2

DiFrisco, J., Love, A. C., & Wagner, G. P. (2023). The hierarchical basis of serial homology and evolutionary novelty. *Journal of Morphology*, *284*(1), 1–18. https://doi.org/10.1002/jmor.21531

Ferry, B., Roozendaal, B., & McGaugh, J. L. (1999). Role of norepinephrine in mediating stress hormone regulation of long-term memory storage: A critical involvement of the amygdala. *Biological Psychiatry*, *46*(9), 1140–1152. https://doi.org/10.1016/S0006–3223(99)00157–2

Figdor, C. (2018). *Pieces of Mind—The Proper Domain of Psychological Predicates*. Oxford University Press.

Figdor, C. (2022). What could cognition be, if not human cognition?: Individuating cognitive abilities in the light of evolution. *Biology and Philosophy*, *37*(6), 1–21. https://doi.org/10.1007/s10539-022-09880-z

Francken, J. C., Slors, M., & Craver, C. F. (2022). Cognitive ontology and the search for neural mechanisms: Three foundational problems. *Synthese*, *200*(5), 378. https://doi.org/10.1007/s11229-022-03701-2

Fusco, G. (2022). Serial homology. *Biological Theory*, *17*(2), 114–119. https://doi.org/10.1007/s13752-021-00395-6

García, C. L. (2010). Functional homology and functional variation in evolutionary cognitive science. *Biological Theory*, *5*(2), 124–135. https://doi.org/10.1162/BIOT_a_00036

Glenberg, A. M., & Hayes, J. (2016, January). Contribution of embodiment to solving the riddle of infantile amnesia. *Frontiers in Psychology*, *7*, 1–6. https://doi.org/10.3389/fpsyg.2016.00010

Greenough, W. T., Black, J. E., & Wallace, C. S. (1987). Experience and brain development. *Child Development*, *58*(3), 539–559. https://doi.org/10.1111/j.1467–8624.1987.tb01400.x

Griffiths, P. E. (2007). The phenomena of homology. *Biology and Philosophy*, *22*(5), 643–658. https://doi.org/10.1007/s10539-007-9090-x

James, W. (1890). *The Principles of Psychology*. Henry Holt and Co. https://doi.org/10.1037/10538–000

Johnson, F., & Whitney, O. (2005). Singing-driven gene expression in the developing songbird brain. *Physiology & Behavior*, *86*(3), 390–398. https://doi.org/10.1016/J.PHYSBEH.2005.08.009

Khalidi, M. A. (2023). *Cognitive Ontology*. Cambridge University Press. https://doi.org/10.1017/9781009223645

Klein, C. (2012). Cognitive ontology and region- versus network-oriented analyses. *Philosophy of Science*, *79*(5), 952–960. https://doi.org/10.1086/667843

Krickel, B. (2024). The new mechanistic approach and cognitive ontology—or: What role do (neural) mechanisms play in cognitive ontology? *Minds and Machines*, 1–19. https://doi.org/10.1007/s11023-024-09679-9

Louder, M. I. M., Kuroda, M., Taniguchi, D., Komorowska-Müller, J. A., Morohashi, Y., Takahashi, M., Sánchez-Valpuesta, M., Wada, K., Okada, Y., Hioki, H., & Yazaki-Sugiyama, Y. (2024). Transient sensorimotor projections in the developmental song learning period. *Cell Reports*, *43*(5), 114196. https://doi.org/10.1016/j.celrep.2024.114196

Love, A. C. (2007). Functional homology and homology of function: Biological concepts and philosophical consequences. *Biology and Philosophy*, *22*(5), 691–708. https://doi.org/10.1007/s10539-007-9093-7

McCaffrey, J., & Wright, J. (2022). 14 neuroscience and cognitive ontology: A case for pluralism. In F. De Brigard & W. Sinnott-Armstrong (Eds.), *Neuroscience and Philosophy* (pp. 427–466). MIT Press.

Miller, E. K., & Cohen, J. D. (2001). An integrative theory of prefrontal cortex function. *Annual Review of Neuroscience*, *24*(1), 167–202. https://doi.org/10.1146/annurev.neuro.24.1.167

Moore, T., & Zirnsak, M. (2017). Neural mechanisms of selective visual attention. *Annual Review of Psychology*, *68*, 47–72. https://doi.org/10.1146/ANNUREV-PSYCH-122414–033400

O'Reilly, R. C., & Frank, M. J. (2006). Making working memory work: A computational model of learning in the prefrontal cortex and basal ganglia. *Neural Computation*, *18*(2), 283–328. https://doi.org/10.1162/089976606775093909

Pampush, J. D., & Daegling, D. J. (2016). The enduring puzzle of the human chin. *Evolutionary Anthropology*, *25*(1), 20–35. https://doi.org/10.1002/evan.21471

Poldrack, R. A. (2010). Mapping mental function to brain structure: How can cognitive neuroimaging succeed? *Perspectives on Psychological Science*, *5*(6), 753–761. https://doi.org/10.1177/1745691610388777

Poldrack, R. A., Kittur, A., Kalar, D., Miller, E., Seppa, C., Gil, Y., Stott Parker, D., Sabb, F. W., & Bilder, R. M. (2011, September). The cognitive atlas: Toward a knowledge foundation for cognitive neuroscience. *Frontiers in Neuroinformatics*, *5*, 1–11. https://doi.org/10.3389/fninf.2011.00017

Price, C. J., & Friston, K. J. (2005). Functional ontologies for cognition: The systematic definition of structure and function. *Cognitive Neuropsychology*, *22*(3–4), 262–275. https://doi.org/10.1080/02643290442000095

Quartz, S. R., & Sejnowski, T. J. (1997). The neural basis of cognitive development: A constructivist manifesto. *Behavioral and Brain Sciences*, *20*(4), 537–596. https://doi.org/10.1017/S0140525X97001581

Saalmann, Y. B., & Kastner, S. (2011). Cognitive and perceptual functions of the visual thalamus. *Neuron*, *71*(2), 209–223. https://doi.org/10.1016/J.NEURON.2011.06.027

Saper, C. B., & Lowell, B. B. (2014). The hypothalamus. *Current Biology: CB*, *24*(23), R1111–R1116. https://doi.org/10.1016/J.CUB.2014.10.023

Sirois, S., & Karmiloff-Smith, A. (2009). Ontogenetic development matters. In *Cognitive Biology* (pp. 321–334). The MIT Press. https://doi.org/10.7551/mitpress/9780262012935.003.0293

Spool, J. A., Macedo-Lima, M., Scarpa, G., Morohashi, Y., Yazaki-Sugiyama, Y., & Remage-Healey, L. (2021). Genetically identified neurons in avian auditory pallium mirror core principles of their mammalian counterparts. *Current Biology*, *31*(13), 2831–2843.e6. https://doi.org/10.1016/J.CUB.2021.04.039/ATTACHMENT/D39BC6B3-6587-4FEC-8676-70BEB6B01E7B/MMC2.MP4

Striedter, G. F. (2002). Brain homology and function: An uneasy alliance. *Brain Research Bulletin*, *57*(3–4), 239–242. https://doi.org/10.1016/S0361–9230(01)00692-X

Tchernichovski, O., & Marcus, G. (2014). Vocal learning beyond imitation: Mechanisms of adaptive vocal development in songbirds and human infants.

Current Opinion in Neurobiology, *28*, 42–47. https://doi.org/10.1016/J.CONB.2014.06.002
Tulving, E. (1993). What is episodic memory? *Psychological Science*, *2*(3), 67–70.
Vanderwolf, C. H. (2007). The evolving brain: The mind and the neural control of behavior. *The Evolving Brain: The Mind and the Neural Control of Behavior*, 1–104. https://doi.org/10.1007/978-0-387-34230-6/COVER
Viola, M. (2017). Carving mind at brain's joints. The debate on cognitive ontology. *Phenomenology and Mind*, *0*(12), 162–172. https://doi.org/10.13128/Phe
Viola, M., & Zanin, E. (2017). The standard ontological framework of cognitive neuroscience: Some lessons from Broca's area. *Philosophical Psychology*, *30*(7), 945–969. https://doi.org/10.1080/09515089.2017.1322193
Wagner, G. P. (1989). The biological homology concept. *Annual Review of Ecology and Systematics*, *20*, 51–69. https://www.jstor.org/stable/2097084
Wagner, G. P. (2007). The developmental genetics of homology. *Nature Reviews Genetics*, *8*(6), 473–479. https://doi.org/10.1038/nrg2099
Weiskopf, D. A., & Adams, F. (2015). *An Introduction to the Philosophy of Psychology*. Cambridge University Press.
Yeo, B. T. T., Krienen, F. M., Eickhoff, S. B., Yaakub, S. N., Fox, P. T., Buckner, R. L., Asplund, C. L., & Chee, M. W. L. (2015). Functional specialization and flexibility in human association cortex. *Cerebral Cortex*, *25*(10), 3654–3672. https://doi.org/10.1093/cercor/bhu217
Zagaria, A., Ando', A., & Zennaro, A. (2020). Psychology: A giant with feet of clay. *Integrative Psychological & Behavioral Science*, *54*(3), 521–562. https://doi.org/10.1007/S12124-020-09524-5

5 Representational vehicles, from regions to cells

Adina L. Roskies

A group of blind men heard that a strange animal, called an elephant, had been brought to the town, but none of them were aware of its shape and form. Out of curiosity, they said: "We must inspect and know it by touch, of which we are capable". So, they sought it out, and when they found it they groped about it. The first person, whose hand landed on the trunk, said, "This being is like a thick snake". For another one whose hand reached its ear, it seemed like a kind of fan. As for another person, whose hand was upon its leg, said, the elephant is a pillar like a tree-trunk. The blind man who placed his hand upon its side said the elephant, "is a wall". Another who felt its tail, described it as a rope. The last felt its tusk, stating the elephant is that which is hard, smooth and like a spear.

—*Ancient Indian parable*

5.1 Representation

Representation talk is rife in the philosophy of mind literature. Computational theories of mind are committed to the notion of representation (Egan, 2018; Fodor, 1980). Classical computational theories of mind are realist theories about computation, holding that computations are operations over representations. Such theories are also, *a fortiori*, realist about representations. For example, Fodor's Representational Theory of Mind holds that mental states such as propositional attitudes are constituted in part by the tokening of symbols that are part of a system of representation and whose tokening is dependent upon formal properties which correlate with semantic properties of the things represented. In contrast, other positions consistent with a computational approach eschew literalism about symbols, positing that mental representations are likely to be distributed patterns of neural activity across nodes in a network, where the semantic value of these nodes is subsymbolic (Churchland, 1989; Clark, 2014). Still others are reluctant to take clear sides in the realism debate, opting instead to view representation talk about mind or brain as an explanatory or pragmatic strategy aimed at understanding various practices, for instance, the

DOI: 10.4324/9781003458531-5

attribution of folk psychological states and use of these in explanation and prediction of behavior (Dennett, 1987), or as a way of describing the scientific practices of seeking and explaining biological phenomena in terms of representations (Egan, 2014, 2018; Cao, 2022). Cao, for example, says:

> What makes something a representation just is that we can identify and re-identify it as such, and moreover, that we can manipulate it effectively to do whatever it is that we think representations ought to do for us in that particular context.
>
> (Cao, 2022, 151:1)

In order to identify representations, Cao introduces the notion of a probe, which is anything that helps us pick out a representation from the enormous complexity of causal interactions that generate behaviors of interest.

As such, according to Cao, a representation is probe-relative, a pragmatic construct whose reidentification is, in some respects, in the eye of the scientist. Egan, on the other hand, is a realist about representational vehicles and deflationary about representational content (Egan, 2020). Regardless of one's stance on the question of realism in the brain or mind, there is widespread consensus that invoking representations is warranted only if one can satisfy a number of commitments.

Classical views of representation differentiate between representational vehicles and representational content (Egan, 2020). The representational vehicle is the physical entity (state or structure) upon which causal processes of computation operate and which carries the representational content. Such a distinction was important in the context of early attempts to understand mental representation, in which it was thought that mental processes were operations over symbolic tokens that could be deployed and redeployed in different combinations in order to support inferences, linguistic behavior, etc. The distinction in more contemporary thinking appears less important, but the notion of a vehicle still plays a role in another criterion for vindicating representational ascription: causal importance. In order to make an argument for the existence of a representation, that representation must have a causal role in the generation of behavior (Shea, 2018). Ultimately, it must play a role in a system of representations that has a function for the organism. Thus, a putative representation must be used, transformed, or interpreted by the system. Moreover, with regard to content ascription, there must be a story of how representational vehicles get their contents, one that is naturalizable. Finally, a representational story should allow for a possibility of misrepresentation: there should be, at a minimum, norms for content ascription (see Egan, 2018; Cao, 2022).

In previous work (Roskies, 2021), I have argued that functional neuroimaging provides evidence for representation in the brain, that neuroimaging data gives us insight into representational structures at a macro level, and that there is indication that the desiderata for representations are satisfied or satisfiable. However, I have argued that because the inferences about content are indirect, brain imaging falls short of providing a clear demonstration of representations in the brain. More specifically, on almost any conceivable theory, the vehicles of brain representation supervene on neural activity. While the fMRI signal is dependent on neural activity, it is only indirectly so. First, the relation between neural activity and fMRI signal involves a number of indirect inferences (Roskies, 2008, 2010). Most functional MRI experiments use the Blood Oxygen Level Dependent (BOLD) signal, which measures the ratio of deoxygenated to oxygenated blood in a spatial region. This ratio is related to neural activity in a complicated way that depends on many factors.[1] More importantly, the information-carrying elements in neuroimaging are voxels, and voxels do not play a role in causal explanations of the transitions between neural states. Voxels are merely proxies for representational vehicles, and thus structures that appear in voxel-wise analyses are suggestive of representations but fall short of meeting the causal requirements for representations. Finally, because of the indirect relationship between the BOLD signal and underlying neural activity, the causal contributions to the computational transformations in the underlying neural substrate cannot be adequately inferred. Consequently, the deliverances of neuroimaging identify only proxy representations.

One may suspect that the reason we are finding only proxy representations is that the methods we use yield information too far removed from the actual causally relevant vehicles, the neurons themselves. If so, the caveats in Roskies (2021) are perhaps peculiarities of neuroimaging techniques. Since the BOLD signal is so far removed both in spatial and in temporal resolution from the relevant neural representations themselves, we are only able to access neural representations' "shadows."

If so, perhaps more direct techniques can give us access to the actual vehicles of brain representations. In particular, recordings from individual cells in the cortex seem well suited to give precise information about the firing patterns and response properties of the neurons ostensibly doing the representing. Thus, let's examine a case in which single-cell recordings allow not only a plausible understanding of what the neurons are representing but one which allows for prediction and manipulation of the response of individual neurons. This may be currently the best scenario for finding representations in the brain.

5.2 A case study: face and object recognition in monkey inferotemporal cortex

Doris Tsao and colleagues' groundbreaking series of studies explore the basis of visual recognition in the monkey cortex (Bao et al., 2020; Bao & Tsao, 2018; Chang & Tsao, 2017). The first studies investigated neural activity in small regions in the inferotemporal (IT) cortex of macaque monkeys with a high proportion of neurons that are sensitive to faces (Tsao et al., 2006). These regions appear to be the monkey homolog of regions in the human brain which are important for facial recognition. Much previous work had suggested that these "face patches" (identified as anterior medial [AM], middle fundus [MF], and middle lateral [ML]) were involved in coding properties of faces, and it had even been suggested that neurons in one of them, AM, encoded facial identity (Freiwald & Tsao, 2010).

In these experiments (Chang & Tsao, 2017), Tsao and colleagues recorded from monkeys while they were shown visual stimuli of faces. The stimulus set consisted of a set of parametrized face stimuli created in the following way. The researchers began with a face image database and averaged those faces to compute an average face shape. Landmarks were placed on each face stimulus at canonical locations, and for each face a transform function was computed to characterize the difference between each individual face and the average, yielding a function that warps each face into the average shape. Once shape-based differences are accounted for, the information that remains to distinguish faces from each other can be thought of as "shape-free" appearance information. Both the shape transform and the appearance image can be treated as mathematical descriptors. Principal component analysis (PCA) was performed on the library of shape-transform and shape-free descriptors separately, and the first 25 principal components from each PCA were used to define a 50-dimensional "face space." Finally, 2000 points were randomly sampled in this space to construct a collection of face stimuli, each fully specified by 50 principal components. Images reconstructed from these components were shown to awake monkeys while single cells were recorded in face patches, areas of cortex that are known to show high selectivity for faces.[2]

A summary of their findings while recording from individual neurons follows. Different neurons showed different firing profiles for each stimulus, and it was found that most neurons were sensitive to a small number (approximately 6–7) principal components of the face space. In order to characterize the stimulus sensitivity of individual neurons, a spike-triggered average (STA) face was computed for each neuron: this is the stimulus that would most effectively cause that neuron to fire. To determine the STA stimulus, each stimulus image shown to the monkey (which by design is encoded by a 50D vector) was scaled by the degree to which it caused the neuron to fire. These scaled vectors were averaged across the corpus, thus producing a stimulus image with component axes scaled to reflect their importance in activating that particular neuron. Analysis of the STA face stimuli across different areas revealed different functional properties. Tsao and colleagues found that in the ML/MF face patch, most neurons were tuned primarily to shape dimensions, whereas in AM, neurons were more tuned to appearance dimensions, comporting with previous findings that ML/MF neurons were tuned to viewpoint and AM neurons to face identity (Freiwald & Tsao, 2010).

In summary, the neuronal populations in the face-sensitive areas of monkey IT span a high-dimensional representational space (Kriegeskorte & Kievit, 2013; Poldrack, 2021), and at least some of the dimensions of this space constitute a "face space." From the neural data, Tsao and colleagues inferred that a cell's activity can be roughly approximated by a linear combination of dimensions in face space and that therefore it should be possible to linearly decode a presented stimulus from the population response. Indeed, they showed over 80% accuracy for decoding individual faces in the test set from sampling responses of only 200 cells (Chang & Tsao, 2017). Thus, relatively small populations of neurons suffice to encode the shape and appearance features of arbitrary face stimuli.

Given the above results, we can describe individual neurons as tuned to an axis (a direction) in a multidimensional face space. Each neuron is sensitive to a direction in this face space, and the approximate direction in this face space can be empirically determined by computing the neuron's spike-triggered average face (see above), the face image that will maximally activate that cell. For any incoming stimulus, the neuron will respond to the projection of the face image along its STA (or, in other words, compute a dot product between the direction defined by its STA and a particular face stimulus). While the increasing sensitivity to faces along a neuron's STA axis is guaranteed by the methods of determining the STA, this model makes a nontrivial prediction that can be empirically tested: if the axis theory is correct, then changes to a stimulus that lie on the hyperplane orthogonal to the neuron's STA will not change the neuron's firing rate, despite significant differences in appearance of the face stimulus. Thus, a cell's response should not change with a stimulus change corresponding to a vector orthogonal to the cell's STA.[3]

In order to provide an intuitive way to think about this result, consider the space of acoustic stimuli. Frequency (pitch) and amplitude (loudness) are two orthogonal axes in that space, and we can build detectors that are sensitive to one axis of the space and not another. One could manipulate a stimulus's pitch without affecting the response of a volume detector and vary the loudness of a stimulus without affecting the reading on the pitch detector. Axis-tuned face cells operate in a similar way: they are sensitive to a direction in face space (albeit one that might be rotated with respect to the PCA-derived axes) and insensitive to variations in dimensions orthogonal to that axis. Thus, if an axis coding model of face-sensitive cells was correct, it would predict that alterations of a stimulus along face-space dimensions orthogonal to the cell's STA axis should not change the cell's response.

This is exactly what researchers found. They constructed faces that differed from the ones presented along dimensions of the constructed face space orthogonal to the neuron's STA axis. Presenting these images to the monkey did not alter the neuron's firing rate, whereas non-orthogonal image changes did lead to predictable changes in neuronal firing. Thus, the hypothesis that the neurons are tuned to axes in a multidimensional face space enabled the deliberate construction of multiple face images (probes) that are visually dissimilar to the originally presented images yet did not alter the firing of the neuron. Other neurons with different STAs did change their firing, as expected. In other words, the theoretical understanding of that neuron's response properties enabled both the targeted manipulation of the neurons and the prediction of the way in which they would or would not respond to that manipulation.

The face patches in the inferotemporal (IT) cortex provide a constrained domain in which to investigate neural coding of visual stimuli. In further studies, Tsao's lab applied similar techniques to object recognition, which is widely thought to be a general function of IT, the larger region within which the monkey face patches reside (Bao & Tsao, 2018). Although various subregions of IT have been mapped to various categories, including faces, much of IT has been difficult to characterize. In exploring object recognition, Tsao and colleagues made use of the deep learning networks that have revolutionized image recognition on the web and whose layers have been found to have response properties like those of cells in various parts of the visual cortex (Yamins et al., 2014; Cadieu et al., 2014). Using similar techniques to characterize a putative object space, Tsao derived the first 50 principal components characterizing a deep layer of AlexNet, a deep learning network for object recognition. The first two principal components of this object space corresponded to shape-related features (that she has characterized as spikiness-stubbiness) and animacy-related features (animate-inanimate). Tsao showed that these two features predict the topographic layout of many of the parts of IT that have characterizable functions, such as selectivity for faces (i.e., subsuming the areas in which her original studies were done), for hands, and for bodies. In addition, she was able to use the principal components to generate visual stimuli of nonexistent objects with visual characteristics that she hypothesized would activate regions of IT that had not yet been functionally characterized. As predicted, Tsao and colleagues showed activity in those areas of IT that they predicted would be activated with the nonsense stimuli, organized according to the first two principal components of the object space. Thus, they describe IT cortical surface topography as organized according to the first two principal components of the object space.

This brief and somewhat simplified description of the work reported by Tsao and colleagues is sufficient to get across the main takeaways from their body of work (for more details and to understand the finer points and many other controls that the lab did, I recommend looking at the original papers). To summarize, first, a cortical area at an advanced stage in the visual processing hierarchy (that, on the basis of many studies of physiology and behavior, appears to be important for object recognition) appears to encode a multidimensional visual object space. IT comprises a general coding scheme for both faces and objects, and the general topography of the surface IT is organized in two dimensions according to the first two principal components describing variation across the visual features of objects. Because of featural similarities, various subregions are particularly sensitive to classes of stimuli, such as faces, animals, body parts, and artifacts. Thus, on the scale of millimeters to centimeters, IT encodes a representational geometry that is spatially arrayed. Second, the coding schemes for objects and faces are analogous, in that neurons are tuned to axes in a multidimensional space that spans the dimensions of variation found in the world of natural visual stimuli. Evidence suggests that there are cells that are sensitive to combinations of visual features that can exist in that space, even if nothing in the natural world embodies characteristics representative of that part of object space. What differentiates one region of IT cortex from another are the dimensions represented. That cells with different constellations of representational properties should be spatially segregated from other groups of cells with other tuning is perhaps not surprising. Aside from biological reasons to keep neurons involved in the same task together (e.g., energetic reasons, timing reasons, to name a few), it has been demonstrated that networks with different tasks naturally segregate and that performance on two tasks is superior with a segregated network versus a network trained to perform both with the same hardware (Dobs et al., 2022). Indeed, recent deep network models in which a cost function is imposed for spatial distance develop topography and response properties that better model cortical structure and organization than competing models (Margalit et al., 2023).

5.3 Lessons for neural representation

Thus far, I have talked about the neurons and regions in IT representing dimensions in face and object space without much defense. I have taken the ability of Tsao and colleagues to characterize the firing of neurons in these areas along axes in a space characterized by the first *n* PCs of the incoming image space and to predict on that basis not only how various actual stimuli will activate the cells but also how to design stimuli to activate cells unresponsive to normal stimuli or to alter stimuli in a preferred class in ways that won't affect responses of individual cells to indicate that they have a grasp on both the actual and counterfactual properties of the cell's representational scheme. But, the critical reader may ask, why am I licensed to talk about this as representational at all?

As you will remember, I began with the hypothesis that single-cell data could give us a handle on neural representations that was not available to us with coarse-grained measures, like fMRI. I have argued that recordings from single cells enable us to understand much about the visual features that individual neurons encode. Indeed, on the face of it, it seems that we can easily speak of the neurons (or perhaps the neuronal firing rates) as the vehicles and the content being those axes in PC-face space that they most respond to. The neurons thus represent those axes in PC-face-space.[4]

However, we quickly run into a problem. If a neuron is a vehicle, then what it represents is a low-dimensional set of visual features, but one that does not map happily into any object in the world. (For example, each neuron in one of the face patches represents the image encoded by its STA, but that is a low-resolution, face-like entity composed of a handful of dimensions in face space, not a particular face.) There is no object in the world that is picked out by a single STA. Other neurons, such as those in hard-to-activate areas of IT, are only activated by visual stimuli specifically designed to activate their particular collection of feature sensitivities, so if they represent, they represent nonexistent objects, another indication that there is not any easy relationship between purported single-cell vehicles of representation and objects in the world.

In order to draw a closer connection to how neurons can represent real objects, it makes more sense to think not of single neurons but of populations of neurons as the representational vehicle. Recall that Tsao et al. demonstrated that an individual image (out of a set of 2000) can be identified or decoded from the activity of on the order of 200 neurons from the face patches, so it may make more sense to think of ensembles of neurons as representational vehicles, with the population that is activated by a particular face as encoding the facial characteristics of that face, or even perhaps face identity. Moreover, any given population instantiating a different pattern of firing may represent a different face stimulus, so a representation's identity may depend not just on a collection of neurons but on their particular activity levels. But then one can ask, which neurons make up that population? Is it the subset of cells strongly activated by a particular face probe? Is it the subset of cells active above baseline or at all? Or is the *entire population* of cells that represent face space relevant to the representation of a face, since the inactivity of some of those cells (e.g., those representing features of face shape or appearance that are *not* exhibited by some stimulus) also could provide information regarding the stimulus to downstream neurons and thus may also be considered part of the vehicle? In both theoretical and practical terms, it is difficult to clearly define what subset of neurons acts as a vehicle for a given content.

The difficulty in identifying vehicles of representation also spells problems for the notion of *redeployment*. If vehicles are epistemically indeterminate, how can we say whether the same vehicle is redeployed at a later time? It does not seem to make sense of this set of neurons as portable or redeployable in other contexts as the classical view of the mind as a formal symbol system held. The classical notion of redeployment literally allowed the same formal symbolic representation to play a role in different parts of the cognitive system so that vehicles could move and multiply with impunity. There is no way in which neural populations can move or multiply, but they can reactivate in different processing streams or different (or similar) circumstances—that is, they may redeploy with the same or similar stimulus, or in recollection, etc., but they are not portable as a symbol which may be tokened in various contexts may be. Thus, redeployment of neural populations must mean only reactivation, but without being able to delineate the vehicle, redeployment can only be used in a metaphorical sense. The idea of vehicles which "carry" content and can move or be redeployed in different parts of the system may also be an outmoded way of approaching neural representations. In modern neuroscience, the conception of vehicles of representation must be altered (see, e.g., Burnston, 2021; Shea, 2013, 2018). For example, the distinction between vehicles and their realization may be elided, and the importance of determining strict identity conditions is much diminished. The difficulties in distinguishing between vehicles and their realizations also put pressure on the autonomy of the psychological level of explanation (Drayson, 2018). For neuroscience, what really matters is the functional contribution of a group of neurons to a capacity or behavior.[5] Even if we have not identified the vehicles of representation through this work, we have learned something significant about the format of representations in these brain areas (Coelho Mollo & Vernazzani, 2024).

Understanding this contribution brings us to the importance of causal relevance. As mentioned, Chang and Tsao's (2017) ability to design a stimulus that *doesn't* change the response of a face-sensitive neuron is a kind of null intervention that demonstrates understanding of the individual neural coding scheme, but it does not show that the active neurons are important for face perception. Showing this kind of causal relevance requires bringing in other evidence than the evidence adduced thus far. Establishing causal import is generally not easy to do in human neuroscience, as intervention is the primary way of demonstrating causation, and few researchers have warrant to intervene in subjects' brains. Even in nonhumans this requirement is not trivial to meet. In the studies discussed so far, Tsao only recorded from neurons while monkeys passively viewed stimuli, and thus her data cannot demonstrate that the neurons she recorded from make a behavioral difference to face perception. However, there is ample evidence that face-sensitive areas of IT cortex are behaviorally relevant. In other studies, Tsao performed interventions by microstimulating monkey face areas and has shown that there are behavioral changes in a delayed match-to-sample task, with response profiles suggesting that perception of face identity is affected by stimulating these neurons (Moeller et al., 2017). Moreover, there is evidence in humans that interventions in the face fusiform area (FFA), a homologous cortical region in humans, selectively affect face perception. Parvizi et al. (Parvizi et al., 2012) had the opportunity to stimulate the cortex of a patient with an epileptic focus in the human fusiform gyrus, which, as in the monkey IT, contains face-selective areas. In the paper reporting this case study, Parvizi et al. describe the response of the patient to stimulation of the fusiform face area as he looks at various things in his hospital room. The patient reports no changes to his perceptual state when he looks at a variety of inanimate objects while two electrodes on an electrocorticography (eCOG) array are used to stimulate regions of the fusiform gyrus. However, when he is looking at a face and his FFA is stimulated, he describes a dynamic morphing of the facial characteristics of the person he is looking at, as if their identity is changing. This effect was specific to faces and only occurred in electrical stimulation trials, not in sham stimulation trials or with stimulation at a nearby location in the cortex (Parvizi et al., 2012). This provides direct evidence of the causal import of FFA activity on face perception specifically. The ability to intervene in these regions and alter face perception in monkey and human is a demonstration that the activity in this small region of the cortex is causally efficacious in the perception of faces, and the finding that stimulation only affects face perception, but not perception of other objects[6] suggests that the function of these areas is quite specific.

However, relying on this information assumes that it is legitimate to bridge between studies, between subjects, and, in this latter case, between species. If we had clear ways of identifying vehicles, this could be done in a principled way. Without an adequate account of vehicles, our inferences are far looser. Are we justified in making them? Perhaps given the totality of evidence we are, though it behooves us to be clear about how shaky these inferences are, with respect to the identity of vehicles. We do not know for certain that the cortical electrodes used in humans were in precisely homologous areas as those identified by Chang and Tsao (2017) in monkey cortex. Even less do we have evidence that the stimulating electrodes were perturbing the same neuronal populations that were measured in the Tsao face coding study, as we lack a clear notion of what "sameness" is in this case. Nonetheless, broadly speaking, given what we know of cross-species functional homology in face-processing regions, it is reasonable to assume that the causal information from the intervention studies can inform us of the causal relevance of the populations recorded from in Tsao's study of face encoding in monkeys.

These intervention studies also give us a handle on misrepresentation. One could argue that what is essential is that it is important that any account of neural representation makes it possible for *an organism* to misrepresent its environment, providing room for slippage between representations and the thing being represented somewhere along the entire processing pathway (and perhaps not discretely), thus lessening the burden of providing an account of misrepresentation for any putative representation. Interference with the normal activity elicited by visual stimuli via microstimulation causes people and monkeys to misperceive faces and face-like stimuli, preventing them from performing simple visual tasks, and in humans, altering their self-reported perceptions. This seems like a clear case of misrepresentation. Moreover, as I and others have argued, a sufficient condition for making sense of misrepresentation is being able to ascribe norms for content ascription, norms that may be flouted.

The account I have developed so far provides an adequate basis for establishing such norms. For example, the layout of IT cortex involves topographic mapping of various representational properties onto the 2D surface of cortex, and the complex but orderly mapping of other kinds of properties into various cortical substructures (Bao et al., 2020; Margalit et al., 2023). These regularities can act as norms. For example, as there is evidence for sparse coding in IT cortex, it is likely that categorical distinctions are made by discerning clusters of activity in relatively densely populated subregions of these sparsely populated high-dimensional representational spaces. Thus, projection to or readout of neuronal activity from part of the map may be taken to indicate categorical information about stimuli, occasionally in cases in which they don't in fact apply. Thus, features in a cloudscape may activate places in feature space congruent with face representations, leading to a percept of a face in a cloud or the judgment that one sees a cow in poor light conditions when the object one sees is a horse. Because the transformations in processing occur over multiple layers in a complex processing hierarchy, there is plenty of room for actual signal or noise to result in the misrepresentation of sensory qualities or the abstraction of information that turns out to be inaccurate. Indeed, models of the ventral visual stream that recapitulate many measurable properties of the cortex perform well on recognition and categorization tasks but occasionally fail to properly categorize particular inputs. Although there is evidence that such models sometimes fail on stimuli that are unlike those that cause biological systems to fail (i.e., so-called adversarial images), it remains to be seen whether human cortex is different in not being subject to such adversarial stimuli or whether we have not developed methods for finding or constructing the relevant stimuli for the particular neural networks we are.

Modern neuroscience has been deeply influenced by work in artificial neural networks (ANNs), and rightly so. Whether it needs to preserve, or would benefit from largely preserving, the notion of vehicles of representation identifiable at an intermediate level between content and realization/implementation remains to be seen. Thinking about representation in the context of high-dimensional vector spaces seems to be fundamentally different from the way in which classical views of representation have approached the problem. It is as yet unclear whether these approaches can or should be ultimately reconciled, whether the classical framework should be altered or abandoned, or whether neuroscience should adopt some version of a deflationary or antirealist position on neural representation. Here I just try to illustrate one element of the traditional framework that seems to fit poorly with our best neuroscience.

5.4 Conclusion

Discussions of representation have traditionally distinguished representational vehicles and representational content. Giving an account of mental representation has been a goal of cognitive neuroscience that has been surprisingly challenging to cash out, at least in classical terms. I have argued that some neuroscientific techniques, such as functional MRI, are suggestive of neural representations, yet due to the nature of the signals they measure, they do not provide access to the vehicles of neural representation (Roskies, 2021). Here I have explored whether more fine-grained brain-measurement techniques fare better. I have argued that while recording from individual neurons allows more direct access to the causal substrate of purported neural representations, the vehicles of representations remain discomfitingly indeterminate. So whereas functional neuroimaging enables us to identify proxy representations, single-cell studies allow us to directly probe the neural representations, but only in the way that the proverbial blind man comes to know the elephant: in any given experiment, we are in contact with only a tail or wrinkle or whisker of the representation we seek to characterize. Granted, over many experiments we can make inferences about the properties of the representation as a whole and about the kinds of transformations or computations that take place in various parts of a processing hierarchy, such as between stages at which face appearance and face identity across different views appear to be computed, that is, at the point at which identity information can be read out by a linear transformation of the neural activity. What is not feasible is to delineate the entire population of neurons that constitute a particular vehicle, or, for that matter, to say whether there is any fact of the matter about what constitutes a vehicle. In contrast to some philosophers who have argued that the notion of representational content in the brain is suspect but accept realism about representational vehicles (Egan, 2020), this discussion problematizes the notion of representational vehicles in the brain (see also Burnston, 2021), as well as recent attempts to support classical representational theories such as the Language of Thought (Quilty-Dunn et al., 2023). Still, while neuroscience continues to put pressure on some of the features of representations, it does not do away with the utility of representational talk. Indeed, the ability to understand the tuning and division of labor of cells in IT provides novel ways to conceptualize brain function, ways that are increasingly aided by novel and more powerful modeling techniques. Whether the indeterminacy of neural representational vehicles should push us toward antirealism about brain representation, or whether it should prompt us instead to rethink traditional notions of representation themselves, remains to be determined.

Notes

1 For example, the BOLD signal reflects neural activity, but it is not closely coupled with it. It is delayed in time and smeared in both time and space; it is influenced by aspects of the vasculature and can produce artifacts; it does not allow one to distinguish excitatory from inhibitory neural activity; it does not support precise quantitative inference about underlying neural activity, although it does vary parametrically with it.

2 Some other manipulations were done to compute viewpoint sensitivity, but I will not go into the details of that here. For more details on this, see Chang and Tsao (2017). To sum up, there is indication that shape-sensitive neurons in ML/MF project to view invariant neurons in AM. In other data reported in this chapter, Tsao et al. report that neurons in AM show highly correlated responses to the same face on appearance dimensions, to frontal and profile faces. They were able to decode faces independent of head orientation on the basis of AM cell activity, suggesting that AM cells project faces in different orientations onto corresponding appearance axes.

3 You might ask on what grounds the face changes in appearance if there is no change in the firing of the cell. The answer is that our perception that the image is different is due to the firing of many cells in face-sensitive visual areas, and most of those cells have STAs with different directions in face space. So the difference in appearance is encoded by the entire population, but not the cells for which the changes are in the null space.

4 It bears mentioning that the PC face space identified by Tsao is a human analytical invention designed to characterize neural properties or a particular *probe* in Cao's (2022) sense. To say that the representational space is 50d is only an approximation; we do not know the actual dimensionality of face or object space

in the brain (or if there is a precise answer to that question), and our characterization of the response properties is only relative to a particular way of measuring that space, that is, the construction of our stimulus set (another probe, a la Cao). Moreover, any set of basis functions that spans the space could do the job. Whether the PCs are the dimensions that best characterize the properties of neurons remains to be seen.

5 There are other ways of conceiving of brain function, such as predictive processing views, that may more easily accommodate notions of representation without relying on a clear vehicle/content distinction.

6 Except for broadly face-shaped objects, like apples (see Moeller et al., 2017).

References

Bao, P., She, L., McGill, M., & Tsao, D. Y. (2020). A map of object space in primate inferotemporal cortex. *Nature*, *583*(7814), 103–108. https://doi.org/10.1038/s41586-020-2350-5

Bao, P., & Tsao, D. Y. (2018). Representation of multiple objects in macaque category-selective areas. *Nature Communications*, *9*(1), 1–16. https://doi.org/10.1038/s41467-018-04126-7

Burnston, D. C. (2021). Contents, vehicles, and complex data analysis in neuroscience. *Synthese*, *199*(1), 1617–1639. https://doi.org/10.1007/s11229-020-02831-9

Cadieu, C. F., Hong, H., Yamins, D. L. K., Pinto, N., Ardila, D., Solomon, E. A., Majaj, N. J., & DiCarlo, J. J. (2014). Deep neural networks rival the representation of primate IT cortex for core visual object recognition. *PLoS Computational Biology*, *10*(12), e1003963. https://doi.org/10.1371/journal.pcbi.1003963

Cao, R. (2022). Putting representations to use. *Synthese*, *200*(2), 151. https://doi.org/10.1007/s11229-022-03522-3

Chang, L., & Tsao, D. Y. (2017). The code for facial identity in the primate brain. *Cell*, *169*(6), 1013–1028.e14. https://doi.org/10.1016/j.cell.2017.05.011

Churchland, P. M. (1989). *A Neurocomputational Perspective*. MIT-Press.

Clark, A. (2014). *Mindware: An Introduction to the Philosophy of Cognitive Science* (2nd edition). Oxford University Press.

Coelho Mollo, D., & Vernazzani, A. (2024). The formats of cognitive representation: A computational account. *Philosophy of Science*, *91*(3), 682–701. https://doi.org/10.1017/psa.2023.123

Dennett, D. C. (1987). *The Intentional Stance*. MIT Press.

Dobs, K., Martinez, J., Kell, A. J. E., & Kanwisher, N. (2022). Brain-like functional specialization emerges spontaneously in deep neural networks. *Science Advances*, *8*(11), eabl8913. https://doi.org/10.1126/sciadv.abl8913

Drayson, Z. (2018). The realizers and vehicles of mental representation. *Studies in History and Philosophy of Science Part A*, *68*, 80–87. https://doi.org/10.1016/j.shpsa.2018.01.005

Egan, F. (2014). How to think about mental content. *Philosophical Studies: An International Journal for Philosophy in the Analytic Tradition*, *170*(1), 115–135.

Egan, F. (2018). The nature and function of content in computational models. In M. Sprevak & M. Colombo (Eds.), *The Routledge Handbook of the Computational Mind*. Routledge.

Egan, F. (2020). A deflationary account of mental representation. In *What are Mental Representations?* Oxford University Press. https://drive.google.com/file/d/1nnbr8QUB-DLqJZR_wO-jzB0DmLSOxtYD/view?usp=drive_open&usp=embed_facebook

Fodor, J. A. (1980). *The Language of Thought* (1 edition). Harvard University Press.

Freiwald, W. A., & Tsao, D. Y. (2010). Functional compartmentalization and viewpoint generalization within the macaque face-processing system. *Science (New York, NY)*, *330*(6005), 845–851. https://doi.org/10.1126/science.1194908

Kriegeskorte, N., & Kievit, R. A. (2013). Representational geometry: Integrating cognition, computation, and the brain. *Trends in Cognitive Sciences*, *17*(8), 401–412. https://doi.org/10.1016/j.tics.2013.06.007

Margalit, E., Lee, H., Finzi, D., DiCarlo, J. J., Grill-Spector, K., & Yamins, D. L. K. (2023). A unifying principle for the functional organization of visual cortex. *bioRxiv: The Preprint Server for Biology*. https://doi.org/10.1101/2023.05.18.541361

Moeller, S., Crapse, T., Chang, L., & Tsao, D. Y. (2017). The effect of face patch microstimulation on perception of faces and objects. *Nature Neuroscience*, *20*(5), 743–752. https://doi.org/10.1038/nn.4527

Parvizi, J., Jacques, C., Foster, B. L., Withoft, N., Rangarajan, V., Weiner, K. S., & Grill-Spector, K. (2012). Electrical stimulation of human fusiform face-selective regions distorts face perception. *Journal of Neuroscience*, *32*(43), 14915–14920. https://doi.org/10.1523/JNEUROSCI.2609–12.2012

Poldrack, R. A. (2021). The physics of representation. *Synthese*, *199*(1–2), 1307–1325. https://doi.org/10.1007/s11229-020-02793-y

Quilty-Dunn, J., Porot, N., & Mandelbaum, E. (2023). The best game in town: The reemergence of the language-of-thought hypothesis across the cognitive sciences. *Behavioral and Brain Sciences*, *46*, e261. https://doi.org/10.1017/S0140525X22002849

Roskies, A. L. (2008). Neuroimaging and inferential distance. *Neuroethics*, *1*(1), 19–30. https://doi.org/10.1007/s12152-007-9003-3

Roskies, A. L. (2010). Neuroimaging and Inferential Distance: The Perils of Pictures. In S. J. Hanson and M. Bunzl (Eds.), *Foundational Issues in Human Brain Mapping*, 195. MIT Press.

Roskies, A. L. (2021). Representational similarity analysis in neuroimaging: Proxy vehicles and provisional representations. *Synthese*, *199*(3), 5917–5935. https://doi.org/10.1007/s11229-021-03052-4

Shea, N. (2013). Naturalising representational content. *Philosophy Compass*, *8*(5), 496–509. https://doi.org/10.1111/phc3.12033

Shea, N. (2018). *Representation in Cognitive Science*. Oxford University Press.

Tsao, D. Y., Freiwald, W. A., Tootell, R. B. H., & Livingstone, M. S. (2006). A cortical region consisting entirely of face-selective cells. *Science*, *311*(5761), 670–674. https://doi.org/10.1126/science.1119983

Yamins, D. L. K., Hong, H., Cadieu, C. F., Solomon, E. A., Seibert, D., & DiCarlo, J. J. (2014). Performance-optimized hierarchical models predict neural responses in higher visual cortex. *Proceedings of the National Academy of Sciences of the United States of America*, *111*(23), 8619–8624. https://doi.org/10.1073/pnas.1403112111

6 Frames of discovery and the formats of cognitive representation

Dimitri Coelho Mollo and Alfredo Vernazzani

6.1 Introduction

A central posit in contemporary cognitive science, broadly understood to include AI, is that cognition is at least in large part a matter of producing and using internal representations (Miłkowski 2013; Neander 2017; Shea 2018). Representations allow cognitive systems to store information about the environment and manipulate it internally, in order to inform their behavior in the world. What representations are about, what they represent, is their content. A map of Umeå represents the spatial layout of a nice little university town in Northern Sweden, a description of the city's winter on a tourist guide is about the features of Umeå winters. The physical states and events that realize representations are representational vehicles: specific combinations of color and shapes on a piece of paper are a common vehicle for maps of Umeå, a sequence of written symbols a vehicle for the description of its winters.

In addition to contents and vehicles, representations also have formats. Public, external representations can have different formats: a map represents Umeå differently from how a picture does, let alone a textual description of the city. It is widely believed that the same point applies to cognitive representations: they can also come in different formats. Accounts of the formats of cognitive representations have tended to focus on analogies with public representations, typically contrasting putative language-like internal representations with putative picture-like representations, and trying to establish which cognitive functions rely on one or the other sort of format (e.g., Sloman 1978; Fodor 2008; Kosslyn 1980, 1994; Paivio 1990).

Our critical target in this chapter is such an approach to formats: investigating representational formats in cognitive science in terms of analogies to public, external representations.

Here is how we proceed in what follows. In Section 6.2, we briefly go into the explanatory role of representational formats in cognitive science. In Section 6.3, we outline how scientific discovery is shaped by scientific

DOI: 10.4324/9781003458531-6

frames and how philosophers and cognitive scientists have used public representational formats as frames for tackling questions around the format(s) of cognitive representations. In Section 6.4, we turn to the shortcomings of such a framing, illustrating them by analyzing two case studies: the compositionality debates of the late 1980s to early 1990s that pitched symbolicists against connectionists; and recent discussion about the nature and explanatory roles of visual demonstratives in understanding human visual perception. Finally, in Section 6.5, we outline a computational approach to cognitive formats that does away with analogies to public representations, and which we take to offer a more promising approach for improving our understanding of cognitive formats.

6.2 Representational formats in cognition

The explanatory value of representational formats in cognitive science and AI is due to the fact that some questions remain unanswered even after we have established what the relevant representational vehicles are in a representational system and what contents they carry. Elsewhere, we have argued that the explanatory purchase of representational formats comes from two main questions that they help answer (Coelho Mollo & Vernazzani 2024):

- *Transformation-based explanations* help explain the workings and behavioral effects of cognitive states and processes in terms of the specific kinds of transformation or manipulation available and performed over such states and processes.
- *Efficiency-based explanations* help explain why certain cognitive states and processes are more (or less) adequate for a specific task in terms of certain sets of transformations/manipulations being more efficient, powerful, less taxing, and/or temporally advantageous.

In other words, cognitive representational formats capture properties of representational vehicles that are relevant to providing explanations concerning the available transformations of representational vehicles, and the extent to which they contribute to making the solution to cognitive tasks more or less difficult and/or resource-consuming.[1]

It may turn out that there are no principled grounds for grouping representational vehicles in ways relevant to those explanatory roles: it may be, for instance, that representational vehicles have properties that radically vary with context in unsystematic ways, or that are overly fine-grained, such that any grouping into formats ends up being ad hoc, or explanatorily unhelpful. Should that be the case, eliminativism about representational formats would be warranted.

Another possibility, which has played an important role, especially in older debates about cognitive formats, is that there is one or at most a few cognitive formats that cover all or most of cognition. The traditional dichotomous approaches to cognitive formats have largely taken this path, pitting language-like versus iconic formats, digital versus analog, discrete versus continuous, and so on (e.g., Fodor 2008; Sloman 1978).

Yet another possibility, the one that we favor and that has received growing attention in recent years (see, e.g., Quilty-Dunn et al. 2023) is a rather liberal pluralism about cognitive representational formats. Representational vehicles may turn out to have a variety of sets of shared properties that yield a large number of representational formats. Moreover, the properties of vehicles that are relevant for typing formats may vary depending on the explanatory aims and needs of specific research endeavors. In other words, different categorizations of formats, focusing on different sets of properties representational vehicles have, can be scientifically legitimate in light of differing explanatory targets.

Past and ongoing debates about cognitive formats have focused on analogies with the formats of the public representations we are familiar with in our everyday lives, such as sentences, pictures, and maps. This sort of analogy may help motivate monism or a limited pluralism about representational formats: natural language is analogous to the language of thought (LOT); pictures, photos, and paintings are analogous to so-called iconic formats in perception and memory; and cartographic maps are analogous to cognitive maps in entorhinal cortex. As those are the dominant formats we use in everyday life, their cognitive analogues are likely to be the main (if not the only) formats at play in cognition too.

This line of reasoning underpins a specific sort of framing of research on cognitive formats, one that takes as its starting point analogies to public representations. As we will argue, such a framing has several shortcomings, as the two domains being analogized are too different to be mutually informative.

6.3 Formats and frames of scientific discovery

6.3.1 Scientific inquiry and frames of discovery

The process of scientific discovery does not happen in a theoretical vacuum. Scientists inquire into phenomena guided by different epistemic concerns and goals (cf. Vernazzani ms), which derive in part from the state-of-the-art of the specific field of research, and in part from the adoption of specific theoretical frames that provide a characterization and structure to the process of discovery.[2]

A theoretical or scientific frame can be understood as a (set of) conceptual tools that, under an intended interpretation, provides a cohesive and unitary schema for understanding a particular target phenomenon (Camp 2019). As such, frames differ from Kuhnian paradigms (Kuhn 2012)—at least under one sense of "paradigm"—in being less general and coherent: they mold research by suggesting hypotheses, possible interpretations, and explanations, and by embodying specific norms for scientific inquiry within a particular field. A frame used for the purposes of scientific discovery can be seen as a lens or template that directs research, even though a frame may later be dropped for another one or coexist with other frames considered to be useful for the same domain of research.

We can construe past and ongoing research on the formats of cognitive representations as essentially guided by different frames of inquiry. According to what we may call—not without some simplification—the "received view," the formats of cognitive representations are modeled on the formats of public representations, most paradigmatically on natural languages (such as English or French), and on icons, a broad category that includes pictures, paintings, maps, and photographs.

Our aim in this chapter is to show that the traditional, public representation-based frame to understanding formats is deeply problematic, as it can lead our research efforts into unfruitful directions. An alternative, computation-based account of formats (Coelho Mollo & Vernazzani 2024), we will argue, is preferable.

6.3.2 *Public representations as frames*

Current research on the format of cognitive representations has often used public representations as frames of discovery. Public representations include a broad range of representational artifacts: paintings, photographs, maps, linguistic scripts, and many other kinds of signs. Philosophers and cognitive scientists typically identify two broad categories or families of public formats: icons and language-like representations (e.g., Fodor 2008; Kosslyn 1980, 1994; Paivio 1990). Applied to cognitive representations, this distinction has been taken to undergird important differences in cognitive architecture. For example, Carey (2009), Burge (2010), and Block (2023) argue that the difference between perception and cognition is to be cashed out in terms of different formats. More specifically, they argue that while perceptual representations are typically iconic or image-like, conceptual cognitive representations are instead language-like.

The strategy typically adopted by philosophers and cognitive scientists using public representations as frames of discovery is to hypothesize functional and/or structural similarities between relevant properties of a public format and the properties of the cognitive representations putatively at play in a cognitive function of interest. This is often motivated by apparent similarities between the kinds of tasks we solve by using external representations and the tasks cognitive systems have to tackle (Coelho Mollo 2022), as well as by considerations regarding the required expressive power of representational systems, be they external or internal, for them to be appropriate to a task. Such similarities are then used to suggest that the sorts of operations and transformations of vehicles made available to users by a public representational format are analogous to the computations made available to cognitive (sub)systems by its analogous cognitive format. It will be instructive to discuss one example of this strategy.

Fodor (2008) contends that pictures, just like sentences, have a compositional semantics. However, he maintains that pictures and natural languages obey different principles of compositionality. Specifically, he argues that pictures obey what he calls the "Picture Principle": if P is a picture of X, then parts of P are pictures of parts of X. In Fodor's view, this principle undergirds the fact that pictures, unlike sentences, do not admit of a canonical decomposition, each part of a picture of a person represents a part of a person, whereas a sentence can only be decomposed into semantically meaningful units according to specific rules.

A direct application of Fodor's Picture Principle to cognitive representations can be found in Quilty-Dunn (2016).[3] Quilty-Dunn argues that the segmentation of the visual scene into discrete objects requires a canonical decomposition into distinct constituents, each requiring a distinct vehicle. This is apparent in the case of multiple object tracking (MOT) (see also Section 6.4), where test subjects are shown different items on a display, among other confounds, and are tasked to keep track of a selected number of objects. Studies indicate that subjects are typically capable of easily tracking up to four or five distinct objects (Pylyshyn 2007). This limitation may be due to the fact that the visual system has a limited number of simultaneously active vehicles to represent and keep track of objects. Now, if the visual scene were represented iconically—so Quilty-Dunn argues—it would be impossible for the visual system to represent objects as segregated from other objects and the scene; this because the segmentation of objects requires different representational vehicles, which in turn is strongly indicative of a canonical decomposition into discrete objects. However, this is the sort of thing that should be impossible were the representation iconic, since, in virtue of the Picture Principle, icons do not admit of canonical decomposition.

This sort of case nicely illustrates the strategy discussed above. The operating assumption is that there is a structural similarity between some alleged properties of icons (public representations), that is, their non-canonical decomposition, and the cognitive representations that undergird visual object perception and scene perception. The kind of compositional semantics operating in the case of object perception does not seem to obey the Picture Principle, hence, allegedly, such representations cannot be iconic.

Although we are about to criticize adopting public formats as frames of discovery for the format of cognitive representations, such a strategy is not unmotivated. Arguably, one reason for adopting public formats as frames of discovery is the widespread use of analogies in science (Hesse 1966). A well-known example is the wave equation, which can be used to study mechanical waves such as water and sound waves, but also electromagnetic waves. The underlying assumption is that all these different phenomena display a wave-like dynamic that can be aptly captured by the same equation.

A similar line of thinking may underlie the use of public formats to model the format of cognitive representations. On the assumption that the two families of representations may have interesting analogies, one may study the computational properties of public formats to gather insights into the format of cognitive representations. In doing so, however, it is crucial to keep in mind the abstractions and idealizations involved in such a process, as the two domains being analogized are likely to have crucial differences, and only partially share interesting features. Failure to recognize such abstractions and idealizations may lead to improper extension of features of one domain to the other, leading researchers astray (Coelho Mollo 2022). As we will argue, framing cognitive formats in terms of public formats exemplifies such a risk.

6.4 The limits of the public representation frame

Our strategy to assess the adequacy of taking public representations as a frame to better understand representational formats in cognition is two-pronged. First, we will focus on theoretical arguments about what public and cognitive representations have in common, and what sets them apart. As we will argue, the similarities between these two domains of representational phenomena are limited, while the differences are substantial. This suggests that using one to help understand the other is epistemically risky.

Second, we will briefly examine two case studies that, we believe, illustrate the perils of relying on the public representation frame when investigating cognitive formats: the debate in the late 1980s and early 1990s about compositionality and cognitive architecture between Fodor & Pylyshyn (1988), Smolensky (1987), and others; and contemporary work on the nature and roles of "visual demonstratives" in visual perception (Burge 2018, 2023; Green & Quilty-Dunn 2021; Pylyshyn 2007).

6.4.1 *Public versus cognitive representations*

Let us take a closer look at what public and cognitive representations have in common, and what distinguishes them. Properties that they share include the following:

1. being about, or meaning, or having as content something, be it concrete or abstract, vague or definite, actual or imaginary;
2. having some physical vehicle(s) that implement the representation, that is, that "carry" the content, be they a sheet of colored paper, strings of digits in a solid-state memory drive, or the activity of neurons in brains;
3. standing in appropriate, often complex sets of relationships to something, as described in (1), which make it so that the vehicle, as described in (2), comes to have the content that it does (shedding light on what these relations are is the core mission of the project of naturalizing representation, see Neander 2017; Millikan 2017; Shea 2018);
4. playing a representational functional role, that is, being used by some system to represent something else.

While these are important similarities, they are features that are essential to being a representation at all. Public representations and cognitive representations, in brief, have contents, are implemented by physical vehicles, and come to be representations courtesy of both appropriate relationships between vehicles and what they represent, and appropriate functional roles. Insofar as they share these features, both phenomena of interest count as representational. If any of them lacked one or more of those features, they would not count as representations. In other words, the features that public and cognitive representations have in common are the minimal ones that make them both cases of representation.

A closer look at their more specific features, over and beyond those that make them into representations, reveals crucial differences.

First, there are important differences between the sorts of vehicles used for public representations—ink on paper, light on screens, paint on surfaces, etc.—and the sorts of vehicles thought to be relevant for cognitive representations, which ultimately involve aspects of the activity of brain cells, especially neurons in the neocortex. While this may seem like a relatively trivial difference, having to do "merely" with implementational details, it is not so. Implementational details constrain the sorts of transformation that representations can undergo. Ink on paper cannot typically be erased or overwritten without loss of clarity, while Unicode in computer memory can. Moreover, implementational details help determine which systems can access and make use of a representation: paint on the surface of neurons is unlikely to be something that downstream cognitive systems can use, as they are not sensitive to the color of neurons.

More generally, the implementation of public representations hinges on communicative intentions, and complex socio-historical processes that led to the maintenance of certain conventions over others, with one guiding constraint being interpretability by the humans target of the representation—which in its turn depends on their familiarity with the chosen conventions.[4] The implementation of cognitive representations, on the other hand, is guided and constrained by rather different factors. They depend on mind-independent evolutionary and developmental constraints, as well as on the fact that the consumer systems are not full-fledged interpreters, but ultimately "dumb homunculi" (Dennett 1991).

Second, the relationships between vehicles and what they are about that partly endow representations with content are fundamentally different when it comes to external public representations versus internal, cognitive ones. In the former, a combination of individual propositional attitudes (e.g., one's intention to convey a certain content to an interlocutor), socially established conventions (e.g., crosses for Christian churches in city maps), and social practices (e.g., Gricean norms of communication) all play a role in establishing the contents of a representation. None of these factors, however, are part of (naturalistic) accounts of cognitive representation. Cognitive representations are supposed to underlie propositional attitudes and, more indirectly, conventions and social practices. Thus, on pain of explanatory circularity, cognitive representations must stand in other kinds of relation to what they represent.

Naturalistic theories of content have settled on a combination of causal-informational and teleofunctional relations as the ones that determine the contents of cognitive representations (Millikan 2017; Neander 2017; Shea 2018; Coelho Mollo 2022). These are natural relations that do not involve nor presuppose mental states and/or social conventions or institutions. While some theorists take such naturalistic theories to also cover or at least help explain public representations (e.g., Millikan 2017), others are skeptical that there is any straightforward extension of an account of representation that applies to cognitive states to one that also applies to public representations, and vice versa (Neander 2017; Shea 2018; Coelho Mollo 2015).

Finally, when it comes to the functional roles of public and cognitive representations, we also find substantial differences. As we pointed out, public representations are typically means of communication between two or more full-fledged intelligent, cognitive agents (including the same agent at different times, as in personal notes and reminders). In most cases of public representation, there are intentions to communicate something to someone else, often with a specific valence, and relying on an assumption that the communicative intention and the representation's "code" are grasped by the receiver.

Typical cases of cognitive representation, on the other hand, are infra-agent and do not involve communicative intentions. They are produced by cognitive subsystems and are "read" or "consumed" by other cognitive subsystems. Such a reading involves no process of interpretation—at least in its common sense—but merely a mindless, "mechanical" responsiveness by downstream subsystems to certain features of the representation.

In brief, public and cognitive representations share only those features that make them both representations. When it comes to the central details about how they come to be representations, the stories that need to be told are very different. The kinds of vehicle, content-fixing relations, and functional roles at play in public and cognitive representations importantly differ. This suggests that public representations are a poor frame for understanding the formats of cognitive representations.

6.4.2 *Two case studies*

Far from being of merely theoretical interest, framing cognitive representational formats in terms of analogies to public representations can actively hinder progress in understanding cognition. We will briefly present two such cases: one in which analogies to public representations led to overstrict views about how a certain capacity need be implemented (the compositionality of thought debate between classicists and connectionists); and one in which allegiance to the public representation frame led to positing cognitive components that may not really exist (visual demonstratives).

6.4.2.1 *The compositionality debates*

In the late 1980s and early 1990s, the connectionist, neural network-based approach to artificial intelligence and cognitive science gained in (short-lived) popularity with the growth of what came to be known as the PDP (Parallel Distributed Processing) program (Rumelhart et al. 1986). Some of its proponents argued that connectionism offered an alternative account of cognitive architecture to the classical paradigm. The latter posits that cognitive processes are a matter of symbolic computations over atomic representations (or molecular compositions thereof), with a syntax similar to that of natural (and artificial) languages (Fodor 1975)[5]; a hypothesis that has come to be known as the Language of Thought Hypothesis or LOTH. The former, on the other hand, has it that cognitive computation is non-symbolic, being rather numerical, or at most "subsymbolic," and that representations are distributed and partially overlapping across nodes in the network, instead of being discrete (e.g., Smolensky 1987).

Fodor and Pylyshyn (1988), central proponents of LOTH, provided a detailed attack of connectionism as a candidate cognitive architecture.

Their core argument was simple. Thought is productive—that is, it can produce an indefinite number of different thoughts—and systematic—that is, being able to think certain thoughts is intrinsically tied to the ability to think other thoughts: a mind that can think "John loves Mary" can also think "Mary loves John." But productivity and systematicity require thoughts to have constituent structure, that is, discrete parts that retain their identity throughout and that can be combined in rule-governed ways. Thought must thereby be compositional. However, representations in connectionist networks lack constituent structure, and their computations are based on stochastic relations, rather than strict symbolic rules that govern how representations can and cannot be combined. Therefore, connectionist networks cannot underpin a compositional representational system, and thus cannot account for the productivity and systematicity of thought. Connectionism thereby fails as a claim about cognitive architecture.

Interestingly, Fodor and Pylyshyn's (1988) attack against connectionism is almost purely theoretical—they make little to no recourse to empirical evidence, and they occasionally rely on intuitions about what kinds of minds can or cannot exist.[6] Crucially, they endorse two central undefended assumptions: (a) sentences in natural language express thoughts; (b) sentences and thoughts have similar structures (the underlying idea being arguably that assumption (b) is needed if sentences are to adequately express thoughts).

Unsurprisingly, given such assumptions, Fodor and Pylyshyn's arguments move seamlessly between features of language and putative features of thought. As natural languages are productive, systematic, and compositional, so must thought be, as the former expresses the latter.

Assumption (b), however, indicates a serious "lack of imagination" on the part of Fodor and Pylyshyn, as Chalmers (1993) puts it. While natural language provides a paradigmatic case of a productive, systematic, and compositional representational system, other kinds of representational system may also realize, or at least approximate, those features so as to lead to appropriate behavior.[7] Those representational systems may have no analogue, however, among the public representations we are familiar with. Indeed, connectionist networks have been shown to have the representational means to, at a minimum, approximate compositional competence, despite lacking the sort of explicit constituent structure and symbolic composition rules that Fodor and Pylyshyn take as characteristic of language-like formats (Smolensky 1987, 1990; Lepori et al. 2024; see Coelho Mollo 2025 for extended argument and illustration).[8]

At any rate, Fodor and Pylyshyn (1988) display a lack of imagination also because they do not consider that expressing thoughts via natural language may involve a transformation of format from a potentially non-LOT cognitive format to the natural language format. Natural language, rather than being the analogue of the cognitive format, may actually be just an approximate means to express thoughts that do not share the same strictures of a language-like format (Smolensky 1987). Classical compositionality may be an artifact of language use, rather than a feature of the cognitive architecture of minds.[9]

In sum, framing cognitive representational formats by analogy with public representational formats seems to have worked as a limit on imagination for the defenders of LOTH within that old debate. The constraints on possible cognitive formats are different from the historical, socio-cognitive constraints on public representation systems. To frame or model the former in terms of the latter risks constraining our scientific imagination, dismissing options that are worth exploring, such as the possibilities allowed by large numbers of numerical computations in artificial neural networks (ANNs).[10]

6.4.2.2 Visual demonstratives

A *Perceptual Object Representation* (or POR, for short) is the representation of a single visual object, a coherent visual whole (e.g., Feldman 2003; Green 2018a; Rubner & Vernazzani ms; Vernazzani 2022). PORs are postulated to explain different feats of our visual system. First, they are posited as a solution to the synchronic binding problem, that is, the problem of unifying different surface and spatiotemporal features (like color, shape, and trajectory) into a single coherent whole. Second, they are supposed to provide a solution to the correspondence problem, that is, the problem of explaining how the visual system re-identifies an object as the same in a dynamic context, despite featural and spatiotemporal changes (Flombaum et al. 2009; Quilty-Dunn & Green 2023; Scholl 2007).

It is generally agreed that PORs include representations of an object's features. In addition, some researchers contend that an additional component, a *visual demonstrative* or *pointer*, must be added (e.g., Alvarez & Franconeri 2007; Echeverri 2017; Pylyshyn 2003, 2007; Scholl et al. 2001). The main motivation for the introduction of a demonstrative component in PORs comes from studies on dynamic object tracking, as exemplified by experiments on MOT.

MOT studies require test subjects to lock reference to a limited number of targets—e.g., one or more white circles—displayed on a screen while the targets move among other figures, which work as confounds (Pylyshyn & Storm 1988). In some studies, the targets do not simply move but also change some of their surface features, such as form and color. Typically, subjects can easily target four or five such objects (Pylyshyn 2007; Green 2018b; cf. also Alvarez & Franconeri 2007). Since objects are tracked despite spatiotemporal and featural changes, it has often been argued that reference is secured to the object itself via a demonstrative-like component and not via the object's properties:

> Unless at some point one can think the equivalent of "*This* has property P", one cannot refer to a particular object token in a way that will allow it to be bound to the arguments of a visual predicate or to serve as the basis for action (e.g. to point toward or grasp the token object in question).
>
> (Pylyshyn 2003, p. 254; cf. also 2007, p. 95)

The visual demonstrative may be construed in different ways. Burge construes it as a complex demonstrative embedded into a proto-propositional structure such as "This *F*" (2018, 2023), where the property *F* is not predicated of the object but merely attributed to it. Alternatively, one may combine the complex demonstrative account with a sentential and predicative structure—such as "This *F* is *G*" (e.g., Quilty-Dunn & Green 2023)—or cash out the visual demonstrative as a simple demonstrative plus a predicate, such as "This *a* is *F*" (e.g., Fish 2009; Fodor & Pylyshyn 2015).

Another way of understanding the visual demonstrative component of PORs is modeled after pointer architecture, a computational structure that represents a reference to a feature stored in memory (Green & Quilty-Dunn 2021). This latter way of construing the visual demonstrative may be coupled with object files, where the pointer plays the twofold role of anchoring reference to a worldly object, while also relating it to information stored in short-term memory (Kahneman et al. 1992).

For our present purposes, it is critical to see that, however visual demonstratives are construed, they suggest a language-like format of PORs. This is very explicit in Pylyshyn's account of the structure of visual objects within his broadly Fodorian approach, which embraces both modularity and LOTH (Fodor 1983). More recently, Green and Quilty-Dunn (2021) have similarly suggested that object files have a propositional structure encoding discrete symbols that can be syntactically combined in different ways.[11]

In this context, the format of a public representational structure, that is, language, works as a frame for scientific inquiry and discovery, structuring and guiding our understanding of PORs' nature and organization, as well as of the way they refer to worldly objects. Furthermore, adopting such a frame of discovery makes some questions more salient: for example, how should we understand visual demonstratives, as complex or simple demonstratives? Such research questions invite a way of treating the problem that draws on the resources provided by linguistics and the philosophy of language, at the same time downplaying other possible ways of thinking about it in nonlinguistic terms.

As we have seen, visual demonstratives are typically introduced to explain dynamic object tracking and to provide a solution to the correspondence problem. However, there are reasons to be skeptical about whether they are needed for these explanatory tasks. In work in progress,

Rubner and Vernazzani (ms) argue that there can be a satisfactory solution to the correspondence problem without appeal to visual demonstratives. The argument unfolds in two steps.

First, they point out that empirical evidence strongly suggests that properties of objects are all that is needed to keep track of objects in a dynamic context. Studies have shown that spatiotemporal properties, such as topological properties or an object's trajectory, play an important role in object tracking (Mitroff & Alvarez 2007; Zhou et al. 2010). More recently, it has been shown that surface properties such as color and shape are also critical for object tracking in some contexts (Jiang 2020; Hein & Moore 2012).

There is thus ample empirical evidence suggesting that properties are all that is needed in keeping track of items, where some properties of the object are used for the purpose of reference guidance or tracking, while other properties take a subordinate role, as they belong to the POR, although they are not used for the purpose of tracking (Green & Quilty-Dunn 2021). Which properties the visual system uses for tracking depends on context. For instance, Tremoulet et al. (2000) show that infants track objects through shapes and colors when spatiotemporal information does not provide sufficient criteria of individuation. Thus, the visual system seems to adopt flexible strategies to the problem of object tracking.

Once we acknowledge the critical role of properties in object tracking, we can articulate a solution to dynamic object tracking tasks without resorting to visual demonstratives. As we have seen, visual demonstratives are introduced to account for the fact that the visual system can keep track of an object despite featural change. Supporters of visual demonstrative have taken this as evidence that object tracking is not achieved via properties. Rubner and Vernazzani, however, advance an alternative solution to the correspondence problem that does not require the postulation of visual demonstratives, relying on what they call the Continuity Principle:

> For any objects O_1, O_2, sets of properties F_1, F_2, and times t_1, t_2, where t_1 and t_2 occur sequentially, if F_1 *guides reference to* O_1 and F_2 guides reference to O_2, then O_1 at t_1 may be perceived as the same as O_2 at t_2 only if $F_1 \cap F_2 \neq \emptyset$.

The key to solving the correspondence problem according to the Continuity Principle is that the visual system does not track only one single property of the object, but rather a set of properties. As long as the intersection of an object's properties used for tracking at two different times is non-empty, reference is secured and the correspondence problem solved, without any need to introduce visual demonstratives in PORs architecture.[12]

It is helpful to illustrate how the Continuity Principle works in practice. Take a case of dynamic object tracking with featural change, for example, an object moving from the left-hand side of the visual scene at t_1 to the right-hand side of the scene at t_2 while changing color (say, from red to green) and shape (say, from rectangular to triangular). According to the Continuity Principle, the visual system is able to keep track of the object and solve the correspondence problem, that is, to identify the object as the same in spite of featural change, because some of the object properties from t_1 to t_2 are preserved—in this case, the spatial trajectory—where the spatial trajectory belongs to the set of properties *F* that the visual system uses for reference guidance or tracking. Notice that the Continuity Principle also enables us to make empirically testable predictions, for it follows from it that object tracking would be disrupted just in case from t_1 to t_2 all properties used for tracking were lost and replaced by other properties. In real-life contexts, this condition virtually never obtains, and this illustrates how the Continuity Principle captures the robustness and flexibility of object tracking.

Accepting the Continuity Principle provides an empirically sound solution to the correspondence problem without resorting to (natural) language as a frame for modeling object tracking.[13] At first, direct implication is that the Continuity Principle makes the postulation of visual demonstratives unnecessary. By removing visual demonstratives from the structure of PORs, further questions directly linked with their posited architecture are removed as well: there is no need to draw on the research on linguistic demonstratives to best capture the nature of visual demonstratives, and no need to think of them as simple or complex demonstratives.

Second, once the need for visual demonstratives is removed, one reason for thinking of PORs as articulated in a language-like format (see Green & Quilty-Dunn 2021) is undermined as well. There is no need to construe them as either proto-attributional structures—as in Burge (2023), "That *F*" where *F* is a visual attributive—or predicational—as in "That *F* is G" or "That *a* is *F*" (Fish 2009).

We do not wish to suggest that the entire question of the language-likeness of the format of PORs rests on visual demonstratives. Rather, our discussion was meant to highlight two important lessons. First, defenders of visual demonstratives read the empirical studies and interpret the correspondence problem through the lens of the public representational format frame, thus inviting specific further research questions, for example, about the nature of visual demonstratives and their roles in PORs architectures. But an empirical and conceptual solution to the correspondence problem does not necessitate a language-like format for PORs.

This brings us to the second lesson. Framing the issue through the public representational format frame downplays alternative ways of interpreting the evidence and construing the correspondence problem. This may obfuscate other empirically viable approaches, such as the one offered by the

Continuity Principle. Embracing the Continuity Principle and dismissing visual demonstratives as a viable solution to the correspondence problem leads to a reconfiguration of the problem, highlighting some issues—e.g., whether there is a hierarchy of property-types for purposes of tracking—while discarding others as conceptual artefacts—e.g., how best to construe visual demonstratives.

In brief, the public representational format frame leads to hypotheses and lines of research that may be less fruitful than alternatives, but that seem natural due to the explicit or implicit adoption of that intuitive frame. In the next section, we put forth an alternative frame for thinking about the formats of cognitive representations that draws, instead, on foundational notions in the cognitive sciences.

6.5 A computational account of formats

In a recent paper, we have put forward and defended a view of representational formats that takes as its starting point the explanatory needs of the cognitive sciences (Coelho Mollo & Vernazzani 2024). Such a starting point casts suspicion on the idea that categories that are useful for understanding the formats of external, public representations should also be applicable to internal, cognitive representations. Indeed, while in both cases we are dealing with representations, they are of such different types that it is implausible to think that a single account, or even two very similar accounts, can capture them both.

If public representations, as we have argued, are indeed ill-suited as a frame for understanding the formats of cognitive representations, we must look for alternatives that are more likely to suit the explanatory needs of cognitive science. We have developed a theory of representational formats for the cognitive sciences that relies on an approach that has proven to be enormously scientifically fruitful: namely, the use of the notion of concrete computation in accounting for cognitive processing—a foundational insight of mainstream cognitive psychology and cognitive neuroscience that underpins much of the progress these fields have made in the past few decades (Miłkowski 2013; Piccinini 2020; Colombo & Piccinini 2023).

The resulting computational view of formats that we suggest identifies cognitive representational formats with what we call "computational profiles." Computational profiles are features of cognitive vehicles. More specifically, they are the set of features of computationally individuated cognitive vehicles that constrain and govern which computational operations and transformations vehicles can undergo. When computational vehicles in cognitive systems are also representational vehicles, their computational profiles are their representational formats.

We distinguish two types of constraints on vehicles, which we dub "inner constraints" and "outer constraints." Briefly, inner constraints capture the ways in which parts of a computational-representational vehicle can vary: how many of its parts play computational roles and which values those parts can take. Outer constraints, on the other hand, are those constraints that come from other computational-representational vehicles within the same representational (sub)system: that is to say, how values taken by parts of vehicles $V_2, \ldots, V_n$ in the same (sub)system influence the values that can be taken by a vehicle V_1.

A helpful, though rather rough, analogy is with action figure toys. If we take each limb of the toy to be a vehicle, the inner constraints are the positions each limb part can take, which tend to vary with the design and quality of the toy. The outer constraints are instead the mutually constraining relations that may exist between limbs: for instance, in some such toys, having an arm at a certain position constrains the positions that the other arm can take (in the simplest case, both arms must always be in the same relative position).

While the inner and outer constraints in action figure toys help determine their degrees of freedom—an important selling point—the computational profiles of representational vehicles in cognitive systems help determine which computational transformations those vehicles can undergo, and thus how cognitive representations can be manipulated. This, we claim, is just what cognitive representational formats are, and why they are scientifically useful notions: they allow us to capture features of cognitive processing that shape how representations can be transformed by downstream systems to contribute to furthering the systems' goals.

The computational view of formats makes use of conceptual tools that are not only widely employed but also foundational to cognitive psychology, neuroscience, and cognition-focused AI. This makes it possible to inform and assess hypotheses about representational formats by appeal to theoretical and empirical results from those fields—as our case studies illustrate. This stands in stark contrast with the extra-scientific origin of the public representation frame, which comes from everyday representational practices of human individuals and societies, rather than the posits of the cognitive sciences.

While extra-scientific concepts and analogies can be heuristically useful in some cases, employing them comes with risks (Coelho Mollo 2022). Indeed, as we have pointed out in Section 6.4, the public representation frame has several shortcomings, some theoretical and some that can be seen in how certain debates in the field have been (unhelpfully) structured. Looking at the case studies we examined, does the computational view fare any better? We believe that it does.

With regard to the issue of the compositionality of thought that exercised Fodor, Pylyshyn, Smolensky, and many others in the 1980s and 1990s, and which has seen a rebirth in recent times with the advent of Large Language Models, the computational view does not share the limits of the public representation frame. As it is not wedded to analogies to natural language, the computational view does not invite a view of compositionality that hinges on how language comes to have that feature. It thus opens the way to seeing the compositionality of thought in a different light, as a phenomenon that may, but need not, be akin to linguistic phenomena. Moreover, by directing focus toward computational profiles and computational transformations of representational vehicles, the computational view invites the idea that similar capacities may be computationally realized in different ways. As much as there are different algorithms for division, there can be different computational means to realize behaviorally appropriate compositionality of thought, with Smolensky's suggestion being one such alternative, and current research on Large Language Models revealing yet others (e.g., Lepori et al. 2024; Merullo et al. 2024).

Similarly, positing visual demonstratives hinges on a rather clear analogy to the role played by demonstratives in natural language. Indeed, the notion comes from researchers committed to the idea that cognitive systems employ mostly or exclusively language-like formats. The computational view works under no such assumptions. Cognitive formats are likely not to resemble anything we are familiar with when it comes to public representations, for the reasons we suggested in Section 6.4. A language-like demonstrative is one potential way in which the correspondence problem in visual perception can be tackled, but far from the only one. We should therefore actively investigate alternatives, such as the one sketched above and in Rubner and Vernazzani (ms), which may be more in line with what we know about the workings of the brain.

6.6 Conclusions

In this chapter, we have presented two competing strategies for tackling the nature and roles of the format(s) of cognitive representations. The received view uses a scientific frame that draws from the formats of public representations, such as maps, pictures, drawings, and natural languages. Researchers adopting such a framing typically start by singling out some computational properties of the former, and then they compare such properties to the properties of a target cognitive system or subsystem. We have questioned the fruitfulness of this strategy, pointing to its theoretical limits, and illustrating its shortcomings with two case studies. We can make headway into thinking about the format of cognitive representations once we sever, or at least considerably weaken, associations with public representational formats, and focus instead on the computational profiles of representational vehicles, and the explanatory roles they can play in our theories of mind and brain.

Acknowledgments

The authors would like to thank participants at the symposium on "The Formats of Cognitive Representations" at the Second Conference of the International Society for the Philosophy of the Sciences of the Mind. Alfredo Vernazzani would like to thank the German Research Foundation DFG, Research Training Group "Situated Cognition" (GRK 2185/2) (Gefördert durch die Deutsche Forschungsgemeinschaft [DFG]—Projektnummer GRK 2185/2).

Notes

1 Other approaches to representational formats focus instead on types of contents (e.g., Peacocke 2019) or content-encoding schemes (e.g., Lee et al. 2023). We are open to the idea that different approaches to representational formats can be useful for different explanatory purposes. At any rate, we believe that the main point of this chapter, namely the claim that there are serious risks with framing cognitive representational formats in terms of public representations, applies to those other approaches as well. Detailed defense of this point is beyond the scope of this chapter. We thank an anonymous referee for suggesting this clarification.

2 See Haziz (2023), for an overview of the epistemology of inquiry; and see Barack (2024), for a biologically inspired perspective on scientific inquiry.

3 Quilty-Dunn (2016) is interested in what he dubs "Iconicity," whose formulation is very close to Fodor's Picture Principle.

4 An illustration of the role of socio-historical conventions in public formats is provided by Wilson (2003), which found that children's drawings vary significantly across cultural backgrounds. For instance, over 70% of children's drawings from Islamic countries in the Middle East tend to represent human torsos with a rectangular shape, whereas only less than 4% of children's drawings from the United States, Australia, European countries, and New Guinea exhibited that feature. In addition, English children in the late 1800s tended to represent bottle-shaped bodies (Wilson 2003; Cohn 2013). According to Wilson (2003), this indicates that children's drawings are not directly drawn from perception, but rather rely on culture-specific schemata.

5 Despite being one of the most forceful proponents of LOTH, Fodor accepted the existence of iconic representations, understood in terms of his Picture Principle mentioned above, even though he ascribed to them rather limited roles in comparison with LOT representations. It is also worth pointing out that Fodor was rather skeptical of the capacity of LOTH to explain global, non-modular cognitive processes (see Fodor, 2001, 2008).

6 A recent defense of LOTH, instead, prefers to focus on empirical evidence and downplays Fodor and Pylyshyn's (1988) compositionality-based arguments (Quilty-Dunn et al. 2023).

7 Given this latter option, defending LOTH by appealing to the competence/performance distinction borders on begging the question (cf. Fodor & Pylyshyn 1988; Quilty-Dunn et al. 2023).
8 There is ongoing research on how current Large Language Models (LLMs) tackle tasks that involve systematicity and compositionality. Some of this incipient research indicates that LLMs may be fruitfully described, with some idealization, as approximating some of those features by non-symbolic means (see, e.g., Merullo et al. 2024). This seems to be along the lines of Smolensky's (1987) suggestion that LOT may be a useful abstraction, but does not capture the nature of the non-discrete representations and non-symbolic computations at play in cognition. As such, LOT should rather be seen as a highly idealized scientific model of linguistic cognition, rather than as a hypothesis about the underlying computational-representational architecture. We thank an anonymous reviewer for pressing us on this point.
9 There are other live options that Fodor and Pylyshyn (1988) fail to consider: formats other than language, like maps and tree diagrams, allow for compositionality (Camp 2009) so that the putative compositionality of thought can be accounted for by appeal to non-LOT representations; and the compositionality of human thought may depend on the "cognitive scaffolding" enabled by learning natural languages (cf. Sterelny 2010). Moreover, Fodor and Pylyshyn seem to maintain that connectionist networks can implement and/or simulate compositional processes, as they agree that connectionist networks can implement LOT architectures, which are compositional. They seem not to appreciate the import of this point for the tenability of their arguments against connectionism, as Chalmers (1993) notices.
10 Contemporary proponents of LOT have indeed abandoned both the earlier claims that LOT is the only cognitive format, and that it should be modeled on natural language (Quilty-Dunn et al. 2023, pp. 2–3).
11 Green and Quilty-Dunn (2021) contend that a POR just is an object file, while Block (2023) argues that PORs are different from object files (see also Rubner & Vernazzani ms).
12 The Continuity Principle is here phrased in terms of properties, but one may phrase it in terms of property-instances or tropes.
13 One may have other reasons to posit a demonstrative, however. For instance, Green and Quilty-Dunn (2021) contend that the demonstrative should be more appropriately understood as a pointer—a notion familiar from computational architectures—and that the pointer plays a twofold role in PORs: connecting an object file to the worldly target, and referring to information stored in a multiple-slot memory structure (see Rubner & Vernazzani ms).

References

Alvarez, George A. & Steven L. Franconeri (2007) "How Many Objects Can You Track? Evidence for a Resource-Limited Attentive Tracking Mechanism" *Journal of Vision* 7(13), pp. 1–10.

Barack, David (2024) "What is Foraging?" *Biology & Philosophy* 39. https://doi.org/10.1007/s10539-024-09939-z

Block, Ned (2023) "Let's Get Rid of the Concept of an Object File" in B. McLaughlin & J. Cohen (eds.) *Contemporary Debates in Philosophy of Mind* (pp. 494–516). New York: John Wiley.

Burge, Tyler (2010) *Origins of Objectivity*. New York: Oxford University Press.
Burge, Tyler (2018) "Iconic Representation: Maps, Pictures, and Perception" in Shyam Wuppuluri & Francisco Antonio Doria (eds.) *The Map and the Territory* (pp. 79–100). Cham: Springer.
Burge, Tyler (2023) *Perception: First Form of Mind*. New York: Oxford University Press.
Camp, Elisabeth (2009) "A Language of Baboon Thought" in R.W. Lurz (ed.) *The Philosophy of Animal Minds* (pp. 108–127). New York: Cambridge University Press.
Camp, Elisabeth (2019) "Perspectives and Frames in Pursuit of Ultimate Understanding" in S. Grimm (ed.) *The Varieties of Understanding* (pp. 17–45). New York: Oxford University Press.
Carey, Susan (2009) *The Origin of Concepts*. New York: Oxford University Press.
Chalmers, David J. (1993) "Connectionism and Compositionality: Why Fodor and Pylyshyn were Wrong" *Philosophical Psychology* 6(3), pp. 305–319.
Coelho Mollo, Dimitri (2015) "Being Clear on Content—Commentary on Hutto and Satne" *Philosophia* 43, pp. 687–699.
Coelho Mollo, Dimitri (2022) "Deflationary Realism: Representation and Idealisation in Cognitive Science" *Mind & Language* 37(5), pp. 1048–1066.
Coelho Mollo, Dimitri (2025) "AI-as-Exploration: Navigating Intelligence Space" THEORIA. *An International Journal for Theory, History and Foundations of Science*. https://doi.org/10.1387/theoria.25837
Coelho Mollo, Dimitri & Alfredo Vernazzani (2024) "The Formats of Cognitive Representations: A Computational Perspective" *Philosophy of Science* 91(3), pp. 682–701. https://doi.org/10.1017/psa.2023.123
Cohn, Neil (2013) *The Visual Language of Comics*. London: Bloomsbury.
Colombo, Matteo & Gualtiero Piccinini (2023) *The Computational Theory of Mind*. New York: Cambridge University Press.
Dennett, Daniel C. (1991) *Consciousness Explained*. Boston: Little, Brown & co.
Echeverri, Santiago (2017) "Visual Reference and Iconic Content" *Philosophy of Science* 84(4), pp. 761–781.
Feldman, Jacob (2003) "What is a Visual Object?" *Trends in Cognitive Science* 7(6), pp. 252–256. https://doi.org/10.1016/s1364-6613(03)00111-6
Fish, William (2009) *Perception, Hallucination, and Illusion*. New York: Oxford University Press.
Flombaum, Jonathan, Scholl Brian J., & Laurie R. Santos (2009) "Spatiotemporal Priority as Fundamental Principle of Object Persistence" in B. M. Hood & L. Santos (eds.) *The Origins of Object Knowledge* (pp. 135–164). New York: Oxford University Press.
Fodor, Jerry (1975) *The Language of Thought*. New York: Thomas Crowell.
Fodor, Jerry (1983) *The Modularity of Mind*. Cambridge, MA: MIT Press.
Fodor, Jerry (2001) *The Mind Doesn't Work That Way*. Cambridge, MA: MIT Press.
Fodor, Jerry (2008) *LOT2*. New York: Oxford University Press.
Fodor, Jerry & Zenon Pylyshyn (1988) "Connectionism and Cognitive Architecture: A Critical Analysis" *Cognition* 28(1–2), pp. 3–71.
Fodor, Jerry & Zenon Pylyshyn (2015) *Minds Without Meanings*. Cambridge, MA: MIT Press.
Green, E.J. (2018a) "A Theory of Perceptual Objects" *Philosophy and Phenomenological Research*. https://doi.org/10.1111/phpr.12521
Green, E.J. (2018b) "What Do Object Files Pick Out?" *Philosophy of Science* 85, pp. 177–200.

Green, E.J. & Jake Quilty-Dunn (2021) "What is an Object File?" *British Journal for the Philosophy of Science* 72(3), pp. 665–699.

Haziz, Eliran (2023) "Norms of Inquiry" *Philosophy Compass*. https://doi.org/10.1111/phc3.12952.

Hein, Elisabeth & Cathleen M. Moore (2012) "Spatio-Temporal Priority Revisited: The Role of Feature Identity and Similarity for Object Correspondence in Apparent Motion" *Journal of Experimental Psychology: Human Perception and Performance* 38(4), pp. 975–988.

Hesse, Mary (1966) *Models and Analogies in Science*. Indiana: University of Notre Dame Press.

Jiang, Hao (2020) "Effects of Transient and Nontransient Changes of Surface Feature on Object Correspondence" *Perception* 49(4), pp. 452–567. https://doi.org/10.1177/0301006620913238.

Kahneman, Daniel, Anne Treisman & Brian J. Gibbs (1992) "The Reviewing of Object-Files: Object Specific Integration of Information" *Cognitive Psychology* 24(2), pp. 175–219.

Kosslyn, Stephen M. (1980) *Image and Mind*. Cambridge, MA: Harvard University Press.

Kosslyn, Stephen M. (1994) *Image and Brain*. Cambridge, MA: MIT Press.

Kuhn, Thomas S. (2012) *The Structure of Scientific Revolutions*. Chicago: University of Chicago Press.

Lee, Andrew Y., Myers Joshua & Gabriel Oak (2023) "The Structure of Analog Representations" *Noûs* 57(1), pp. 209–237.

Lepori, Michael, Thomas Serre & Ellie Pavlick (2024) "Break It Down: Evidence for Structural Compositionality in Neural Networks" *Advances in Neural Information Processing Systems* 36, pp. 42623–42660.

Merullo, J., C. Eickhoff & E. Pavlick (2024) *Circuit Component Reuse Across Tasks in Transformer Language Models*. ICLR.

Miłkowski, Marcin (2013) *Explaining the Computational Mind*. Cambridge, MA: MIT Press.

Millikan, Ruth G. (2017) *Beyond Concepts*. New York: Oxford University Press.

Mitroff, Stephen R. & George A. Alvarez (2007) "Space and Time, Not Surface Features, Guide Object Persistence" *Psychonomic Bulletin & Review* 14(6), pp. 1199–1204.

Neander, Karen (2017) *A Mark of the Mental*. Cambridge, MA: MIT Press.

Paivio, Allan (1990) *Mental Representations: A Dual Coding Approach*. New York: Oxford University Press.

Peacocke, Christopher (2019) *The Primacy of Metaphysics*. New York: Oxford University Press.

Piccinini, Gualtiero (2020) *Neurocognitive Mechanisms: Explaining Biological Cognition*. Oxford: Oxford University Press.

Pylyshyn, Zenon (2003) *Seeing and Visualizing*. Cambridge, MA: MIT Press.

Pylyshyn, Zenon (2007) *Things and Places*. Cambridge, MA: MIT Press.

Pylyshyn, Zenon & Ron W. Storm (1988) "Tracking Multiple Independent Targets: Evidence for a Parallel Tracking Mechanism" *Spatial Vision* 3(2), pp. 179–197.

Quilty-Dunn, Jake (2016) "Iconicity and the Format of Perception" *Journal of Consciousness Studies* 23(3–4), pp. 255–263.

Quilty-Dunn, Jake & E.J. Green (2023) "Perceptual Attribution and Perceptual Reference" *Philosophy and Phenomenological Research* 106(2), pp. 273–298.

Quilty-Dunn, Jake, Nicolas Porot & Eric Mandelbaum (2023) "The Best Game in Town: The Reemergence of the Language-of-Thought Hypothesis Across the

Cognitive Sciences" *Behavioral and Brain Sciences* 46(e261), pp. 1–75. https://doi.org/10.1017/S0140525X22002849.

Rubner, Andrew & Alfredo Vernazzani (ms) "A Bundle Theory of Perceptual Object Representation" Unpublished Manuscript.

Rumelhart, David E., James McClelland & The PDP Research Group (1986) *Parallel Distributed Processing: Explorations in the Microstructure of Cognition. Vol. 1: Foundations*. Cambridge, MA: MIT Press.

Scholl, Brian J. (2007) "Object Persistence in Philosophy and Psychology" *Mind & Language* 22(5), pp. 563–591.

Scholl, Brian J., Zenon Pylyshyn & Jacob Feldman (2001) "What is a Visual Object? Evidence from Target Merging in Multiple Object Tracking" *Cognition* 80, pp. 159–177.

Shea, Nicholas (2018) *Representation in Cognitive Science*. New York: Oxford University Press.

Sloman, Aaron (1978) *The Computer Revolution in Philosophy*. Hassocks, England: Harvester Press.

Smolensky, Paul (1987) "The Constituent Structure of Connectionist Mental States" *Southern Journal of Philosophy Supplement* 26, pp. 137–160.

Smolensky, Paul (1990) "Tensor Product Variable Binding and the Representation of Symbolic Structures in Connectionist Systems" *Artificial Intelligence* 46, pp. 159–216.

Sterelny, Kim (2010) "Minds: Extended or Scaffolded?" *Phenomenology and the Cognitive Sciences* 9(4), pp. 465–481.

Tremoulet, P., A.M. Leslie & D.G. Hall (2000) "Infant Individuation and Identification of Objects" *Cognitive Development* 15, pp. 499–522.

Vernazzani, Alfredo (2022) "Do We See Facts?" *Mind & Language*. https://doi.org/10.1111/mila.12336

Vernazzani, Alfredo (ms) "Animating Scientific Understanding: Objectualism, Explanationism, and the Plurality of Science's Epistemic Concerns" Unpublished Manuscript.

Wilson, Bret (2003) "The Artistic Tower of Babel: Inextricable Links between Culture and Graphic Development" *Visual Arts Research* 29(57), pp. 25–38.

Zhou, Ke, Huan Luo, Tiangang Zhou, Yan Zhou & Lin Chen (2010) "Topological Change Disturbs Object Continuity in Attentive Tracking" *PNAS* 107(50), pp. 21920–21924.

7 Structural representation as complexity management

Manolo Martínez

7.1 Introduction

Representations are entities that *stand for* or *are about* other entities. They are a central theoretical posit in cognitive science and the main ingredient in one of its foundational hypotheses: the idea that cognition can often be seen as consisting of sequences of computations over representations—what is sometimes called the *computational theory of mind*.

While original formulations of the computational theory of mind (e.g., Fodor 1980) conjured a mental model in which quasi-linguistic entities (the representations) are subjected to formal manipulation (the computations), contemporary versions take more seriously the fact that cognitive systems (e.g., brains, or brain plus bodies plus environment, depending on who you ask) are complex systems, where clean, symbol-based processes are, to put it mildly, hard to identify. One popular reaction to this fact consists in seeking analyses of the notion of representation that may be applied to complex systems, with the hope that they will guide us in the discovery of the computational-representational underpinnings of cognition (Millikan 1984; Papineau 1987; Neander 2017; Shea 2018).

Many of these analyses rely, implicitly or explicitly, on information as it is understood in the Shannonian tradition (Shannon and Weaver 1998–1949; Cover and Thomas 2006). In this approach to information, we first consider a target system, which can be in different states at different moments in time. Having information about that system consists in being in a position to make better-than-chance predictions about the state it is in. Representations are often taken to be, among other things, entities that carry information, in this sense, about what they represent.[1] In a popular elaboration of this idea, representations are *trackers* (Bourget and Mendelovici 2014) or *detectors* (Gładziejewski and Miłkowski 2017; Facchin 2021): entities that are in the business of tracking (detecting) the states of a target system.

DOI: 10.4324/9781003458531-7

One shortcoming of this conception is that many, perhaps most, representations postulated in cognitive science and neuroscience are not of this sort: cognitive maps (O'Keefe and Nadel 1978), theories (Rehder 1999), and in general entities whose structure resemble that of some target system fit awkwardly, if at all, in the detector mold. Recently, philosophers of cognitive science have been paying a fair amount of attention to these other, so-called *structural representations* (also *SRs* henceforth: Piccinini 2020, sec. 12.2; Shea 2024, chapter 2; Artiga 2023; Gładziejewski 2015, among many others).

It is sometimes suggested that informational connections and structural resemblance are just two different ways in which representations can make themselves useful[2]; and therefore, that one should be a pluralist about the grounds of representational status. I show here that detectors and SRs are more closely integrated than that. Consideration of what information (or, as I will say, *signal*) processing involves reveals a place for SRs alongside detectors in a unified whole: the signal processing in which much cognition consists depends both on the very *signals* that are processed and transformed (some of which, I suggest, are closely related to the trackers/detectors of the philosophical literature) and on the devices, the *coders*, that effect those transformations—and these will be closely related to the SRs of the philosophical literature.

Another important idea in cognitive science is that many interesting cognitive phenomena and strategies are the result of resource management: it is better (and sometimes necessary) to implement target transformations, or approximations thereof, in a shorter time, in less space, with processes of lower complexity. In particular, what makes SRs useful, and sometimes necessary, is the need to keep the complexity of transformations manageable. See Section 7.2 for more on signal processing as a model for cognition; and Section 7.4 for more on signals and coders. Section 7.3 introduces some key ideas in the formal study of complexity.

I spell out how complexity management gives rise to SRs in Sections 7.5 and 7.6. These sections exploit the idea (which I first encountered in MacKay 2003) that one can reduce the complexity of some transformations by spreading them out across a graph of very simple processes that, crucially, is structurally similar to some extra-cognitive domain to which the transformation applies. I develop this idea for two kinds of processes often studied in cognitive neuroscience. The first is cognitive maps (Section 7.5): all efficient path-finding algorithms that we know of rely on SRs. I then do it again for hierarchical processing (Section 7.6): the efficient algorithms for object recognition that we know of (that is to say, deep neural networks and their ilk) also work by relying on a data structure that mirrors how visual objects are constructed.[3]

Section 7.7 builds on the previous discussion to intervene in two debates in the philosophical study of representations: first, the argument, popular in 4E cognitive science, that representations are less useful than usually supposed because they can be substituted by the world itself (as "its own model," as the phrase goes); second, the debate surrounding the perceived liberality of some accounts of SR. Section 7.8 offers some concluding remarks.

7.2 The signal-processing aim of cognition

Two ideas on which there is broad consensus in cognitive science and philosophy are, first, that cognition is somehow related to the generation of adaptive behavior; and, second, that what counts as adaptive depends on the circumstances of the agent, information about which is provided by sensory input, broadly conceived. As Valerie Hardcastle puts it: "[t]he very organization of the brain is more intelligible if we take the perspective that the nervous system exists fundamentally for creating motor output in response to sensory input" (1995, 102f).

We can give this idea of "motor output in response to sensory input" a straightforward signal-processing gloss. We model sensory input as a time series, *S*, and motor output as another one, *M*. Cognition then becomes the process of transforming *S* into *M*—see Fig. 7.1. In what follows, I will call Hardcastle's transformation of sensory input into motor output the *signal-processing aim of cognition.* I will assume, without much argument, that the signal-processing aim is indeed an important (but not necessarily *the*) aim of cognition.

Let me reassure readers about a couple of possible worries. First, one does not need to believe that the brain is a computer in order to accept that cognition aims at signal processing in this sense. Many theorists in the anti-representationalist, anti-computationalist tradition accept it too: for a couple of examples, Rodney Brooks claims that agents "must cope appropriately and in a timely fashion with changes in its dynamic environment" (1991, 145); the very idea of *enaction*, as presented by Varela and colleagues, is that perception amounts to "perceptually guided action" (Varela et al. 2016, 173). One may take exception with describing perceptually guided action as a transformation from one variable to another, but that would be confusing the math and the territory (Andrews 2021): transformations between variables are just models of the processes that the enactivist and the embodied-cognitive scientist already accept. In particular, this transformation is carried out by a *cognitive system*, *C*, which is just that: whatever it is that performs the transformation between *S* and *M*. We need not assume that *C* is a brain (as opposed to a brain-in-a-body, or some even more encompassing system) and we need not assume that

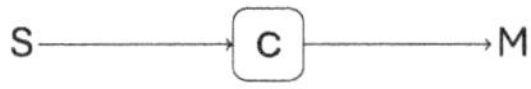

Figure 7.1 Cognition as signal processing—the embarrassingly simplistic picture: a cognitive system *C* transforms input, *S*, into output, *M*.

C doesn't change across time, as a result of its exposition to *S*. It is also perfectly possible for the value of *S* at some time to depend on the value of *M* at earlier times. While Figure 7.1 presents the model in its most general form (and its least useful) I will be more interested in particular instances of variable transformation: a signal gets in and another gets out, as part of the performance of some specific cognitive task.

Second, accepting the signal-processing aim does not involve a commitment to a now implausible feedforward-only picture of the brain (*cf.* Catani and Ffytche 2005, 2232, but see below on object recognition): *C* might involve as many feedback loops as necessary, and as I said above, there might and often will be feedback loops between *S* and *M* themselves. The main commitment one takes when accepting the signal-processing aim is that there is an asymmetry between *S* and *M* (again, recognized by most): we take one aim of cognition to be to generate behavior that is adapted to the agent's circumstances, and not to generate circumstances that are adapted to the agent's behavior. This is compatible with niche construction, scaffolding, and other processes in which behavior feeds back into the agent's circumstances (Stotz 2010).

Apart from these two time series: *S*, which encodes sensory input; *M*, which encodes motor output; and *C*, which we think as a possibly evolving, possibly stateful, device that transforms the former into the latter, the signal-processing perspective also requires some way of cashing out the "adaptive" in "adaptive behavior," or the "appropriately" in "cope appropriately." I'll assume that these vague claims of appropriateness can be given (equally vague, if necessary) numerical expression: given a certain run of the *S* and *M* series up to time *t*, some motor responses at *t* are better than others. More concretely, I'll assume a *loss function* that takes runs of the *S* and *M* series up to *t* as inputs, and a score from 0 to 1 as output, where 0 is maximally appropriate (adaptive, or apt) and 1 minimally so. Typically, our models only consider a few turns into the past, and not the whole history of the time series. Sometimes these series are continuous, and there will be no turns, strictly speaking.

Optimally meeting the signal-processing aim of cognition amounts to always producing the best or one of the best possible motor outputs given some history of circumstances and behaviors. It's unlikely that there is a nomologically possible *C* able to implement this optimal transformation in a reasonable time; in any event, naturally occurring *C*s will typically fall significantly short of that. Much computational cognitive science aims at reconstructing interesting aspects of cognitive function as strategies to adequately meeting the signal-processing aim in the face of, for example, computational or information-processing shortcomings and constraints (Kirby et al. 2015; Sims 2016; Rosch 1999, among many others). In particular, in this chapter I will be interested in *computational* shortcomings and how sometimes they give rise to structural representations.

One final word for the brain is not a computer camp so that they can keep reading with a clear conscience: exploring the computational shortcomings of a certain device, and the strategies by which these shortcomings are negotiated, does not imply that this device is a computer. Most results in complexity theory and information theory are mathematical in nature and are true of every device, independently of whether we regard them as computers, information processors of any other sort, or none of the above. So, for example, NP-complete problems are (in all likelihood) not solvable in polynomial time. That is to say, not solvable by computers, *or any other means*, in polynomial time (Aaronson 2005). Similarly, the fact that something is not a computer or an information-processor of some sort does not somehow allow it to losslessly compress some source to a size smaller than its entropy. That this cannot be done is a mathematical theorem (Shannon and Weaver 1998–1949). These are well-known facts about computability and information processing, on which I expand in Section 7.3. My current point is simply that the bar for some phenomenon to fall under their purview is very low—it just needs to be modelable as a transformation of some variables into other variables. Cognition, as conceived by classical-representationalist and 4E cognitive science alike, qualifies.

7.3 The complexity of transformations

Despite occasional proclamations to the contrary (e.g., Floridi 2013), there is basically one, relatively well understood theory of information transmission, for which Claude E. Shannon (1948) laid the foundations (Cover and Thomas 2006; MacKay 2003 are two excellent textbook expositions). In this body of work, information (or entropy) is uncertainty about the value taken by a *source*: a random variable, V, or a time series, that can take different values with different probabilities, at different times.[4] The entropy of V is notated $H(V)$. Another variable, $\hat{V}$, "carries information" about the source (in information-theoretic parlance, the mutual information between these two variables is nonzero) to the extent that the probability distribution of the source changes when we conditionalize on values of the second variable. Mutual information is notated $I(V,\hat{V})$. This makes entropy and mutual information useful quantities in cognitive science, whenever we are studying a changing system (a source) that a certain cognitive agent is interacting with and keeping track of, via signals that carry some information about that source (Rieke et al. 1999; Dimitrov et al. 2011).

Agents don't keep track of changing systems just because. The point is to act back on them. Above I suggested that we can think of this cycle of world change and behavioral response, in full generality, as signal processing: turning variables informative about the world into variables that encode behavioral plans. One important shortcoming of Shannon's approach to information is that it is blind to crucial aspects of this sort of transformation. I will first present the intuitive idea and then discuss ways to formalize it.

Suppose that we have a random variable, V, whose values are lists of five random numbers from 1 to 100. For example: [54, 60, 28, 24, 13]. We are tasked with finding out, for values of V, if a certain number, say, 42, is higher than any number in the list. According to the signal-processing gloss, we are transforming lists of five numbers into a single binary digit: 1 if 42 is higher than all five numbers, and 0 otherwise. Doing this implies computing five comparisons.

Suppose now that another random variable, $\hat{V}$, spits a tuple: its first member is a list of the same numbers in the same order but additionally, helpfully, its second member is their maximum. For the example above, one of the values of $\hat{V}$ would be ([54, 60, 28, 24, 13], 60). Now our task is five times simpler: we only need to compare 42 with the second member in the tuple. $\hat{V}$ makes our life easier in a clear, tangible sense, but one that cannot be captured by information theory. Both sources are *informationally equivalent* in the sense that one can convert losslessly from any one of them into the other: they have the same entropy, and the mutual information between them is equal to that entropy (cf. Simon 1978; Larkin and Simon 1987): random variables V and $\hat{V}$ are informationally equivalent iff $H(V) = H(\hat{V}) = I(V;\hat{V})$.

V and $\hat{V}$ are informationally equivalent, but not *computationally equivalent*: some computations are more easily done on one than on the other. In particular, in this case, if each list of numbers in V and $\hat{V}$ has n numbers, computing if 42 is larger than the largest in the list requires n comparisons if we are dealing with V and only 1 if we are dealing with $\hat{V}$.

Unfortunately, the crucial notion in this rough characterization of computational equivalence, "easily done," does not lend itself to as clean a formalization as the one afforded by information theory for informational equivalence. One of the main ideas here[5] is to find a function of some key parameter of the problem (which we take to be a yes/no decision, such as "is 42 bigger than all numbers in the list?") whose value grows as slowly as possible yet is always bigger than the number of steps it takes to reach a correct decision. In the example above those functions are n and 1, respectively: the number of steps needed (here, comparisons to be made) grows with the size of the list for V, and doesn't grow at all for $\hat{V}$. We say that the *time complexity* of the respective decision problems are $O(n)$ (pronounced "big oh of n") and $O(1)$ ("big oh of 1").[6] When the function bounding the number of steps has the form n^k for $k > 0$, we say that the problem in question is in complexity class TIME(n^k), better known as P—these are the tractable problems. If *verifying* a given proof of the answer to a decision problem can be done in $O(n^k)$ we say that the problem is in NTIME(n^k), better known as NP (Papadimitriou 1993)—these can be intractable. Analogous *space complexity classes* can be defined when what we bound is the space, or memory, necessary to reach a solution to the decision problem. There are many, *many* complexity classes—the *Complexity Zoo* catalogues many of them (Aaronson et al. 2005).

This approach to complexity is not without problems.[7] For one, complexity classes are very coarse-grained. For example, complexity class P doesn't care if the problem can be solved in n or n^{500} steps. They also only make sense for decision problems—those with binary, yes/no answers. This is the price one pays for being able to prove nontrivial theorems about the complexity of procedures. One consequence of this situation is that when neuroscientists and cognitive scientists study complexity they often rely on computationally feasible measures and heuristics. For example, transformations are taken to be easy if linear, and hard if nonlinear (DiCarlo et al. 2012); easy if low dimensional and hard otherwise (Pang et al. 2016); or complexity is somehow identified with patterns of connectivity (Tononi et al. 1994; Reijneveld et al. 2007).

As I said above, constrained optimization and trade-offs of various kinds are central to computational cognitive science. It's unfortunate that computational complexity remains imperfectly integrated with other formal tools. Still, the approach summarized in this section will allow me to formulate the main idea in this chapter: at least sometimes, it is illuminating to see structural representations as the result of computational-complexity management.

7.4 Signals and coders

There is at least one natural role for representations under the signal-processing view of cognition: they are the signals. Information needs to flow downstream, and something needs to take it there. Figure 7.2 shows

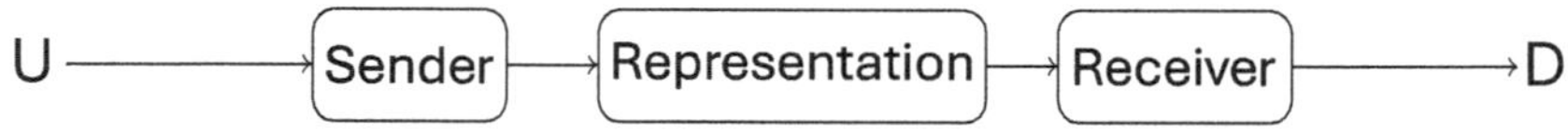

Figure 7.2 The main sender–receiver model of representation.

a bare bones version of how this goes: in the process of transforming some upstream variable, *U*, into a downstream variable, *D*.[8] at least two kinds of entities are involved: *signals* (the representation node) and *coders* (the sender and receiver nodes). The representation node is, like *U* and *D*, a time series: it has values that change across time, as a response to changes in *U*—and, perhaps, indirectly, to changes in *D*. These representations have the same approximate temporal profile as the changes in *U* that they are tracking, and therefore they are comparatively short-lived. The encoder and decoder nodes, on the other hand, are longer-lived entities (although they may be stateful, and may change), tasked with the transformation of *U* into *D* with the help of intervening signals such as *Representation*.

There is another, perhaps less obvious role for representations under the signal-processing view: they are the coders. Indeed, one of the main messages of this chapter is that many of the representations postulated in cognitive science that are thought of as signals should be thought of as coders. This mischaracterization has caused some confusion.

Many examples in discussions of the nature of representations by cognitive scientists and neuroscientists (but only lately by philosophers of cognitive science) present representations as what we could call *active maps*: "systematically organized patterns of activity that are isomorphic with the external world" (Poldrack 2021, 1308). Hippocampal place cells (O'Keefe and Nadel 1978) are a very common example of this kind of representation: as usually described, these are cells such that (i) their activity correlates with the place occupied by the agent and (ii) cells corresponding to contiguous places are preferentially interconnected. It is often claimed that this kind of arrangement helps agents navigate their immediate environment. Peter Godfrey-Smith, for example, discusses a model of hippocampal maps by Allison Reid and John Staddon (1997, 1998) and argues, correctly in my view, that a Reid-Staddon map is not read by anything, but is rather "a self-reading map, or a map that works without being read at all" (Godfrey-Smith 2012, 57). In the following section I show how thinking of Reid-Staddon and other active maps as coders and not signals, and in particular coders aiming at reducing the computational complexity of minimal-path calculations, removes much of the puzzlement surrounding maps without reader. After that I discuss hierarchical representations, under the same light.

7.5 Cognitive maps

I have argued that one important constraint on variable transformations is their complexity: for some variable *S*, the best possible transformation (according to whatever loss function is currently relevant) might be *M*, but still, possible coders *C* that transform *S* into *M* may be too complex for this to be a feasible solution to the problem at hand. Other possible outcome variables M', not optimal but still satisficing, and which are more easily decoded from S, may be preferable.

So, what makes for simpler transformations? An important insight here is that one can do "complicated calculations using simple distributed hardware" (MacKay 2003, chapter 16). The "distributed hardware" in question is composed of simple nodes (capable of, for example, adding, averaging, or calculating the maximum or minimum of its neighbors, but little more) connected in a way such that their composition results in the target calculation. Crucially, this often requires connecting them in a way that mirrors some relevant structure. Hence SRs.

Suppose that some reward (food, perhaps) is sitting somewhere on a certain region of space. It will sometimes be useful to learn how to reach this reward as quickly as possible from elsewhere in that space. One can naturally express this task in terms of a coder C: its input is a signal S that says, "We are <here>," and it is to output another signal M that says "Go <there>," where "there" is a small step in a direction that takes us closer, ideally closest, to the goal. More precisely, C is governed by a loss function that takes pairs of values of S and M to real numbers, such that the smaller the number, the better the value of M is as a next step to the value of S, in the process of reaching the reward.

One, absolutely awful, way of going about this task is to walk that space at random from some starting point until we reach the reward, keeping track of how much distance we have walked, then starting again and again until we are satisfied with the minimum distance in all of those trials. Another, better but still pretty awful, way is to do this systematically: we start from some location and go right, ahead, left, back, etc., in an ordered sequence, keeping track of which combination of moves takes the least time to reach the reward. Awful as they are, these two methods have the advantage of not needing any internal map. The coder C just needs to keep track of sequences of moves—a very, very big number of sequences of moves indeed (see Section 7.7.2) but no map.

It turns out that this task can be solved much more efficiently if we equip the coder with a map-like data structure, such as a *graph*: a collection of *nodes* (think of them as standing for locations) plus a collection of *edges*, each connecting two nodes (think of edges as saying that two locations are directly reachable from one another). These edges, furthermore, are *weighted*, or associated with a real number, which we can think of as the distance between the two nodes the edge connects. Dijkstra (1959–2022) describes an algorithm that solves the above shortest-path task[9]: we start from the node that holds the reward and move outwards. At each step, we visit the node that is closest to the current node and update the distances from all its neighboring nodes to the reward, storing them in the nodes

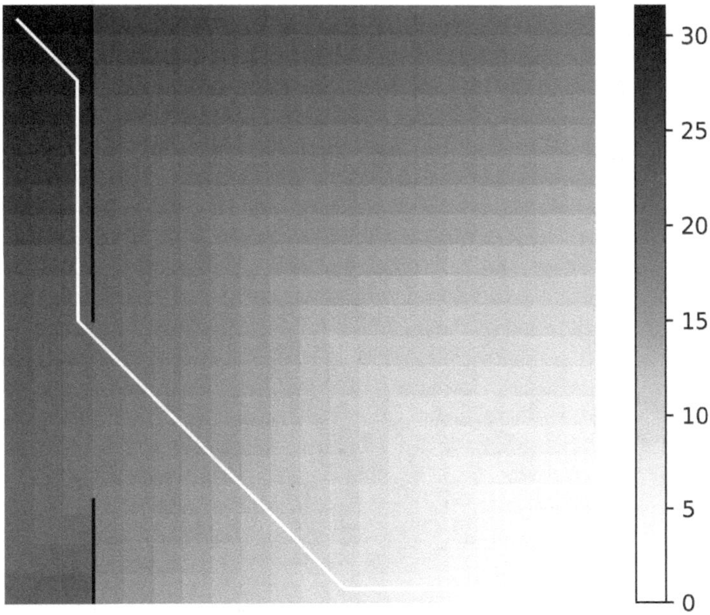

Figure 7.3 A "maze" with one single wall (the black lines). The shade of gray at each node corresponds to its distance to the reward node (at the lower right corner). The white line shows one optimal path from the upper left corner to the reward node.

themselves (this is important—see Section 7.7.1). Once there are no unvisited nodes, or no unvisited node is a neighbor of the current one, we are done. As a result we have, at each node, the shortest distance to the reward, and now a shortest path can be found from any node simply by always moving to the neighbor with the shortest associated distance to the reward (see Figure 7.3).

In a series of papers in the 1990s, Reid and Staddon (1997, 1998) developed a model of how cognitive maps are exploited that can be seen as a variation on Dijkstra's algorithm[10]: instead of a systematic process of visiting nodes, they allow an "activation" (a value that is initially 1 at the reward node and 0 elsewhere) to diffuse through the graph following a simple averaging process (Figure 7.4). If the space is not too complicated (not too many barriers) and diffusion is allowed to happen for long enough, *and* the agent is able to detect small activation gradients (as they diminish exponentially as one gets away from the reward node), then spaces can be navigated efficiently by allowing a Reid-Staddon map to calculate next node from previous node, until the reward node is reached. One can see this as a sort of "neuromorphic Dijkstra": diffusion is, presumably, easier

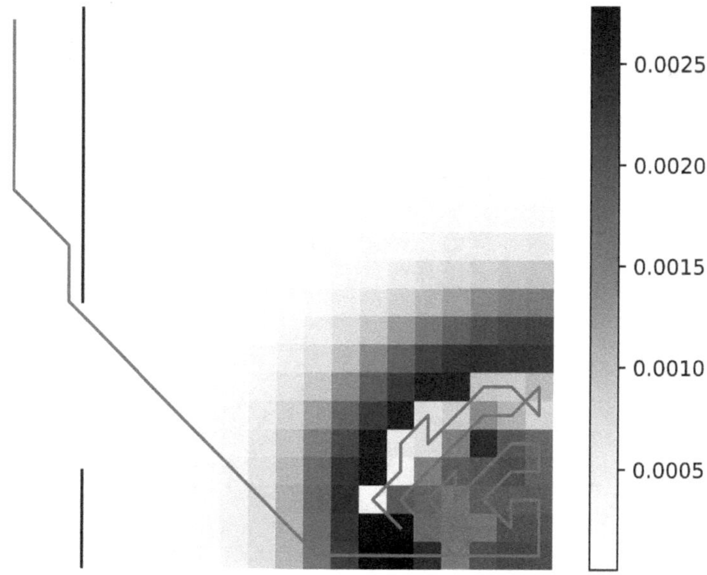

Figure 7.4 The same maze as in Figure 7.3. The erratic behavior of the agent once it reaches the vicinity of the reward node is by design: the Reid-Staddon map-constructing algorithm has quirks that mimic area-restricted search.

to implement by neural populations than Dijkstra updating. While there is no algorithmic guarantee that Reid-Staddon diffusion will result in optimal paths, it will often result in good enough (satisficing) ones at somewhat lower complexity: a loss-complexity trade-off.

In both Dijkstra and Reid-Staddon algorithms, efficient calculation of optimal, or satisficing, paths to a reward is achieved by *message passing* between nodes in a graph that is structurally similar to the region of space where the reward is. Each node does precious little: they just compare distances to the reward among their immediate neighbors and add their own distance to their minimum (for Dijkstra) or average activations among immediate neighbors (for Reid-Staddon). Maps, or at any rate these maps, are useful because almost all the computational heavy lifting is done by the graph being wired up the way it is.

The question of map reading does not even arise. These maps are just data structures that afford easy calculation of some target quantities. Nothing is supposed to read them; they are just ways of computing go-there from you-are-here.

7.6 Hierarchical processing

It is perhaps not terribly surprising that maps are useful in solving spatial tasks—although how precisely they are useful is interesting in its own right. In any event, as it happens, the very same kind of ideas (about passing messages encoding simple computations across distributed processing that is laid out according to a structurally significant graph) are useful in many other, non-spatial tasks. Visual object recognition (the ability to correctly label an object presented visually, DiCarlo et al. 2012) is a case in point. DiCarlo and Cox (2007, 337) propose that we think of the processing that underlies this ability as "untangling object manifolds." That is to say, the information necessary to identify an object is (by hypothesis) in the visual stimulus; it "just" needs to be transformed ("reformatted," DiCarlo et al. 2012, 417) in a way that can be easily turned into a label-assigning decision. This is a clear example of signal processing, the way I described it in Section 7.2.

It is sometimes claimed that invariance is "*the* computational crux of recognition" (DiCarlo et al. 2012, 417, original emphasis): object recognition must somehow be specific enough, for example, to tell different faces from one another; yet invariant to irrelevant transformations, such as pose, rotation, apparent size (e.g., objects depicted closely or further away) or illuminant. It is not entirely clear how cognitive systems solve this problem, but there is broad consensus that, in mammals, this happens in the visual ventral stream through a hierarchical (Güçlü and van Gerven 2015; Poggio and Ullman 2013; Cao and Yamins 2024), mostly feedforward process sandwiched between V1, which detects spatial frequencies and orientations (e.g., Hubel and Wiesel 1968); and the inferotemporal cortex, which seems to support "core object recognition" (DiCarlo et al. 2012), with representations of increasingly complex features in between.

The main algorithmic idea here (dating at least from Hubel and Wiesel 1962; see DiCarlo et al. 2012, 425) is that the ventral hierarchy alternates between linear, AND-like stages (such as the summation of different inputs); and nonlinear, OR-like stages (such as taking the maximum of those inputs). The intuitive idea is that ANDs help with specificity (only accept entities that have this feature *and* that, *and* that other one, etc.), and ORs help with invariance (accept entities that have this feature *or* that one, *or*, etc.). This intuition aside, a lasagna of alternating linear and nonlinear operations is, in essence, a neural network (Riesenhuber and Poggio 1999); and it is well-known that neural networks are universal function approximators (e.g., Mhaskar 1996). In object recognition, the function they'd be approximating is one from pixels (or some other way of capturing retinal input) to labels ("car," "chair," or "face").

The connection with active maps comes at this point. Poggio et al. (2020) show that, while shallow neural networks (ones with just one, arbitrarily

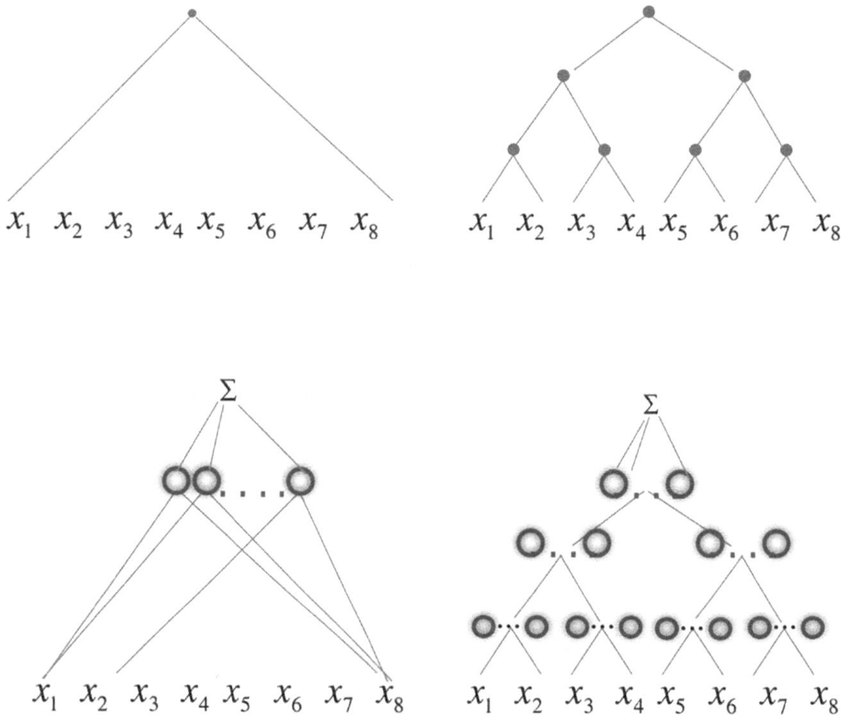

Figure 7.5 Above we have two ways in which a certain collection of worldly variables, $x_1, \ldots, x_8$, could be constructed. On the left, the general case (they are generated somehow or other), and on the right, the computationally tractable case (they are generated via a hierarchical process, in this case a binary graph of simpler transformations.) Below we have a shallow neural network (left) and a deep one (right). The shallow network needs to have exponentially more nodes than the deep one to match the latter's accuracy of reconstruction, if the deep one mirrors the structure of the variable-constructing graph, above.

Source: Reproduced from Poggio et al. (2020, p. 2)

wide, hidden layer) are indeed universal function approximators, one can approximate functions with the same precision with exponentially fewer neurons, *if the neural-network graph is the same as the graph that generates the target function*. The main idea in this result is to think of the function to be approximated as a composition of simpler functions, along a binary tree (see Figure 7.5). If the neural network that is to approximate it has the same width and depth, then it can be much smaller than if it simply crams all of its computational complexity into a single hidden layer. Again here, we are trading computational complexity for distributed processing; and, again, this works because we allow our neural network to pass messages in a way that emulates how the actual worldly feature is constructed.[11]

7.7 Some lessons about structural representation

I have offered two examples (cognitive maps and hierarchical processing) where structural representations help reduce the complexity of computations associated with a certain task (finding shortest paths and object recognition, respectively). In both cases, the mechanisms involve the same principles: (i) nodes in a graph that bears task-relevant similarity to some worldly structure; (ii) very simple computations carried out by each of these nodes; (iii) such that their results are passed along and aggregated across the graph.

Beyond message passing, there are surely other ways in which structural representations can make themselves useful to a cognitive system (see, e.g., Section 7.8), but this is a particularly clear, central way. It is also a useful model to have in mind when evaluating claims about SRs. In what follows, I draw some lessons about structural representations that rely on this mental model.

7.7.1 Usefulness

One important source of 4E-style skepticism about the need for structural (or any other) representation in cognition is captured by Rodney Brooks in a characteristically eloquent way:

> [The physical grounding hypothesis] states that to build a system that is intelligent it is necessary to have its representations grounded in the physical world. Our experience with this approach is that once this commitment is made, the need for traditional symbolic representations soon fades entirely. The key observation is that *the world is its own best model*. It is always exactly up to date. It always contains every detail there is to be known. The trick is to sense it appropriately and often enough.
>
> (Brooks 1990, 5, emphasis mine)

I find the idea that the world is its own best model appealing and useful. Why, indeed, would you bother constructing a model of a place, say, and rely on *that*, if you can just get your informational input, all the informational input, from the place itself? Philosophers often talk of decouplability in this context: Gładziejewski, for example, claims that "we are not dealing with a truly representational system or mechanism unless the structures we want to treat as representations can be used off-line" (Gładziejewski 2015,

76). But, as Brooks suggests, it might be that the trick to cognition lies precisely in insisting on sensing the world appropriately and often enough, rather than on retreating from it and working with a surrogate.

An alternative reply to Brooks is that *the world does not always support message passing*. The shortest-path algorithms I reviewed above rely on being able to store a "message" (an accumulated distance or, more or less interchangeably, a measure of diffused activation) at each node that can be then passed onto neighboring nodes. The world sometimes affords this: some rewards give out, for example, semiochemicals that diffuse more or less according to the Reid-Staddon model—at least so far as the diffusion doesn't have to go around corners. But very often it does not: the reward can be buried, or not be a reward at all, just an important location. In those cases, calculating shortest paths in an efficient manner, at least insofar as we know how to do it, requires having a slot per location in which an intermediate result (provisional minimal distances to the reward or activation coefficients, depending on the model) can be inscribed. Again, perhaps sometimes the world itself can play this role: we can bath it in semiochemicals, or cover it in PostIt notes. Often, though, this too will be impossible or inconvenient. Hence, the need for a map. The case of hierarchical representations is even worse for the world-as-its-own-model strategy: in object recognition nodes do not represents places, but something like stages in the generation of whole visual objects from features such as edges or blobs, through increasingly complex features. The notion that we may store messages on real-world edges or on Gabor patterns is close to nonsensical.

There may be other algorithms that solve the above tasks without relying on maps or hierarchies—ones we haven't come up with yet. It is also possible that natural cognitive systems "satisfice" by relying on worse (higher complexity) algorithms to solve them. Finally, it is possible that the world can afford extramental message passing in many more situations than I have considered. All these possibilities are worth considering; and in particular 4E cognitive scientists have made great strides in showing everyone else that the world is more accommodating to various extended forms of cognition than traditionally supposed. Still, the above considerations provide substantial support to the claim that some SRs are useful, and how.

7.7.2 *The liberality challenge*

Another important worry about structural representations concerns whether there is a principled way to tell useful from useless structure: it is easy to make any extramental system and any cognitive system "structurally correspond" to each other, by taking one graph, and then labeling some of the entities and relations of each system (there will be many candidates) as nodes and edges of that one graph: now both systems "instantiate the graph." Shea's (2018, chapter 5) response, as far as it goes, is perfectly sensible: not all structural correspondences count, but only those that are *exploitable* by the system.

Shea unpacks exploitability in terms of the claims that the extramental structure is "of significance" to the system, and the structurally corresponding cognitive structure is one that the system is "systematically sensitive" to (Shea 2018, 120). I don't think that much more can be said about the exploitability of structure that provides fully general demarcation criteria between useful and useless. On the other hand, the notions of significance and systematic sensitivity risk being too unspecific to inform substantial theorizing about SRs. For example, Facchin (2021) has argued that what he calls *receptors* (systems the states of which reliably correlate with some extramental variable: the signals of Section 7.4, or close enough) count as SRs, among other things because at least sometimes cognitive systems *exploit* structural similarity between receptors and the world (Facchin 2021, sec. 4.1). This is a problem because, Facchin claims, receptors have been independently shown to fail Ramsey's (2007) "job description challenge": we do not need to think of them as representations to understand how they contribute to the system that hosts them.

Regardless of how one feels about Facchin's argument, this line of attack is opened by tying structural-representation status to nonspecific notions of exploitability. Being more specific helps. The idea I have been developing here is that the structural homomorphisms that "count" are those that meaningfully reduce the complexity of a target transformation—they are exploitable in that precise sense, and to that precise degree. For example, Dijkstra's algorithm can be completed in a number of steps that grows as the square of the number of nodes (Skiena 2020, 260): its complexity is $O(n^2)$. The complexity of a naive brute-force exhaustive search over all paths is much higher. For illustration, consider only "self-avoiding walks" (those that do not cross themselves) on a 10 × 10 grid, using a Manhattan neighborhood (only nodes up, down, left, or right count as neighbors, not those across a diagonal). The problem of calculating the optimal path from the top left node to the bottom right node with a Dijkstra-style approach takes $10^2 = 100$ steps, but enumerating all paths between them and then finding the shortest takes involves considering 1,568,758,030,464,750,01 3,214,100 different paths (Bousquet-Mélou et al. 2005, table 2).[12] This is a very striking computational difference; one that clearly makes cognitive maps useful. While we don't know that brains implement Dijkstra-style algorithms over these maps, "it is difficult to imagine solutions of the planning problem which do not involve this core idea" (Khajeh-Alijani et al. 2015, 2). For hierarchical representations, as we have seen, the difference in complexity is also very large: the number of nodes in a shallow net that is able to approximate a certain function *F* to some precision is exponentially larger than for a network that approximates it to the same precision, but has the same structure as the compositional graph of *F*. Here, though, it is size and not time that grows. The job description challenge is met just as explicitly: you can do object recognition with three layers, but you can do it vastly more efficiently if your processing graph mimics the generative structure of the objects you are recognizing.

I am claiming that an SR is exploitable if it reduces the complexity of a target transformation, but of course the usual provisos apply. Exploitability doesn't entail actual exploitation (Rathkopf 2017). For many problems, there will be various different, similarly useful SRs, and complexity considerations will not allow us to pick one among them.

7.8 Concluding remarks

Cognition can often be interpreted in terms of variables being transformed into other variables. These transformations are sometimes dramatically simplified by offloading part of the requisite computations to a data structure, homomorphic in some relevant respects to the real-world phenomenon the transformations in question are related to. I have suggested that these structures are a central example of what philosophers call structural representation.

One potentially awkward aspect of seeing some "variable transformers," aka coders, as representations is that coders are not there to be read (interpreted, consumed, etc.), which is what the mainstream sender–receiver approach claims representations are for. Coders are there to transform variables into other variables. Sometimes, as in the examples discussed in this chapter, they need to *be* structurally similar to some worldly entity in order to do this efficiently, but nothing needs to be there to take note of that similarity. So, this seems to result in a pluralistic metaphysics of representation: there are signal-like representations and coder-like representations, and no nontrivial genus they are both species of. Shea (2018, chapter 2) agrees, for different reasons: he sees structural correspondence and correlational information as two different kinds of exploitable relation, perhaps among others. This is surely true enough, but, as I have argued above, we should not make our theories of representation rest on notions as general as exploitability or significance. Another, and I suggest better, way of integrating signal-like and coder-like representations in a meaningful whole is, precisely, as the two main types of nodes in the sender–receiver architectural motif (Figure 7.2): sometimes signals need to carry world-related information from one variable to another (this may result in a representational role), *and* in order to do this efficiently coders may need to be structurally similar to some entity relevant to the task at hand. This latter fact also commonly elicits attributions of representational role in neuroscience and cognitive science (Poldrack 2021). Our task as philosophers of cognitive science is not to correct this usage, but to understand it.

I have spent some time here spelling out how distributed processing helps explain computationally efficient variable transformation. I have focused on message passing, but surely there are instances of complexity management that do not fit the message-passing mold or do so imperfectly. Imperfect fit will sometimes invite thinking of the phenomenon in question as a peripheral, or borderline, case of structural representation. Some other times, it will invite looking for other complexity-management strategies that still rely on implementing a world-homomorphic graph (e.g., as a way to keep the volume of axonal tracts as small as possible, Kriegeskorte and Diedrichsen 2019, sec. 3.4). On the other hand, I suspect that the usefulness of modeling SRs after public maps is sometimes overblown.

I regard the complexity-management argument for SRs as continuous with, and complementary to, other defenses of the role of representations in cognition. For example, Piccinini's "argument from complex control":

1. Nervous systems perform complex control functions in a computationally tractable way.
2. Performing complex control functions in a computationally tractable way requires processing structural representations.

Therefore, nervous systems process structural representations (Piccinini 2020, 264).

Here, I have provided some complexity-theoretic detail in favor of Piccinini's second premise.

Similarly, Clark and Toribio famously characterize a "representation-hungry" problem as one that "involves reasoning about absent, nonexistent, or counterfactual states of affairs" or "requires the agent to be selectively sensitive to parameters whose ambient physical manifestations are complex and unruly (for example, open-endedly disjunctive)" (Clark and Toribio 1994, 419). Nothing I have said contradicts the importance of representation-hungry problems in the emergence of representations. On the other hand, at least some of the tasks that I have discussed here are not representation-hungry in this sense. A laboratory mouse trying to navigate a region of space can be fully "coupled" to its surroundings and still need representations for the simple reason that the space in question does not have slots in which to store partial distances. Nor is navigation open-endedly disjunctive, complex, or unruly. Indeed, it is a perfectly tractable problem, solvable in polynomial time. Focusing on the complexity of transformation (when this analysis is at all feasible) sometimes provides a more explicit spelling out of the need for reliance on representations than appeals to representation-hungriness.

Finally, nothing of what I have said should be taken to imply that SRs are everywhere useful. The "distributed hardware" trick may work when the target variable transformation (i) can be approximated by many sufficiently simpler transformations that are (ii) hooked up in a world-mirroring way. Even if the existence of universal function approximators perhaps means that the first condition is universally met, there's no reason to think that every cognitively relevant transformation also meets the second condition.

Notes

1 Some accounts make information an essential ingredient in the analysis of representation (Dretske 1986; Isaac 2019; Martínez 2019; see also Mann 2023). Others, most prominently Millikan's biosemantics (1984, 2004) take informational links to be, not representation-constituting relations themselves, but typical consequences of representation-constituting relations.
2 This seems to be Shea's (2018) position.
3 These sections can be seen as developing the idea the representational status and computational status are, at least sometimes, tightly intertwined. See, for example, Lee (2019); Piccinini (2022); Ramsey (2023).
4 Throughout the chapter I use "variable" for both random variables and time series (sequences of random variables.)
5 What follows is a very crude sketch of some ideas in classical complexity theory. An excellent introduction to these ideas for cognitive scientists and philosophers is Rooij et al. (2019). One standard textbook exposition is Papadimitriou (1993).
6 Slightly more precisely, $f(x)$ is $O(g(x))$ if $f(x) \leq a \cdot g(x)$ for some positive constant a, for big enough values of x.
7 Another important approach to complexity is so-called *algorithmic information theory*. Shannon's information theory only makes sense in the context of probabilistic sources, but we also want to make sense of the idea of the informational content of a static object, such as a book, or an algorithm that transforms one variable into another. A suggestion made independently by Kolmogorov (1965), Chaitin (1966), and Solomonoff (1964) equates this informational content with the length of the shortest program (for a designated universal Turing machine, perhaps, Li and Vitányi 2008) that reproduces the object in question. This is called the *algorithmic information content*, or the *Kolmogorov complexity* of the object. I won't be relying on algorithmic information theory here.
8 More complicated models, for example, with two or more senders or receivers, or relays between them, are studied in *network information theory* (El Gamal and Kim 2011).
9 If all the edges weigh the same then an even simpler *breadth-first search* can find shortest paths. Dijkstra's algorithm is the way to solve it for graphs with arbitrary (nonnegative) weights (Skiena 2020, sec. 8.3). This makes no difference for my current purposes: breadth-first searches are also implemented on a graph that is homomorphic to the space to be navigated.

10 While Reid and Staddon do not seem to be aware of Dijkstra's algorithm, or at least do not make reference to it in their papers, other theorists of hippocampal maps, such as Muller et al. (1996) or Khajeh-Alijani et al. (2015), are, and do.

11 One thing to bear in mind here, though, is that this allusion to the construction of worldly features is handwavy and potentially misleading. A neural network that classifies photographs into cars and not cars obviously does not recapitulate the construction of actual cars, assembly-line style. It recapitulates the construction of something like *cars as perceived*: from textures and edges, to simple shapes, to more complicated ones (see, e.g., Zeiler and Fergus 2014, figure 2). Chollet (2021, 205, my emphasis) remarks that it is "the *visual world* [that] is fundamentally spatially hierarchical." Vision scientists tend to find talk of, for example, edges detection unproblematic, but there is some awkwardness to claiming that the visual world *really* is constructed like this—in some flat-footed sense, it really is not.

12 Shea (2024, 36), discussing cognitive maps, sketches an optimal-path algorithm that

> starts by activating a target location to be reached and runs through sequences of co-activation that trace back to the current location. It [then] picks the shortest of the simulated sequences and, switching to online mode, follows that sequence in reverse in order to reach the target.

In this section of his book, Shea is not thinking about complexity questions, but simply offering an intuitive algorithm over a cognitive map. It is still interesting to note that the time that this search would take, unless suitably constrained, grows super-exponentially with the number of place cells.

References

Aaronson, Scott. 2005. "NP-Complete Problems and Physical Reality." *SIGACT News* 36 (1): 30–52. https://doi.org/10.1145/1052796.1052804.

Aaronson, Scott, Greg Kuperberg, and Christopher Granade. 2005. *The Complexity Zoo*. https://cse.unl.edu/~cbourke/latex/ComplexityZoo.pdf.

Andrews, Mel. 2021. "The Math is Not the Territory: Navigating the Free Energy Principle." *Biology & Philosophy* 36 (3): 30. https://doi.org/10.1007/s10539-021-09807-0.

Artiga, Marc. 2023, November. "Understanding Structural Representations." *The British Journal for the Philosophy of Science* 728714. https://doi.org/10.1086/728714.

Bourget, David, and Angela Mendelovici. 2014. "Tracking Representationalism: William Lycan, Fred Dretske and Michael Tye." In *Philosophy of Mind: The Key Thinkers*, 209–235. London: Bloomsbury. https://books.google.com/books?hl=es&lr=&id=SFGpAgAAQBAJ&oi=fnd&pg=PT171&dq=tracking+representationalism&ots=UKVXCLqHl0&sig=bJrGjhsTM5XH6F80CRH-ndgCQDw.

Bousquet-Mélou, M., A. J. Guttmann, and I. Jensen. 2005, September 15. *Self-Avoiding Walks Crossing a Square*. https://doi.org/10.48550/arXiv.cond-mat/0506341.

Brooks, Rodney A. 1990. "Elephants Don't Play Chess." *Robotics and Autonomous Systems* 6 (1–2): 3–15.

Brooks, Rodney A. 1991. "Intelligence without Representation." *Artificial Intelligence* 47 (1–3): 139–159. https://doi.org/10.1016/0004-3702(91)90053-m.

Cao, Rosa, and Daniel Yamins. 2024, September. "Explanatory Models in Neuroscience, Part 1: Taking Mechanistic Abstraction Seriously." *Cognitive Systems Research* 87: 101244. https://doi.org/10.1016/j.cogsys.2024.101244.
Catani, Marco, and Dominic H. Ffytche. 2005. "The Rises and Falls of Disconnection Syndromes." *Brain* 128 (10): 2224–2239. https://doi.org/10.1093/brain/awh622.
Chaitin, Gregory J. 1966. "On the Length of Programs for Computing Finite Binary Sequences." *Journal of the ACM (JACM)* 13 (4): 547–569.
Chollet, François. 2021. *Deep Learning with Python*. Manning.
Clark, Andy, and Josefa Toribio. 1994. "Doing without Representing?" *Synthese* 101 (3): 401–431. https://doi.org/10.1007/BF01063896.
Cover, T. M., and J. A. Thomas. 2006. *Elements of Information Theory*. New York: Wiley.
DiCarlo, James J., and David D. Cox. 2007. "Untangling Invariant Object Recognition." *Trends in Cognitive Sciences* 11 (8): 333–341. https://doi.org/10.1016/j.tics.2007.06.010.
DiCarlo, James J., Davide Zoccolan, and Nicole C. Rust. 2012. "How Does the Brain Solve Visual Object Recognition?" *Neuron* 73 (3): 415–434. https://doi.org/10.1016/j.neuron.2012.01.010.
Dijkstra, E. W. 1959–2022. "A Note on Two Problems in Connexion with Graphs." In *Edsger Wybe Dijkstra*, edited by Krzysztof R. Apt, and Tony Hoare, 1st ed., 287–290. New York, NY, USA: ACM. https://doi.org/10.1145/3544585.3544600.
Dimitrov, Alexander G., Aurel A. Lazar, and Jonathan D. Victor. 2011. "Information Theory in Neuroscience." *Journal of Computational Neuroscience* 30 (1): 1–5. https://doi.org/10.1007/s10827-011-0314-3.
Dretske, Fred. 1986. "Misrepresentation." In *Belief: Form, Content, and Function*, edited by Radu J. Bogdan, 17–36. Oxford University Press. https://philarchive.org/rec/DREM.
El Gamal, Abbas, and Young-Han Kim. 2011. *Network Information Theory*. Cambridge university press.
Facchin, Marco. 2021. "Structural Representations Do Not Meet the Job Description Challenge." *Synthese* 199 (3–4): 5479–5508. https://doi.org/10.1007/s11229–021-03032–8.
Floridi, Luciano. 2013. *The Philosophy of Information*. Oxford University Press.
Fodor, Jerry A. 1980. *The Language of Thought*, 1st ed. Cambridge, MA: Harvard University Press.
Gładziejewski, Paweł. 2015. "Explaining Cognitive Phenomena with Internal Representations: A Mechanistic Perspective." *Studies in Logic, Grammar and Rhetoric* 40 (1): 63–90. https://doi.org/10.1515/slgr-2015–0004.
Gładziejewski, Paweł, and Marcin Miłkowski. 2017. "Structural Representations: Causally Relevant and Different From Detectors." *Biology and Philosophy* 32 (3): 337–355. https://doi.org/10.1007/s10539-017-9562-6.
Godfrey-Smith, Peter. 2012. "Signals, Icons and Beliefs." In *Millikan and Her Critics*, edited by D. Ryder, and J. Kingsbury. Blackwell.
Güçlü, Umut, and Marcel A. J. van Gerven. 2015. "Deep Neural Networks Reveal a Gradient in the Complexity of Neural Representations Across the Ventral Stream." *Journal of Neuroscience* 35 (27): 10005–10014. https://doi.org/10.1523/JNEUROSCI.5023–14.2015.
Hardcastle, Valerie Gray. 1995. "A Critique of Information Processing Theories of Consciousness." *Minds and Machines* 5 (1): 89–107. https://doi.org/10.1007/BF00974191.

Hubel, D. H., and T. N. Wiesel. 1962. "Receptive Fields, Binocular Interaction and Functional Architecture in the Cat's Visual Cortex." *The Journal of Physiology* 160 (1): 106–154. https://doi.org/10.1113/jphysiol.1962.sp006837.

Hubel, D. H., and T. N. Wiesel. 1968. "Receptive Fields and Functional Architecture of Monkey Striate Cortex." *The Journal of Physiology* 195 (1): 215–243. https://doi.org/10.1113/jphysiol.1968.sp008455.

Isaac, Alistair M. C. 2019. "The Semantics Latent in Shannon Information." *The British Journal for the Philosophy of Science* 70 (1): 103–125. https://doi.org/10.1093/bjps/axx029.

Khajeh-Alijani, Azadeh, Robert Urbanczik, and Walter Senn. 2015. "Scale-Free Navigational Planning by Neuronal Traveling Waves." *PLoS One* 10 (7): e0127269. https://doi.org/10.1371/journal.pone.0127269.

Kirby, Simon, Monica Tamariz, Hannah Cornish, and Kenny Smith. 2015, August. "Compression and Communication in the Cultural Evolution of Linguistic Structure." *Cognition* 141: 87–102. https://doi.org/10.1016/j.cognition.2015.03.016.

Kolmogorov, Andrei N. 1965. "Three Approaches to the Quantitative Definition of Information." *Problems of Information Transmission* 1 (1): 1–7.

Kriegeskorte, Nikolaus, and Jörn Diedrichsen. 2019, July. "Peeling the Onion of Brain Representations." *Annual Review of Neuroscience* 42: 407–432. https://doi.org/10.1146/annurev-neuro-080317–061906.

Larkin, Jill H., and Herbert A. Simon. 1987. "Why a Diagram is (Sometimes) Worth Ten Thousand Words." *Cognitive Science* 11 (1): 65–100. https://doi.org/10.1111/j.1551-6708.1987.tb00863.x.

Lee, Jonny. 2019. "Structural Representation and the Two Problems of Content." *Mind & Language* 34 (5): 606–626. https://doi.org/10.1111/mila.12224.

Li, Ming, and Paul Vitányi. 2008. *An Introduction to Kolmogorov Complexity and Its Applications. Texts in Computer Science*, Vol. 9. New York: Springer.

MacKay, David J. C. 2003. *Information Theory, Inference and Learning Algorithms*. Cambridge University Press.

Mann, Stephen Francis. 2023, December. "The Relevance of Communication Theory for Theories of Representation." *Philosophy and the Mind Sciences* 4. https://doi.org/10.33735/phimisci.2023.10992.

Martínez, Manolo. 2019. "Representations are Rate-Distortion Sweet Spots." *Philosophy of Science* 86 (5): 1214–1226. https://doi.org/10.1086/705493.

Mhaskar, H. N. 1996. "Neural Networks for Optimal Approximation of Smooth and Analytic Functions." *Neural Computation* 8 (1): 164–177. https://doi.org/10.1162/neco.1996.8.1.164.

Millikan, Ruth Garrett. 1984. *Language, Thought and Other Biological Categories*. Cambridge, MA: The MIT Press.

Millikan, Ruth Garrett. 2004. *Varieties of Meaning*. The MIT Press.

Muller, R. U., M. Stead, and J. Pach. 1996. "The Hippocampus as a Cognitive Graph." *Journal of General Physiology* 107 (6): 663–694. https://doi.org/10.1085/jgp.107.6.663.

Neander, Karen. 2017. *A Mark of the Mental: In Defense of Informational Teleosemantics*. MIT Press.

O'Keefe, John, and Lynn Nadel. 1978. *The Hippocampus as a Cognitive Map*. Oxford; New York: Clarendon Press; Oxford University Press.

Pang, Rich, Benjamin J. Lansdell, and Adrienne L. Fairhall. 2016. "Dimensionality Reduction in Neuroscience." *Current Biology* 26 (14): R656–R660. https://doi.org/10.1016/j.cub.2016.05.029.

Papadimitriou, Christos. 1993. *Computational Complexity*, 1st ed. Reading, MA: Pearson.

Papineau, David. 1987. *Reality and Representation*. Oxford: Basil Blackwell.
Piccinini, Gualtiero. 2020. *Neurocognitive Mechanisms: Explaining Biological Cognition*. Oxford; New York: Oxford University Press.
Piccinini, Gualtiero. 2022, April. "Situated Neural Representations: Solving the Problems of Content." *Frontiers in Neurorobotics* 16: 846979. https://doi.org/10.3389/fnbot.2022.846979.
Poggio, Tomaso, Andrzej Banburski, and Qianli Liao. 2020, June. "Theoretical Issues in Deep Networks." *Proceedings of the National Academy of Sciences*. https://doi.org/10.1073/pnas.1907369117.
Poggio, Tomaso, and Shimon Ullman. 2013. "Vision: Are Models of Object Recognition Catching Up with the Brain?" *Annals of the New York Academy of Sciences* 1305 (1): 72–82. https://doi.org/10.1111/nyas.12148.
Poldrack, Russell A. 2021. "The Physics of Representation." *Synthese* 199 (1–2): 1307–1325. https://doi.org/10.1007/s11229-020-02793-y.
Ramsey, William M. 2007. *Representation Reconsidered*. Cambridge: Cambridge University Press. https://books.google.com?id=GW59h6Qs5xcC.
Ramsey, William M. 2023, December. "The Hard Problem of Content Is Neither." *Review of Philosophy and Psychology* 1–22. https://doi.org/10.1007/s13164-023-00714-9.
Rathkopf, Charles. 2017. "Neural Information and the Problem of Objectivity." *Biology & Philosophy* 32 (3): 321–336. https://doi.org/10.1007/s10539-017-9561-7.
Rehder, Bob. 1999. "A Causal Model Theory of Categorization." In *Proceedings of the 21st Annual Meeting of the Cognitive Science Society*, 595–600. Psychology Press.
Reid, Alliston K., and J. E. R. Staddon. 1997. "A Reader for the Cognitive Map." *Information Sciences* 100 (1): 217–228. https://doi.org/10.1016/S0020-0255(97)00042-X.
Reid, Alliston K., and J. E. R. Staddon. 1998. "A Dynamic Route Finder for the Cognitive Map." *Psychological Review* 105 (3): 585–601.
Reijneveld, Jaap C., Sophie C. Ponten, Henk W. Berendse, and Cornelis J. Stam. 2007. "The Application of Graph Theoretical Analysis to Complex Networks in the Brain." *Clinical Neurophysiology* 118 (11): 2317–2331. https://doi.org/10.1016/j.clinph.2007.08.010.
Rieke, Fred, David Warland, Rob de Ruyter Van Steveninck, and William S. Bialek. 1999. *Spikes: Exploring the Neural Code*. Cambridge, MA: The MIT Press.
Riesenhuber, Maximilian, and Tomaso Poggio. 1999. "Hierarchical Models of Object Recognition in Cortex." *Nature Neuroscience* 2 (11): 1019–1025. https://doi.org/10.1038/14819.
Rooij, Iris van, Mark Blokpoel, Johan Kwisthout, and Todd Wareham. 2019. *Cognition and Intractability: A Guide to Classical and Parameterized Complexity Analysis*. Cambridge; New York, NY: Cambridge University Press.
Rosch, Eleanor. 1999. "Principles of Categorization." In *Concepts: Core Readings*, edited by Eric Margolis, and Stephen Laurence, 189–206. The MIT Press.
Shannon, Claude E. 1948. "A Mathematical Theory of Communication." *The Bell System Technical Journal* 27 (3): 379–423.
Shannon, Claude E., and Warren Weaver. 1998–1949. *The Mathematical Theory of Communication*. University of Illinois Press.
Shea, Nicholas. 2018. *Representation in Cognitive Science*. Oxford: Oxford University Press.

Shea, Nicholas. 2024. *Concepts at the Interface*, 1st ed. Oxford: Oxford University Press. https://doi.org/10.1093/9780191997167.001.0001.
Simon, Herbert A. 1978. "On the Forms of Mental Representation." In *Perception and Cognition*, edited by W. Savage, 9–3. University of Minnesota Press.
Sims, Chris R. 2016. "Rate–Distortion Theory and Human Perception." *Cognition* 152: 181–198.
Skiena, Steven S. 2020. *The Algorithm Design Manual. Texts in Computer Science*. Cham: Springer International Publishing. https://doi.org/10.1007/978-3-030-54256-6.
Solomonoff, Ray J. 1964. "A Formal Theory of Inductive Inference. Part I." *Information and Control* 7 (1): 1–22.
Stotz, Karola. 2010. "Human Nature and Cognitive–Developmental Niche Construction." *Phenomenology and the Cognitive Sciences* 9 (4): 483–501. https://doi.org/10.1007/s11097-010-9178-7.
Tononi, G., O. Sporns, and G. M. Edelman. 1994. "A Measure for Brain Complexity: Relating Functional Segregation and Integration in the Nervous System." *Proceedings of the National Academy of Sciences* 91 (11): 5033–5037. https://doi.org/10.1073/pnas.91.11.5033.
Varela, Francisco J., Evan Thompson, and Eleanor Rosch. 2016. *The Embodied Mind: Cognitive Science and Human Experience*, Revised ed. Cambridge, MA; London, England: The MIT Press.
Zeiler, Matthew D., and Rob Fergus. 2014. "Visualizing and Understanding Convolutional Networks." In *Computer Vision—ECCV 2014*, 818–833. Cham: Springer. https://doi.org/10.1007/978-3-319-10590-1_53.

8 The mind-brain is a computer, but what is (neural) computation?

Corey J. Maley and Oron Shagrir

8.1 Introduction

A major hypothesis in the cognitive and brain sciences is that the brain is a computing system and that the computations it performs play a key role in producing cognitive function and behavior.[1] But what sort of computer is the brain? It is widely acknowledged that the rise of artificial neural networks (ANNs) in the 1980s, then again in the 2000s, has resulted in a paradigm shift in artificial intelligence and in the cognitive and neural sciences. Attending to the details of how artificial and biological neural networks compute reveals that, implicit in the main premise of *neural computationalism*—the view that the *brain* is literally a computational system—is that the brain is a different kind of a computer than we might have thought.[2]

More specifically, the view that the mind is computational relies on a view of computation defined by the implementation of Turing machines (TMs) or other abstract automata. The idea that there may be an important connection between computation and mentality was developed soon after Turing's (and others') work on the mathematical theory of computation. Philosophers of mind developed the idea in detail, resulting in the computational theory of mind (CTM): the view that the mind is an implementation of a TM, and mental operations are explained via the computations that the mind performs.[3]

The development of learning algorithms for multilayer perceptrons, or ANNs, in the 1980s led to the view that *mental* operations were not computations based on the implementation of TMs, but more akin to the type of computation performed by ANNs. This view—connectionism—led to a large literature in the philosophy of mind about whether mental computation is computation via rules and representations (i.e., an implementation of a TM or other abstract computational specification, like a software program) or computation via neural-network computation.

We will call the claim that cognitive function and behavior produced by the brain is a result of neural-network computation *neural*

DOI: 10.4324/9781003458531-8

computationalism. To a certain extent, this claim is an empirical one, and we will leave it to the neuroscientists to debate it. Our focus is *philosophical*. A philosophical theory of computation aims to explicate the notion of computation at play in these (and other) investigations. It asks: What distinguishes physical computing systems, such as nervous systems and laptops, from other non-computational physical systems, such as stomachs, mousetraps, and hurricanes? Is the difference between computing and non-computing a matter of fact, or something that is in the eye of the beholder, i.e. observer relative? And how exactly do computational facts contribute to the explanation of cognitive function and behavior? Our claim is that the rise of neural computation has also challenged the notion of computation that dominated the philosophy of mind until about a decade ago. In other words, our claim is that neural computationalism asserts not only that the brain is a different kind of computational system but also that a clear understanding of neural computation *itself* challenges the core notion that was at the heart of philosophical theories of computation.

Our aim in this chapter is twofold. First, we show the ways in which neural computation—both artificial and biological—is fundamentally different from computation as understood in logic and the theories of computation (proof theory, classical computability, automata theory and formal languages, computational-complexity theory, and so on). In other words, whatever neural computation is, it is not as simple as the implementation of a TM or other abstract automata. Second, we argue that this fact reveals some dogmas about the nature of physical computation that must be challenged if we are to take seriously—as we do—neural computationalism.

The chapter proceeds as follows: Section 8.2 highlights the conceptual underpinnings of early accounts of physical computation from the 1960s to the 2000s. We highlight two conceptual assumptions which we call the *logical dogma* and the *architectural dogma*. Section 8.3 shows how the re-appearance of neural computation in the 1980s has shaken the dogmas and forced us to rethink the nature of physical computation. Section 8.4 turns to the more recent accounts of computation—mechanistic, semantic, and pragmatic—that have abandoned the dogmas and advanced core notions of computation that can arguably accommodate classical, neural, and other kinds of computation.

8.2 Old-style accounts of computation (the 1960s to the 2000s)

A theory of physical computation primarily aims to identify the conditions distinguishing computing from non-computing physical systems. A natural place to start the investigation is with the mathematical theory of computation (classical computability theory, automata theory, and related topics) that defines precisely certain types of computing systems. It is thus hardly surprising that the early philosophical theories of physical computation are tightly connected to the formal notions we find in logic and computer science. Scholars described physical computing systems as *syntactic engines* (Dennett 1991) and *automatic formal systems* (Haugeland 1981), or as being "illuminated by the idea of a Turing machine" (Crane 2015: 70), while they described computing processes as executions of *programs* (Cummins 1989), *algorithms* (Copeland 1996), and *effective procedures* (Crane 2015), or as implementations of *automata* (Chalmers 1994). We call this assumption—that there is a strong linkage between these formal notions and physical computation—*the logical dogma*. While these notions are perfectly sound in their original mathematical contexts, the dogma is the assumption that they serve us well when defining physical computing systems. To be sure, we do not here challenge the idea that a physical computing system implements (or at least is described by) some formalism. The dogma, rather, is the requirement that the formalism is one found in logic and classical computability.

Consider, for example, the well-known TM. This formalism[4] is the conceptual basis for the mathematical study of computability. The TM has also been accepted as the conceptual basis for computation itself. However, from our perspective, it is important to remember that the TM was developed—quite explicitly—as a mathematical *model* of human computation.[5] Moreover, it is a mathematical model of only *one* type of computation. Specifically, TMs model humans performing computations by manipulating the individual digits of a digital representation of a number (e.g., the usual, digit-by-digit way that we multiply large numbers by hand). Somewhat more technically, TMs are a mathematical model of an effective procedure: an algorithm that a person can (in principle) use to solve a mathematical problem by performing a finite number of steps using only paper, pencil, and a finite number of symbols, where each step is maximally simple (meaning it can be done without any ingenuity or guessing) and where there is a clear stopping point.

However, human computers also used slide rules and other devices whose use is not a step in an effective procedure in the standard sense: again, effective procedures are those procedures that humans can carry out with only paper and pencil (Copeland 2024). Of course, one can often compute the same *function* (e.g., addition) using either a slide rule or an effective procedure; this is just computing the same function by different methods. But the use of a slide rule, an electronic integrator, or really any device beyond a writing instrument and something to write on, is not an effective method.

Nevertheless, the study of TMs (as well as related mathematical formalisms of effective procedures) became known as the Theory of Computation, and in turn, this family of abstract automata was taken as *the* conceptual foundation for *all* computation. Coupled with the decreased use of analog computers in science, industry, and academia in the last few decades of the 20th century, analog computers were forgotten, and digital computation was often taken to be the only type of computation, or at least the only type of computation relevant to the mind sciences.

The second dogma—*the architectural dogma*—is that the difference between computing and non-computing physical systems involves (at least in part) a distinct (abstract) *causal structure*. This is not meant to imply that non-computing physical systems lack causal structure, but rather that they lack the *right kind* of structure. Another way to put it is that computing systems possess the right kind of architectural or functional profile. This distinct profile is often associated with discrete, digital, or stepwise architectures, which are usually markedly different from more continuous ones. Thus, physical computing systems have been characterized in terms of "specific architectural" properties (Newell and Simon 1976: 117) and, more concretely, as *physical symbol systems* (Newell and Simon 1976). Physical computation processes have been defined as "the generation of output strings of digits from input strings of digits in accordance with a general rule" (Piccinini 2008a: 34), or in terms of *step-satisfaction* (Cummins 1989), *syntactic processes* (Fodor 1980; Stich 1983), or classical architectures (Fodor and Pylyshyn 1988), meaning (roughly) that they traffic "in a languagelike medium" (Fodor 1994: 9). The dogma that is shared by these accounts is that possessing the right kind of architectural profile is a *necessary* condition for computing, and, therefore, systems that lack this profile do not compute. These accounts diverge, however, on the question of whether the right architecture is also a *sufficient* condition. Some say yes, while others identify other (necessary) features. We focus on the necessity of architectural profiles. No one denies, of course, that computational vehicles have some architectural profiles. The dogma is the assumption that physical computation favors certain kinds of architectural profiles (e.g., discrete, or digital) over others (e.g., continuous, or hybrid ones).

Robert Cummins's account is typical of earlier theories of physical computation. He says that virtually every physical system satisfies some input–output (physical) function: a causal process that connects physical inputs to physical outputs. Some of these physical processes compute functions, whereas others do not. As Cummins says, "functions need not be computed to be satisfied. Set mousetraps satisfy a function from tripping to snapping without computing it, and physical objects of all kinds satisfy mechanical functions without computing them" (p. 91). Computation, he continues, is a special way of satisfying functions. "To compute a function g is to execute a program that gives o as its output on input i just in case $g(i) = o$. Computing reduces to program execution" (p. 91). Program execution, in turn, "involves *steps*," and so "program execution reduces to step-satisfaction" (p. 92). The bottom line, then, is that computation is defined by step-satisfaction, and computation occurs when every step in a process is satisfied.

You can see how Cummins embeds and relates the logical and architectural dogmas. First, he associates computation with a certain notion from computation theory—namely, a *program* (this is the *logical dogma*). Specifically, he claims that the difference between a physical system that satisfies a function by computing it and one that satisfies a function without computing it is that the former "executes a program" while the latter does not. But what exactly does it mean that a physical system executes a program? At this point, Cummins associates the theoretical notion—that is, the execution of a program—with a distinct architectural property that is characteristic of computing physical systems (this is the *architectural dogma*). This architectural property is *step-satisfaction*, which apparently refers to processes that proceed in a stepwise fashion. This architectural property is a necessary feature of computing systems: processes that do not possess it are not computing. A non-computational physical process typically proceeds in a non-stepwise, more continuous fashion.

In what follows, we discuss the challenges that neural computation raises with respect to the logical and architectural dogmas and how more recent accounts of computation deal with these challenges.

8.3 Dogmas reconsidered (the 1980s to the 2000s)

Given the centrality of the notion of computation in the philosophy of mind and in the computational sciences, it is astonishing that philosophers assumed—rather than clarified—what "computation" means. But the situation gradually changed in the 1980s due to two sources of pressure. One is the triviality results that were famously advanced by Putnam (1988), Searle (1992), and others.[6] These results imply that virtually every physical system, including rocks, mousetraps, and chairs, implements every automaton or executes every program or algorithm. Even when further constraints are imposed on the notion of computation, we are left with computational indeterminacies that force us to further reshape our philosophical theories of computation (see, e.g., Shagrir 2022). We will not deal with this issue of trivialities and indeterminacies here.

A second source of pressure on philosophical theories of computation was the increasing acknowledgment of many kinds of physical computation, such as analog, biological, quantum, and so-called natural and unconventional computation.[7] These kinds of computation, or at least some of them, do not fit well with the dogmas. The disparity between these kinds of computation and old-style theories of computation did not make a great impression in philosophy, at least in the early days, perhaps because it was assumed that these kinds of computation are irrelevant to the mind/brain context, which philosophers cared about. This situation has changed dramatically with the re-appearance of neural networks in cognitive science (mainly under the name "connectionism") and in theoretical neuroscience in the 1980s. These networks were seen as a new paradigm, challenging the classical computational approaches in cognitive science, and today they dominate the fields of AI, computational neuroscience, and computational cognitive science. More importantly for our purposes, these networks raise serious concerns about the validity of the dogmas. These concerns cannot be ignored if one wants to advance a theory of computation relevant to the sciences of mind and brain.

The emergence of neural computationalism was particularly impactful for CTM. One implication was that the mind is not necessarily a digital TM-style computer, as emphasized by many early supporters of CTM. At the very least, one should somehow bridge the gap between the allegedly digital computations that take place in the mind and the neural computations that take place in the brain. It is thus not surprising that along the years philosophers extended the meaning of computation within CTM. For example, in an early version of the CTM entry in the *Stanford Encyclopedia of Philosophy*, CTM is identified with digital TM-style computation:

> This essay is concerned with a particular philosophical view that holds that the mind literally is a digital computer (in a specific sense of "computer" to be developed), and that thought literally is a kind of computation. This view . . . will be called the "Computational Theory of Mind" (CTM).
>
> (Horst 2003)

In a more recent version of the entry on CTM, however, Rescorla (2020) identifies this statement with the classical computational theory of mind (CCTM):

> According to CCTM, the mind is a computational system similar in important respects to a Turing machine, and core mental processes (e.g., reasoning, decision-making, and problem solving) are computations similar in important respects to computations executed by a Turing machine.

CTM itself is taken here to be the view that mental processes are either classical (TM-style) or nonclassical computations.

Our concern here is with another implication of neural computationalism to CTM. Neural computationalism implies that the notion of computation assumed in CTM—basically the old-style theories of computation—is not tenable. It implies that even if mental processes are digital, TM-style computations (which is itself a coherent claim), the notion of computation that underlies CTM must be refined, since it is in tension with other kinds of computation, most notably with neural computation. Indeed, theorists and philosophers have started to reexamine their assumptions about computation with an eye to the rapid advancements within the computational approaches in the cognitive and neural sciences. In what follows, we will describe some of the tension between neural computation and old-style theories of computation (Section 8.3.1) and then review three strategies that were proposed to deal with this tension (Section 8.3.2).

8.3.1 On the tension between neural computation and old-style theories of computation

The tension between neural computation and the dogmas is not straightforward. In fact, pioneering works linking neural networks with logic and automata theory might suggest there is no tension at all. McCulloch and Pitts (1943) showed how a neural network can implement logic gates and logical systems, which inspired work that defines finite-state automata in terms of these neural cells (Minsky 1967). But there are many other types of neural computation that do not fit well within the dogmas. We will focus here on two of these types, one artificial, the other biological. First, we consider attractor neural networks, a particular type of artificial neural-network architecture that has proven extraordinarily useful for modeling how the brain maintains persistent states of activity. Next, we consider different neural coding schemes in biological neurons as they are currently understood in neuroscience.

Even before going into these more sophisticated examples, it is interesting to see the tension between Cummins's account of computation (outlined above) and even simple two-layer feedforward "connectionist" networks (e.g., Rumelhart and McClelland's renowned past-tense model). It is unclear in what sense these networks satisfy anything like step-satisfaction (thus executing programs) because these networks have only two layers (one input layer and one output layer). Therefore, their dynamics consist of one (input–output) step. Assuming that *single*-step satisfaction is not step satisfaction (to avoid trivializing "computation"), this network does not proceed via steps and thus does not compute.

Now to attractor neural networks. These are fully recurrent networks in which every unit is typically interconnected to all other units, and each unit is updated according to the information it receives from all other units. In a way, all units ("cells") have the same role: there are no designated input, hidden, or output units. As such, these are circuit-level models aimed at characterizing a cognitive/neural phenomenon in terms of the network's overall activity. A state of the network is simply the activity of all its cells at a certain moment. The dynamics of the network involves iteratively updating one or more cells ("asynchronous") or all cells ("synchronous") at once, where these updates are a function of information from other cells. The dynamics can, in principle, proceed *ad infinitum* in indefinitely many different trajectories; Hopfield (1982) showed, however, that if a network satisfies certain constraints, it will always reach a fixed point ("attractor") where its state will never change without new inputs. These attractor networks are used to describe many cognitive phenomena, such as short-term memory (Hopfield 1982; Amit 1989; Khona and Fiete 2022) or how the brain maintains stable eye positioning, even in total darkness (Seung 1996).

How do these attractor networks fare with the logical and architectural dogmas? Let's start with the logical dogma. The behavior of these networks is neither described nor explained in terms of programs or automata. The theory of these neural networks appeals to the principles found in dynamical theory and state-space analysis. Hopfield and those who followed him were physicists who imported the mathematical framework from statistical mechanics. Hopfield himself described the behavior of the network with the notion of *energy landscape* or *function*, which represents the evolution of the network with respect to time and captures all possible trajectories as a function of the initial state of the network. Each point on this landscape represents the total activity of *all* cells in the network at a certain moment (state), and the dynamics of the network is described in terms of movement through this landscape. The theory of attractors neural networks simply does not rely on logic and computability. The description and the explanation of its behavior does not invoke the notions of automaton, program, algorithm, or other familiar notions in logic and computability. However, it does invoke principles of statistical mechanics (and other mathematical theories) applied to describe and explain the behavior of non-computing physical systems.

Considering the architectural dogma, the structures of attractor networks look similar to the architectures of non-computing dynamical systems. This is not surprising. Attractor neural networks were designed with an eye to exploiting the kind of dynamics found in random magnetic systems, such as spin-glass systems (Hopfield 1982). In the spin-glass magnetic system, particles (which are analogous to cells) spin in certain directions (analogous to activities), depending on the magnetic force of other cells (analogous to the weighted input to the cell). Again, the dynamics of the system is described in terms of movement through an energy landscape, where the energy is often the temperature of the system, and the equilibrium points are states in which the system has cooled off. This means that attractor neural networks have structures and dynamics similar to those of spin-glass magnetic systems. If the structure and dynamics of the system conform with their architectural or functional profile, there is not much difference between the architecture of the attractor neural networks that compute and the architecture of magnetic systems that do not.

As noted above, many old-style accounts of computation assume that the architectural profile is only a necessary (but not sufficient) condition for computation. Their supporters could retort that *both* the attractor networks and the magnetic systems have an architectural profile that is necessary for computing, but only the first has some other feature (a sufficient one) that results in computation (e.g., processing representations). The most straightforward reply is to point out that the architecture of Hopfield networks is nonclassical, and, therefore, does not comply with the conditions set by old-style theories of computation. Our argument against the architectural dogma, however, is more nuanced. It starts with the premise that attractor networks and magnetic systems have similar architectures. The next step is to present a dilemma: Does a spin-glass magnetic system possess the right kind of architectural profile, namely, one that is typical of computing systems? If the answer is NO, then the same answer applies to attractor neural systems; namely, these networks do not possess the right kind of architectural profile and, therefore, do not compute. If the answer is YES, then, it seems, very many physical systems hold the right kind of architectural profile, or at least those whose dynamics is described in terms of particles and the forces that apply to them. These include the boiling water in the kettle, changes in the atmospheric pressure, and many other systems that we do not consider as computing. If all these systems hold the right kind of architectural profile, then the architectural profile plays little to no role in distinguishing between computing and non-computing systems. It is more like a dogma. Thus, one must choose between the claim that attractor networks do not compute and the claim that architecture plays little role, if any at all, in the characterization of computation.

A similar tension arises when we attend to the details of how neuroscientists characterize computations performed by biological neurons. In contrast to attractor networks, the focus here is on single neurons, rather than circuits, and a description of the activity of actual biological neurons, rather than an idealized model. Nevertheless, the lesson is the same: the computations performed by attractor networks and single neurons—both of which are used extensively in neuroscience research—present serious challenges to the dogmas mentioned above.

Individual neurons are often taken to be the basic building blocks of computation in the brain. Koch (1999) describes in detail the physical processes that allow for information processing in single neurons. More recently, Rolls (2021) states that "single neurons can be thought of as the computational units of the [nervous] system, and the level at which the information is exchanged by the spiking activity between the computational elements of the brain" (p. 8). Some of the superficial similarities between neural spikes and the bits of digital computers might suggest that neural computation and digital computation are much more alike than they really are, but it is important to be clear about their differences.

Unlike the bits of digital computers, the individual spikes of individual neurons do not typically matter for information processing[8]; rather, it is the frequency that a neuron fires, the location of the neuron relative to other neurons in its network, or the timing of a neural spike's arrival at a neuron relative to another neural spike that carries the relevant information. Flipping a single bit in a digital computer can lead to dramatic misrepresentation of a variable's value, the wrong instruction being processed, or even catastrophic system failure.[9] An errant single spike in a neuron, however, typically does not have a significant effect on the information being carried or processed.

The way that these variables—frequency, location, or timing—carry information is very different than the way it is carried digitally, thus putting a different kind of pressure on the dogmas we mentioned above. While the details of neural computation and representation are still being actively studied, particularly when it comes to ensembles of neurons in circuits and networks, some elementary principles are relatively well understood. We will briefly examine firing frequency in single neurons, but other information-carrying features of neural firing can be similarly characterized.

The frequencies of neural spike trains can often be characterized by a tuning curve, which describes a particular neuron's firing rate as a function of a stimulus that can vary along some dimension. Neurons are often said to have a "preferred" value of the stimulus, which means that the neuron fires with the highest frequency at that value. As the distance of the value increases from that preferred value, the frequency decreases. A more concrete example illustrates this idea.

Many studies, starting with Hubel and Wiesel (1962), have shown that neurons in the visual cortex selectively respond to stimuli oriented in a particular direction. So, for example, a neuron might fire most rapidly when presented with a visual bar oriented at 45 degrees. That neuron will still fire when the bar is 10 degrees away (i.e., 35 degrees or 55 degrees) but less rapidly. And it will fire still less rapidly when the bar is 20 degrees from the preferred orientation (i.e., 25 or 65 degrees). Figure 8.1 shows an example: the left side shows visual stimuli presented at different orientations and the resulting neural firing patterns; the right side shows how firing frequency is a function of the bar's orientation.

This firing rate is a straightforward analog representation of the stimulus: the frequency increases (or decreases) as the stimulus presented gets close to (or further from) the preferred angle (Maley 2011, 2021).

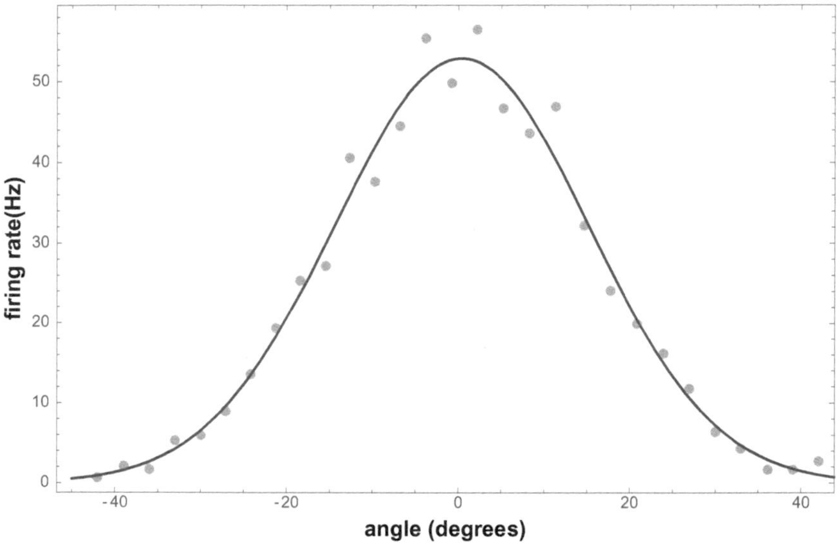

Figure 8.1 Tuning curve from a neuron in monkey primary visual cortex, showing a decrease in firing rate as the magnitude of the angle deviates from the neuron's preferred angle.

Source: Adapted from Dayan and Abbott (2001), from data originally in Henry et al. (1974)

This type of representation is quite common: "most neurons in the cortex respond selectively to different stimuli, and the firing rate increases with the intensity of the stimulus" (Rolls 2021, p. 14). While neural spikes are like bits (in electronic computers) in that they are binary and consist of electrical potentials, that is where the similarity ends. An individual bit is a digit in a larger numerical representation, and that larger numerical representation encodes some variable. For example, one might encode the brightness of a pixel using an eight-bit variable that ranges from 0 to 255. In binary, this range would be from 00000000 to 11111111. Starting at some value, say 10001011, and increasing the brightness by one would cause the variable to change from 10001011 to 10001100. Notice that the last three bits all changed from 0s to 1s (or vice versa). When the intensity (here, brightness) increases, the overall pattern of bits will change, and in turn, the physical electrical potentials that implement the bits will change but not in any systematic way such that those electrical potentials increase as the "stimulus" intensity (the brightness) increases.

Much like the attractor networks mentioned above, nothing about biological neural computation relies on the implementation of automata (the logical dogma) or an old-style computational architecture (the architectural dogma). Additionally, the mathematical theories used to describe biological neural dynamics also apply to other non-computing systems (just as with attractor networks), and we are left with the same dilemma as before.

8.3.2 *Tackling the tension between neural computation and old-style theories of computation*

These and other tensions between neural computation and old-style theories of computation moved philosophers and theorists to reassess the status of the two dogmas vis-à-vis neural computation. We will consider here three strategies that have been taken in response to this challenge. *The strategy of exclusion*, as the name suggests, is to exclude all neural networks and neural "computation" from the proper computational domain. This aim is achieved if we strengthen even further the conditions that are set by the two dogmas in terms closer to logic gates, finite-state automata, or cellular automata, thus excluding neural networks and neural computation. This strategy has been taken by some proponents of CCTM. Pylyshyn (1984: 70–71), for example, draws the line between computers and non-computers along the architectural property of possessing a distinction between *program* and *memory*. A TM, he says, is a computer because it possesses this property, but finite-state automata that do not have (a separate) memory are not computers. In other words, and per the logical dogma, computation is identified with some systems that are studied in computability theory (TMs, pushdown automata) but not all of them (finite-state automata). This strategy excludes the "new connectionist" machines (Pylyshyn's term), but as Pylyshyn himself notes, it also excludes finite-state automata, cellular automata, and other systems that many see as computing. Thus, this strategy of throwing out the baby (finite-state automata) with the bathwater (networks) has not gained popularity.

The splitting strategy is to draw the line between computing and non-computing within neural computation. According to this strategy, we leave the constraints (dogmas) on computation as they are. Those neural networks that satisfy these constraints compute, while the other networks that fail to satisfy the constraints do not compute. Cummins might have adopted this strategy. At one point, Cummins suggests that at least some neural networks do not compute at all. In a paper with Georg Schwarz, they write that a network "whose representational states are real-valued activation vectors and weight matrices, and whose dynamics is given by a set of differential equations" does not compute (Cummins and Schwarz 1991: 69). Presumably, they do not compute because their dynamics does not involve step-satisfaction. At another point, however, Cummins writes: "What I really hope is that the conservative connectionist will turn out to be on the right track" (1989: 155). The hope is that at least the simple two-layer feedforward network can be seen as computing, even though, arguably, its dynamics consist of only a single step.

This idea that some neural networks compute and others don't is put forward more explicitly by Piccinini (2008b). The rough idea is that networks that operate with discrete values typically compute, whereas networks that operate with continuous (or real) values typically do not ("since continuous variables are not strings of digits, their manipulation does not count as computation"). The difficulty with this idea is apparent: replacing the binary-value cells in a network with sigmoid value cells (or cells that satisfy some other continuous function) will make the network continuous. But no one in the sciences thinks that this replacement makes the network go from computing to not computing. Either both networks compute or both do not. Another example is a Hopfield network with discrete point attractors versus one whose fixed points lie along a continuous line ("line-attractor"). There is a sense in which the first is discrete but the second is not. Yet no one thinks that the latter networks are less computational. In any case, Piccinini (2015) himself later abandons this strategy and develops a version of a mechanistic theory that purports to dispense with the dogmas and accommodate analog and neural computation in full.

Lastly, *the strategy of inclusion* embraces all kinds of neural networks as computing systems. This is achieved when you relax the constraints on computation enough that they are satisfied by continuous networks, two-layer networks, attractor networks, and so on. David Chalmers, for example, developed a theory of implementation involving automata in the mid-1990s. But when pressed about other formalisms, he presents an inclusive stance. He declares that "computations are generally specified relative to some formalism, and there is a wide variety of formalisms: these include Turing machines, Pascal programs, cellular automata, and neural networks, among others" (2011: 228). The price of including neural networks of all kinds, however, is the inclusion of many formalisms that are used to describe both the behavior of computing and non-computing systems (such as statistical mechanics used for attractor networks but also for non-computing systems). Chalmers does say that among the other formalisms that are implemented are ones found in dynamic systems, evolution, and artificial life, including continuous mathematics; he says that "the current framework can fairly easily be extended to deal with computation over continuous quantities such as real numbers" (2011: 347). However, it is unclear to what extent this claim is actually true.[10]

Even setting that concern aside, let us assume, with Chalmers, that to compute is to implement some formalism. As Chalmers himself acknowledges, every physical system implements some formalism (often more than one). But many of these systems do not compute. In other words, if you extend the theory to *every* formalism, then it encompasses too much by applying to virtually every physical system. The advantage of the logical and architectural constraints is that they draw a line between physical systems that satisfy the constraints and those that do not (and the criticism is that this line does not separate computing and non-computing systems). Chalmers's proposal, however, takes us back to square one. Since any physical system can be seen as implementing some formalism, we are left with no conditions that distinguish computing and non-computing systems.

To summarize this section: neural computation has raised some concerns about the dogmas that define the core notion of computation. Three strategies have been taken to deal with this challenge, but none of them have resulted in a fully satisfactorily account. The new generation of theories of computation, developed in the 2010s and after, aim to do better and to offer characterizations of computation that can accommodate neural and other kinds of computation, but exclude non-computing systems.

8.4 Computation without dogmas (the 2010s and onward)

Current theories of computation have taken a fourth strategy. They associate, like Chalmers, computation with some formalism; some accounts further suggest that the abstract formalism identifies a "medium-independent" causal structure. But this condition cannot suffice for computation. Every physical system can be associated with, and even can be seen as implementing, a formalism, and often more than one. The old-style theories of computation adopt the logical and architectural dogmas precisely to limit the formalisms and their implementations to the ones found in logic and computability theory and so to their more discrete architectures. But the developments in cognitive science, neuroscience, and AI have shown that the dogmas exclude the wrong kind of systems, including neural computation, and have largely led researchers to abandon the two dogmas. The challenge now is to find a replacement for the logical and architectural dogmas, namely, to supplement or strengthen the medium-independent condition in some way. The dominant theories of computation today aim to do just that. They add another condition that will accommodate, on the one hand, neural and other kinds of computation, but that will exclude, on the other hand, non-computing physical processes. In this section, we will briefly refer to the main philosophical theories (or families of theories) of computation today: the mechanistic, the semantic, and the pragmatic accounts.

Mechanistic accounts locate computation in the mechanistic framework in philosophy of science. They require that the system operate in a medium-independent fashion and that one of the system's teleological functions be to proceed in a medium-independent fashion.[11] The most complete version of this account is provided in Piccinini (2015). According to this account, hurricanes, stomachs, and mousetraps, even if we can describe their processes in abstract, medium-independent terms, do not have the function to proceed in this way; hence, they do not compute. Mechanists mainly disagree on the nature of the function. Piccinini defines the teleological function in dispositional terms (see also Maley and Piccinini 2017). Others think that it should be defined in historical (e.g., etiological) terms (Coelho Mollo 2019; Fresco and Miłkowski 2021); yet others argue that the mechanistic account does not need to appeal to teleological functions at all (Dewhurst 2018; Tucker 2018).

Semantic accounts locate computation within the framework of representation. They say that computing systems represent; mousetraps, stomachs, planets, and rocks represent nothing; thus, they do not compute. More specifically, semantic accounts claim that the contents of physical states constrain the way that the correct abstract (implemented) structure is selected or individuated (Shagrir 2001; Sprevak 2010). The most comprehensive version of this account is advanced in Shagrir (2022). Proponents of this account disagree about the additional ingredients that define computation and about the nature of representational content in computing systems.[12]

Finally, *pragmatic* accounts emphasize the utility of computational models and descriptions over the metaphysics of computation. These accounts differ in the role assigned to computational models: this role can be prediction (Cao 2018; Schweizer 2019), explanation (Dewhurst 2018; Egan 2025; Schweizer 2019), simplification (Chirimuuta 2024), inference (Matthew and Dresner 2017), or something else. Pragmatists might disagree about the properties that play this role, namely, whether computational properties are more formal (Matthews and Dresner 2017; Egan 2025; Schweizer 2019), semantic (Colombo 2021), or mechanistic (Dewhurst 2018). In this respect, pragmatic accounts can be seen as variants of mechanistic, semantic, or other accounts. What is distinctive about pragmatic accounts is the role played by the scientists when employing these properties. *Perspectival* accounts, advanced, for example, by Dewhurst and Schweizer, say that the scientist's perspective is an essential part of the individuation of computation.

Pragmatists differ about the ontological commitments of computational descriptions. Some take a more realist position. They assume that computational descriptions are true descriptions of the physical system, and the fact that the scientist opts for descriptions that are more (e.g., explanatorily) useful than others is not distinctive to computational descriptions; it applies to other mathematical (e.g., measurement schemes) descriptions of physical systems (Dewhurst 2018; Matthews and Dresner 2017; Egan 2025). Chirimuuta (2024) takes a more agnostic approach regarding the ontological commitments behind the claim that the brain is literally a computer and proposes an analogical interpretation. Schweizer (2019) adopts an antirealist approach, which he calls the *computational stance*. He suggests that computational descriptions, in contrast to other mathematical descriptions, are normative; there is something the system *should* implement. But he leaves it to observers to determine what, exactly, that "something" is.

In sum, current philosophical theories characterize computation with minimal appeal to the dogmas. They identify computation neither with notions from logic and computability theory nor with specific architectural profiles. Instead, they appeal to other features that are, arguably, essential to distinguish computing from non-computing systems. Mechanistic accounts appeal to teleological function, semantic accounts appeal to representational content, and pragmatic accounts appeal to the utility of computational descriptions.

8.5 Summary

The chapter outlined the story of how computational approaches in cognitive and neural sciences affected our philosophical theories of computation. In the old times, just a few decades ago, theories of computation identified computation with theoretical notions found in logic, computability, or automata theory (e.g., program) with specific architectural features (e.g., state-transition) and with a specification of algorithms that explain how a mathematical function is *computed*. The re-appearance of neural computation in the 1980s, in psychology, neuroscience, and AI, has led to a gradual abandonment of these logical and architectural dogmas. There is still work to be done: there is not yet consensus on a correct theory of computation that accommodates both extant computational types alongside neural computation in a principled manner. By abandoning different dogmas, conceptual space is opened for new theories of computation that are liberated from our past assumptions about the nature of computation.

To understand neural computation, we must understand the method or algorithm that performs that computation, which in turn depends on how the values of the mathematical operation are represented. This requires understanding the details of neural mechanisms. For example, various details about the structure and function of individual neural components are needed to explain how a single neuron integrates its inputs from dendrites (Koch 1999). Understanding such neural integration requires understanding how the physical features of neural representations are manipulated by physical neural processes (Maley 2021).

Notes

1 Christof Koch, for example, says: "The brain computes! This is accepted as a truism by the majority of neuroscientists engaged in discovering the principles employed in the design and operation of nervous systems" (1999: 1).
2 Notably, claims that relate thinking and computing go back at least to Leibniz and Hobbes (Colombo and Piccinini 2023). Von Neumann (1958) already highlighted the differences between the brain and digital computing systems.
3 Whether all mental operations or only cognitive mental operations are computational is an ongoing debate in the philosophy of mind and cognitive science.
4 Other extensionally equivalent formalisms include Church's lambda calculus, Markov algorithms, and Gödel's theory of general recursive functions. Other less-powerful formalisms, such as pushdown automata and finite-state automata, are also used to study issues of computability.
5 Strictly speaking, the TM model is a letter machine, and as such it is more restricted than a human computer. But Turing (1936) outlines a reduction theorem that shows that the TM model precisely captures the functions that a human computer can compute. See Gandy (1988), Sieg (1994), and Copeland (2004) for further discussion on Turing's analysis of human computation.
6 See Sprevak (2018) for a review.
7 See, for example, Brabazon et al. (2015).
8 Although some recent studies suggest that the precise "shape" of a neural spike does have measurable downstream effects on neurons to which it is connected (Zbili and Debanne 2019).
9 Single bit flips, thought to have been caused by neutrons from cosmic rays, have caused election counts to be off by thousands, and caused aircraft to suffer near loss of control while airborne.
10 Maley (2021) addresses this question in the case of analog computers.
11 Here is the way Piccinini puts the requirement: "One of its functions is to manipulate vehicles based solely on differences between different portions of the vehicles according to a rule defined over the vehicles" (2015: 274).
12 There are middle-ground ("pluralistic") accounts that say that in some contexts (e.g., brains) or levels we take a semantic approach, whereas in other contexts (e.g., computer science) we take non-semantic (e.g., mechanistic) approach (Rescorla 2013; Lee 2021; Curtis-Trudel 2022).

References

Amit, D.J. (1989). *Modeling Brain Function: The World of Attractor Neural Networks*. Cambridge: Cambridge University Press.
Brabazon, A., O'Neill, M., and McGarraghy, S. (2015). *Natural Computing Algorithms*. Berlin: Springer.
Cao, R. (2018). "Computational Explanations and Neural Coding." In *The Routledge Handbook of the Computational Mind*, edited by M. Sprevak and M. Colombo, pp. 283–296. London: Routledge.
Chalmers, D.J. (1994). "On Implementing a Computation." *Minds and Machines* 4(4):391–402.
Chalmers, D.J. (2011). "A Computational Foundation for the Study of Cognition." *Journal of Cognitive Science* 12:323–357.
Chirimuuta, M. (2024). *The Brain Abstracted: Simplification in the History and Philosophy of Neuroscience*. MIT Press.

Coelho Mollo, D. (2019). "Are There Teleological Functions to Compute?" *Philosophy of Science 86*:431–452.
Colombo, M. (2021). "(Mis)computation in Computational Psychiatry." In *Neural Mechanisms*, edited by F. Calzavarini and M. Viola, Vol. 17, pp. 427–448. Springer.
Colombo, M., and Piccinini, G. (2023). *The Computational Theory of Mind.* Cambridge University Press.
Copeland, B.J. (1996). "What is Computation?" *Synthese 108*:335–359.
Copeland, B.J. (2004). *The Essential Turing: Seminal Writings in Computing, Logic, Philosophy, Artificial Intelligence, and Artificial Life Plus the Secrets of Enigma.* Oxford: Clarendon Press; New York: Oxford University Press.
Copeland, B.J. (2024). "The Church-Turing Thesis." In *The Stanford Encyclopedia of Philosophy*, edited by Edward N. Zalta and Uri Nodelman, Spring 2024 Edition. https://plato.stanford.edu/archives/spr2024/entries/church-turing/
Crane, T. (2015). *The Mechanical Mind: A Philosophical Introduction to Minds, Machines and Mental Representation.* London: Routledge.
Cummins, R.C. (1989). *Meaning and Mental Representation.* Cambridge, MA: MIT Press.
Cummins, R.C., and Schwarz, G. (1991). "Connectionism, Computation, and Cognition." In *Connectionism and the Philosophy of Mind*, edited by T. Horgan and J. Tienson, pp. 60–73. Dordrecht: Kluwer.
Curtis-Trudel, A. (2022). "The Determinacy of Computation." *Synthese 200*:43.
Dayan, P., and Abbott, L.F. (2001). *Theoretical Neuroscience.* MIT Press.
Dennett, D. C. (1991). *Consciousness Explained.* Little, Brown, and Co.
Dewhurst, J. (2018). "Computing Mechanisms Without Proper Functions." *Minds and Machines 28*:569–588.
Egan, F. (2025). *Deflating Mental Representation.* MIT Press.
Fodor, J.A. (1980). "Methodological Solipsism Considered as a Research Strategy in Cognitive Psychology." *Behavioral and Brain Sciences 3*:63–73.
Fodor, J.A. (1994). *The Elm and the Expert.* Cambridge, MA: MIT Press.
Fodor, J.A., and Pylyshyn, Z.W. (1988). "Connectionism and Cognitive Architecture: A Critical Analysis." *Cognition 28*:3–71.
Fresco, N., and Miłkowski, M. (2021). "Mechanistic Computational Individuation without Biting the Bullet." *The British Journal for the Philosophy of Science* 72:431–438.
Gandy, R.O. (1988). "The Confluence of Ideas in 1936." In *The Universal Turing Machine*, edited by R. Herken, pp. 51–111. New York: Oxford University Press.
Haugeland, J. (1981). "Semantic Engines: An Introduction to Mind Design." In *Mind Design*, edited by John Haugeland, pp. 1–34. Cambridge, MA: MIT Press.
Henry, G.H., Dreher, B., and Bishop, P.O. (1974). Orientation Specificity of Cells in Cat Striate Cortex. *Journal of Neurophysiology 37*(6):1394–1409. https://doi.org/10.1152/jn.1974.37.6.1394
Hopfield, J.J. (1982). "Neural Networks and Physical Systems with Emergent Collective Computational Abilities." *Proceedings of the National Academy of Science, USA 79*:2554–2558.
Horst, S. (2003). "The Computational Theory of Mind." In *The Stanford Encyclopedia of Philosophy*, edited by E.N. Zalta. https://plato.stanford.edu/archives/sum2003/entries/computational-mind/
Hubel, D.H., and Wiesel, T.N. (1962). "Receptive Fields, Binocular Interaction and Functional Architecture in the Cat's Visual Cortex." *The Journal of Physiology 160*:106–154.

Khona, M., & Fiete, I.R. (2022). Attractor and Integrator Networks in the Brain. *Nature Reviews Neuroscience 23*(12):744–766. https://doi.org/10.1038/s41583-022-00642-0
Koch, C. (1999). *Biophysics of Computation: Information Processing in Single Neurons*. New York: Oxford University Press.
Lee, J. (2021). "Mechanisms, Wide Functions, and Content: Towards a Computational Pluralism." *The British Journal for the Philosophy of Science 72*:221–244.
Maley, C.J. (2011). "Analog and Digital, Continuous and Discrete." *Philosophical Studies 155*:117–131.
Maley, C.J. (2021). "The Physicality of Representation." *Synthese 199*: 14725–14750. https://doi.org/10.1007/s11229-021-03441-9
Maley, C.J., and Piccinini, G. (2017). "A Unified Mechanistic Account of Teleological Functions for Psychology and Neuroscience." In *Explanation and Integration in Mind and Brain Science*, edited by D.M. Kaplan, pp.236–256. New York: Oxford University Press.
Matthews, R.J., and Dresner, E. (2017). "Measurement and Computational Skepticism." *Noûs 51*:832–854.
McCulloch, W.S., and Pitts, W.H. (1943). "A Logical Calculus of the Ideas Immanent in Nervous Activity." *Bulletin of Mathematical Biophysics 5*:115–133.
Minsky, M.L. (1967). *Computation: Finite and Infinite Machines*. Englewood Cliffs, NJ: Prentice-Hall.
Newell, A., and Simon, H.A. (1976). "Computer Science as Empirical Inquiry: Symbols and Search." *Communications of the Association for Computing Machinery 19*:113–126.
Piccinini, G. (2008a). "Computers." *Pacific Philosophical Quarterly 89*:32–73.
Piccinini, G. (2008b). "Some Neural Networks Compute, Others Don't." *Neural Networks 21*:311–321.
Piccinini, G. (2015). *Physical Computation: A Mechanistic Account*. New York: Oxford University Press.
Putnam, H. (1988). *Representation and Reality*. Cambridge, MA: MIT Press.
Pylyshyn, Z.W. (1984). *Computation and Cognition: Toward a Foundation for Cognitive Science*. Cambridge, MA: MIT Press.
Rescorla, M. (2013). "Against Structuralist Theories of Computational Implementation." *The British Journal for the Philosophy of Science 64*:681–707.
Rescorla, M. (2020). "The Computational Theory of Mind." In *The Stanford Encyclopedia of Philosophy*, edited by E.N. Zalta. https://plato.stanford.edu/archives/fall2020/entries/computational-mind/
Rolls, E.T. (2021). *Brain Computations: What and How*. Oxford University Press.
Schweizer, P. (2019). "Computation in Physical Systems: A Normative Mapping Account." In *On the Cognitive, Ethical, and Scientific Dimensions of Artificial Intelligence: Themes from IACAP 2016*, edited by D. Berkich and M.V. d'Alfonso, pp. 27–47. Berlin: Springer.
Searle, J.R. (1992). *The Rediscovery of the Mind*. Cambridge, MA: MIT Press.
Seung, H.S. (1996). "How the Brain Keeps the Eyes Still." *Proceedings of the National Academy of Sciences USA 93*:13339–13344.
Shagrir, O. (2001). "Content, Computation and Externalism." *Mind 110*:369–400.
Shagrir, O. (2022). *The Nature of Physical Computation*. New York: Oxford University Press.
Sieg, W. (1994). "Mechanical Procedures and Mathematical Experience." In *Mathematics and Mind*, edited by A. George, pp.71–117. New York: Oxford University Press.

Sprevak, M. (2010). "Computation, Individuation, and the Received View on Representation." *Studies in History and Philosophy of Science 41*:260–270.

Sprevak, M. (2018). "Triviality Arguments About Computational Implementation." In *Routledge Handbook of the Computational Mind*, edited by M. Sprevak and M. Colombo, pp. 175–191. London: Routledge.

Stich, S.P. (1983). *From Folk Psychology to Cognitive Science: The Case Against Belief*. Cambridge, MA: MIT Press.

Tucker, C. (2018). "How to Explain Miscomputation." *Philosophers' Imprint 18*:1–17.

Turing, A.M. (1936). "On Computable Numbers, with an Application to the Entscheidungsproblem." *Proceedings of the London Mathematical Society 42* (Series 2):230–265.

Von Neumann, J. (1958). *The Computer and the Brain*. London: Yale University Press.

Zbili, M., & Debanne, D. (2019). Past and Future of Analog-Digital Modulation of Synaptic Transmission. *Frontiers in Cellular Neuroscience 13*:1290. https://doi.org/10.3389/fncel.2019.00160

9 Inference in (neuro)cognitive systems

Urte Laukaityte and Matteo Colombo

9.1 Introduction

In artificial intelligence, psychology, neuroscience, and biology, inferences and inferential processes have variously been attributed to different kinds of systems to explain many of their mental capacities. For example, in the psychology of human reasoning, several researchers have argued that deductive reasoning is explained by a mental logic composed of syntactically defined inference rules such as *modus ponens* (Braine & O'Brien 1998; Inhelder & Piaget 1958; Rips 1994). Others disagree that there is a mental logic, arguing that inferences defined over semantic mental models explain human reasoning instead (Johnson-Laird 1983). Still other researchers account for human reasoning by appealing to neither a mental logic nor mental models but to a form of inferential heuristics (Gigerenzer & Brighton 2009) or else to two types of reasoning systems both capable of producing inferences despite their different properties—one system would be deliberative, slow, conscious, and rule-following, while the other intuitive, fast, unconscious, and associative (Colombo 2018; Evans & Stanovich 2013; Kahneman 2011).

Inferential processes have been ascribed to both neural and nonneural cognitive systems to explain mental phenomena beyond human reasoning (Kihlstrom 1987; Nes & Chan 2020, Part I). Researchers interested in sensory processing have posited unconscious inferences to study and elucidate the workings of perceptual systems and the perceptual experiences they generate, including illusions and hallucinations (Gregory 1980; Hatfield 2002). Others have argued that perception can result from rationally evaluable inferential processes, and so perceptual experiences can themselves be more or less rational (Siegel 2017). Researchers interested in the role of uncertainty in behavior have argued that the mind implements probabilistic inferences that can account for a great many mental capacities as well as their disruption, including in areas such as motor control, perception, emotion, learning, memory, decision-making, reasoning, communication,

DOI: 10.4324/9781003458531-9

social behavior, and group decisions in both biological and artificial systems (Bartolo & Averbeck 2021; Colombo 2023; Jardri & Denève 2013; Oaksford & Chater 2007; Pettit 2023; Tenenbaum et al. 2011).

Some have made the case that plants, fungi, and

> cellular collectives . . . carry out, not a rote set of steps, but rather a suite of second-order functions such as active inference [which is the process of acting upon the environment to generate the expected sensory input] to achieve coherent function in a wide range of changing and unpredictable scenarios.
>
> (Levin 2023, 1879)

(See Calvo and Friston (2017) for the hypothesis that plants make inferences and Money (2021) for an argument that fungi possess cognitive capacities.) Even individual cells have been said to rely on inferences to make sense of their world in that "[a] single small agent, like a cell, . . . is able to infer, store, and operate with respect to a small subset of the patterns existing in its environment" (Levin 2019, 11; see also Lyon 2015). Similarly, Reber et al. (2023, 101) note that cells have a particular "information architecture . . ., enabling cellular prediction as inference and anticipation." This broadening of the scope of cognition in the mind and life sciences is part of the so-called *biogenic approach*, or the emerging *basal cognition* research program (Lyon 2006; Lyon et al. 2021—see also the rest of the Philosophical Transactions B Theme Issue "Basal cognition: conceptual tools and the view from the single cell"). In other words, under this view, even comparatively simple (i.e., basal) living systems are characterized by information processing and cognition—and these have also been studied in terms of inference.

So the word "inference" is used in many ways and for a wide range of purposes in cognitive science. Sometimes, it is used to refer to a cluster of distinct *processes*, such as combining multiple sources of information based on their reliability, estimating the value of a variable based on uncertain data, drawing causal conclusions, assigning a given object to a certain category, computing the probability of a hypothesis based on explanatory or statistical evidence, making predictions and generalizations, or drawing deductively valid conclusions. At other times, the term "inference" is used to denote the *output* of a process, including sensory, associative, or more abstract rule-following, deliberative processes. As noted above, researchers ascribe both inferential processes and inferential outputs to entities besides people, including cells, plants, fungi, non-human animals, and group agents. These ascriptions, in turn, inform the study of capacities beyond explicit reasoning and allow for analyses and explanations of those capacities that have normative character.

In spite (or likely because) of this widespread usage across fields, it is challenging to state with precision what inferences are, what various forms of inference in different kinds of (neuro)cognitive systems have in common, why and how (neuro)cognitive systems employ them, and how inferences could be implemented in different physical substrates and identified as such. Ideally, a general account of inference in (neuro)cognitive systems would help us answer these challenges. We see three conditions such an account should satisfy. First, it should do justice to how researchers in cognitive science employ "inference" for disparate purposes and potentially feed back into scientific practice in a useful way. Second, it should allow us to draw fruitful distinctions between different forms of inference, as well as between inferential and non-inferential processes, as needed. And third, it should capture the normative character of inference, whereby inferential operations and inferential outputs can be evaluated as more or less rational. Here, we sketch a philosophically viable, updated account of inference in (neuro)cognitive systems that satisfies these desiderata, affirming the meaningfulness of the sorts of uses seen in the cognitive science literature.

We start, in Section 9.2, by examining four assumptions found within recent philosophical work on inference: (i) inference involves conscious reckoning, (ii) rational inference is best understood in contrast to a-rational associative processes, (iii) inference involves language-like, symbolic representations, and (iv) inferential processes compel an indirect relationship between mind and world. We argue that each of these assumptions is out of touch with scientific practice, and hence current philosophical formulations are not sufficient for the needs of cognitive scientists. In Section 9.3, we address a potential rejoinder against loosening the stringency of current philosophical conceptions of inference. Following this, in Section 9.4, we work out a broader account of inference that can be used to make sense of the nature of inferences in (neuro)cognitive systems across multiple spatiotemporal scales. This notion builds on Susanna Siegel's (2017) theoretical work, combining it with insights from basal cognition (widening the scope of cognition as well as representation) and ecological rationality (extending the scope of rational appraisal). In Section 9.4, we conclude by summarizing the purview and significance of our treatment of inference in (neuro) cognitive systems.

9.2 Inference in philosophy—four common assumptions

Inference has been a foundational notion in philosophy, especially logic and epistemology. As a result, it is reasonable to consider existing philosophical accounts for clarifying cognitive scientists' usages of this concept. However, most philosophical treatments of inference tend to make a number of overly stringent assumptions, which means that, in their current form, these accounts are less helpful than they could be in illuminating and potentially feeding back into scientific practice. The assumptions in question may rest on a history of philosophical uses of "inference" primarily in the context of abstract argument and logical form rather than cognitive or neural processing.

9.2.1 *Inference involves conscious reckoning*

One of these common assumptions is that inference involves conscious reckoning. In fact, by "inference" philosophers often mean a species of reasoned, personal-level, conscious change in view, in which one starts off with some thoughts, and then, following a process of rule-based reasoning, draws some conclusion, provided that the starting point gives adequate epistemic support or reasons for coming to that conclusion (Harman 1986; Adler 2008). While the emphasis on reasons plausibly indicates that inference is a kind of process that can modulate the rational status of the states it produces, one common qualification is that drawing inferences is a conscious activity which only humans can perform.

For example, after distinguishing two species of inference—namely, inferences executed by System 1 unconsciously, automatically, and quickly, and inferences performed by System 2 consciously, deliberately, and slowly (Kahneman 2011)—Boghossian (2014, 2) settles on theorizing about inferences which consist in processes performed by what he calls System 1.5 reasoning. This would be the sort of reasoning process that is conscious and voluntary but need not be slow and effortful. Given this characterization, Boghossian (2014, 5) suggests that "[i]nferring necessarily involves the thinker *taking* his premises to support his conclusion and drawing his conclusion *because* of that fact." Boghossian (2014) is not primarily concerned with accommodating the uses of inference in cognitive science, but any account that construes inference as a personal-level activity that is necessarily conscious cannot do justice to those uses.

To begin, if inference were to require conscious reckoning, then established ways of talking about unconscious inference in cognitive science would be inappropriate or at best metaphorical ways of speaking. But there are in fact clear cases of inference, as studied by cognitive scientists, that do not arise from conscious deliberation. For instance, inferring somebody's emotional state from subtle features of their face is something humans and other animals do all the time without awareness; navigating the environment without bumping into other objects involves inferences about one's speed, current position, and distance from other objects, all of which we are not consciously aware of. Furthermore, there is telling evidence that non-human animals can make social and causal inferences, likely without explicit awareness of making them (Grosenick et al. 2007; Völter & Call

2017). This is because, as Siegel maintains, "reckoning is not necessary for inferring" (Siegel 2017, 78); and so, inferrers need not be able to identify the reasons for why they draw a certain conclusion. Again, Boghossian (2014) may well agree that a broader range of phenomena could fall under "inference," focusing his picture on a narrower subset for other theoretical reasons. But as it stands, his formulation does not meet the desiderata that we have outlined, and hence it is not suitable for empirical research across cognitive science.

Although distinctions between subpersonal and personal-level mental processes can be made and are helpful for some purposes (e.g., Colombo 2013), there is an important sense in which the latter are grounded in the former. Consider visual processing from light waves to having a clear percept of an acacia tree, or speech processing from audio waves to appreciating a 16th-century poem. As Bargh and Hassin (2022, 204–205) put it, "[m]uch preparatory work needs to be done in order to create our conscious [personal-level] experiences, and by logical necessity, all this work is unconscious [subpersonal]." Acknowledging this continuity in different kinds of systems is key for developing an adequate account of inference in (neuro)cognitive systems. Such an account should be based on the idea that differences between so-called higher forms of mental capacities, like conscious reckoning or linguistically guided, rule-following, deliberative reasoning, are elaborations of more basal forms of cognition, which emerge gradually over the course of developmental and evolutionary history, although much cognitive processing remains subpersonal and unconscious (e.g., Allen & Reber 1998). The notion of inference should be available to apply in this context, provided it is scientifically fruitful to do so.

9.2.2 *A dichotomy between a-rational association and rational inference*

Not all philosophical accounts over-intellectualize inference by requiring it to involve conscious reckoning. Quilty-Dunn and Mandelbaum (2018), for example, distinguish "bare inferential transitions" from "rich inferential transitions," where the former do not involve conscious endorsement of the premises that feature in the inference as reasons to draw the conclusion.

According to Quilty-Dunn and Mandelbaum's (2018) account, a system's bare inferential transitions from state A to state B should be defined in terms of three individually necessary and jointly sufficient conditions: (i) the states A and B featuring in the inference have a "discursive format"; (ii) some If-Then rule of a mental logic built into the architecture of the system governs the transition from A to B, such that A satisfies the antecedent of the rule in virtue of A's syntactic structure and B satisfies the consequent in virtue of B's syntactic structure; and (iii) there is no other factor accounting for the transition from A to B (Quilty-Dunn & Mandelbaum 2018, 539). According to this picture, to clarify and precisify the definition, we should contrast inferential transitions with "dumb," or a-rational, associative transitions.

Under this conception, a-rational associative transitions between two representational states of a system are not governed by any mental logic but only depend on some facts about the causal history of the system or some resemblance between the two states. These transitions would consist in a series of states without any normatively appraisable relation between them and without evident asymmetry between the premises and the conclusion in the transition. For instance, consider the transition from the idea that it is cold in the office to the idea that one should play a song by Ice Cube now. In contrast, according to Quilty-Dunn and Mandelbaum (2018; Mandelbaum & Quilty-Dunn 2020), inferences are truth-preserving, asymmetric, rationally appraisable transitions between language-like representational states of a system. Some of these representations constitute premises and a particular representation constitutes a conclusion, all of which are characterized by formal, syntactic relations sanctioned by some system of mental logic.

There are at least two reasons why this view of inference cannot do justice to scientific usage either, nor can it ground an adequate account of inference for (neuro)cognitive systems. One reason is that Quilty-Dunn and Mandelbaum's (2018; see also Mandelbaum & Quilty-Dunn 2020) account is committed to the claim that inferences require a language of thought with logico-syntactic properties. But there is no evidence the neural code could constitute a language of thought (e.g., Maley 2020; Piccinini & Bahar 2013), and, as discussed in the next section, it is unclear what it even means for states featuring in an inference to have a "discursive representational format."

Another reason is that Quilty-Dunn and Mandelbaum's (2018) account assumes a clear-cut dichotomy between rational inference and a-rational association. And yet, there is in fact heterogeneity within both associative and inferential processes, which ultimately lie along a continuum—a-rational, inflexible, stereotypical responses time-locked to a stimulus on the one extreme end and rational, flexible, deliberative language-mediated responses to a set of premises grounded in a formal system of logic on the other end (Papineau & Heyes 2006).

Consider the second issue concerning the association-inference dichotomy first. This dichotomy is ambiguous between the relation among the states involved in a transition and the learning process responsible for creating that relation. Our focus is the former. For Quilty-Dunn and Mandelbaum (2018), both bare and rich inferential transitions—that is, both unconscious and conscious inferences—can flexibly adapt to variation within and across the cognitive tasks at hand, changes in the information available to the system, and shifts in the relevant normative reasons (i.e., considerations that are normatively pertinent to the beliefs, actions, reactions, and attitudes of the system). In contrast, according to Quilty-Dunn and Mandelbaum (2018), associative transitions cannot adjust to these types of changes, since their dynamics would rigidly follow a predetermined sequence of states. And so, while inferences are flexible and sensitive to normative reasons, associations—as stipulated in their account—are rigid and could only be affected by extinction and counterconditioning. Moreover, while inferential transitions can be rationally appraised with respect to their underlying mental logic, associative transitions cannot, as they would not follow any such logic. Yet these stipulations are problematic in four ways.

First, flexibility is a property that comes in degrees. A transition between representational states is flexible to the extent that it is sensitive, within and across tasks, to changes in contextual task demands, motivation, time pressure, and similar factors. The greater the flexibility, the greater a (neuro) cognitive system's ability to achieve a given goal via different means in a context-sensitive manner. For instance, suppose that a system faces the task of recognizing a facial expression. The transition to a representation of the relevant facial expression would be flexible to the extent that the system can base it on a variety of cues (e.g., position of the mouth, eyes, nose, even body posture) across different situations, including new settings (e.g., in the fog, in bright light, in moonlight). Flexibility is also greater to the extent that the relevant transition can occur more quickly when under time pressure or more reliably when a mistake is costly. Consider a toy example to do with the possible range of inputs: a system that can infer a facial expression by applying *modus ponens* but only to information about the mouth (e.g., if the position of the mouth is such and such, then the facial expression is joy) may be less flexible than a system that does not follow a mental logic but uses information about either the mouth or eyes. Not following *modus ponens* or any other rule of "mental logic" would make a transition associative for Quilty-Dunn and Mandelbaum (2018), but it could nonetheless possess a degree of flexibility. More generally, transitions that are not reliant on a mental logic need not comprise a causally brute, purely a-rational sequence of states (cf. Buckner 2019).

The set of cognitive or biological systems whose behavior includes nothing but a small number of mindless, stereotyped, time-locked reactions to triggers in the environment is likely exceedingly small to nonexistent (see Keijzer 2013 questioning the standard upshot of the *Sphex* story).[1] Vertebrates, cephalopods, and arthropods seem to be able to flexibly solve various tasks by recruiting learned associations between a practically unlimited number of states and actions (Birch et al. 2020). Bacteria have been argued to possess cognitive capacities, including "sensory ST [signal transduction], valence, communication, sensorimotor coordination, memory, learning, anticipation, and decision-making in complex and changing circumstances," which are parallel to those of multicellular organisms "not only at the heuristic level of functional analog, but also at the level of molecular mechanism, evolution and ecology" (Lyon 2015, 2). Given the heterogeneity in how associative structures are learned and deployed, it is also empirically implausible to stipulate that counterconditioning and extinction define associative relations between representational states, thus differentiating them from inferential relations. After all, counterconditioning may only change the valence of the states involved in some associative transitions, and extinction might not cause unlearning of existing associations, instead creating other more salient ones to modify the behavioral outcomes (see, e.g., Bouton 2004; Redish et al. 2007).

Second, similar points apply to the proposed dichotomy of rational inferential versus a-rational merely causal transitions. Quilty-Dunn and Mandelbaum (2018) explain that only inferential but not associative transitions can be rational, since inferences follow certain rules constituting a mental logic, while associative transitions are governed by no logical rule. While Quilty-Dunn and Mandelbaum (2018) do not provide us with any details about the actual mental logic supporting inferences in any system, their discussion assumes an internalist model of deductive rationality, where the normative appraisal of an inference is grounded in whether or not the inference follows a rule of deductive logic. In cases where a particular inference does not follow such a rule, the cognitive system makes an inferential mistake, such as, say, committing the fallacy of affirming the consequent or denying the antecedent. Quilty-Dunn and Mandelbaum (2018) take such mistaken inferences to be "performance errors" rather than evidence for logical incompetence or a lack of a mental logic altogether. But these assumptions are overly strict.

From the standpoint of normativity within the empirical mind and life sciences, it is more appropriate to adopt an externalist, *ecological model of rationality*, where a system's rationality is determined not only by factors that are internal to the system, such as how things seem to it from its own perspective or whether it follows certain rules of logic but is also dependent on additional factors external to the system, such as the informational and material structure of the environment in which the system makes choices to achieve certain goals over time. Picture a classic brain-in-a-vat with simulated experiences qualitatively indistinguishable from those of a regular perceiver; it may even draw inferences that comply with the rules of logic. Nevertheless, a brain-in-a-vat secluded from the external environment, as stipulated, would not enjoy ecological rationality—in the classic scenario, it would have no ecology.

The degree of ecological rationality of a system's inferences would depend on the etiology of the inference (e.g., whether the inferential transition was triggered by a disturbance in a system's cognitive processing that did not arise from any interaction with the environment or whether it was triggered by specific demands of its environment), on the epistemic status of the inputs or current representations featuring in the inference (e.g., the extent to which they are well- or ill-founded), on how its outputs are deployed by the system to complete a task or achieve a goal in a given situation, and on properties of the relationship between inputs and outputs, such as its energy efficiency, computational complexity, and time (Gigerenzer et al. 1999). From an externalist perspective, the degree of ecological rationality of an inference should be judged in terms of the kinds of checks and balances that an environmentally and historically situated system producing the inference exhibits in its use of sensory input, current representations, and information in the environment to generate a conclusion that facilitates the attainment of a goal.

Indeed, if an account of inference ought to allow for rational appraisal and accommodate scientific usage, as we suggest, then an ecological model of rationality is a natural assumption for such an account. It makes it possible to appraise the degree of rationality of a system's inferences and fits scientific practice in comparative psychology (e.g., Stevens & King 2013; see Hurley & Nudds 2006), basal cognition (Lyon et al. 2021; Lyon & Cheng 2023), and possibly deep learning approaches to reasoning and decision-making (Buckner 2024, 74–85).

Third, as argued by Dacey (2016; see also Halina 2022), a dichotomy between inferential and associative processes distracts cognitive (neuro)scientists from the more substantial questions about which task a given transition between states is meant to solve, what sort of information is recruited for the transition, how it might be used to reach a goal in a computationally tractable and energy efficient way, and how the transition could be physically implemented and rationally appraised. Researchers often do (and should) reject the claim that associative and inferential transitions are categorically, mutually exclusive kinds of processes as empirically unproductive, and instead adopt a more fine-grained multidimensional approach to study different kinds of transitions between informational states.

Finally, Ginsburg and Jablonka (2019) make the case that there is deep continuity between associative and inferential processes on evolutionary grounds. They present unlimited associative learning (UAL) as open-ended (inferential) learning over a practically unlimited space of possible associations and suggest it serves as a marker for subjective experience. Crucially, they contrast UAL with limited associative learning (LAL), on which it builds. Although in the eyes of many, LAL could be deemed to more properly constitute the domain of associative processes, the term UAL itself stresses a continuum being the relevant conceptions here. This, of course, does not mean there are no notable differences that emerge over evolutionary history—the ends of said continuum do accommodate different sorts of processes. In light of that, a clear-cut distinction between merely causal (a-rational) and more deliberative (rational) types of transitions is coherent in principle. However, we submit it is impracticable to adequately establish as a dichotomy and, even if that could be done, it is often unhelpful to strictly maintain such a dichotomy in empirical practice, whether its focus is synchronic or diachronic.

9.2.3 *Inference involves language-like representations*

Consider now the second problem for Quilty-Dunn and Mandelbaum's (2018) account, namely, the assumption that the representational states involved in an inferential transition must have a discursive, language-like, or symbolic format rather than, say, the format of a map, picture, or tree diagram (see also Fodor and Pylyshyn [1988] on this distinction). As argued by Mollo and Vernazzani (2024), such a putative contrast between discursive and iconic formats takes for granted that external, public representations, such as written and spoken words or pictures and maps, are the best way to model how different patterns of bioelectric or neural activity implement representations in (neuro)cognitive systems.

The problem is that modeling the workings of such systems based on the intuitive categories of discursive versus iconic representational formats, as applied to public, external representations, is misleading. It is certainly true that plenty of experimental data can be modeled using language-like representations, for instance—after all, humans are linguistic creatures and the relevant psychology studies are regularly structured around linguistic prompts (Roskies and Allen 2023). Whether the proposal stands up to empirical scrutiny more generally is an open question (Lupyan 2023). It is unclear how bioelectric or neural signals could have a linguistic or iconic format. And yet, they could comprise representations in virtue of the functional roles they play in the system and their degrees of freedom, such as the timing and rate of occurrence of a certain spike of activity (see Schöner 2020; Baluška & Levin 2016). Indeed, one of the key challenges in contemporary cognitive neuroscience is understanding what degrees of freedom matter for the representations underpinned by bioelectric or neural code, as well as how changes in them support different cognitive functions, including various species of inference—whether deductive, inductive, or abductive.

Although we suggest the bioelectric or neural code that a (neuro)cognitive system relies on is unlikely to be language-like, this does not automatically disqualify it from making inferences. What this means instead is that the representations featured in the system's inferences do not have a syntax similar to the rules governing the combination of words and phrases in natural or constructed languages. Their "syntax" may instead consist in the "evolution rules" governing the dynamics of bioelectrical or neural signals in the system over non-discursive, action-oriented, context-sensitive content, which arguably comprises the more basic and ancient species of inference in (neuro)cognitive systems. Since this form of "syntax" should nonetheless be able to support different kinds of inference, it need not involve language-like representations characterized by rules like *modus tollens* or *disjunctive syllogism*, as defined within a formal system of logic.

Now, suppose one does not find the challenge particularly problematic. After all, some sort of an epistemic gap between various cognitive functions and their neural implementation is arguably not unique to Quilty-Dunn and Mandelbaum's (2018) account. For instance, one could hold that there is some level of (perhaps not neural but still subpersonal) cognitive processing, which is usefully cashed out in terms of inferential transitions characterized by a discursive format. Even so, the onus would be on Quilty-Dunn and Mandelbaum (2018) to demonstrate that all empirically fruitful uses of "inference" rest on that level to justify including discursiveness in its definition. We find this implausible, and, in our view, current scientific practice in cognitive science bears out the need for broadening the scope of inference beyond transitions operating over language-like representations.

9.2.4 *Inference only allows for an indirect relation to the world*

The question regarding the directness (or otherwise) of the mind-world relationship has often been discussed as part of the so-called representation wars in cognitive science. The term "representation" has been contentious and used in different ways in different contexts. To many, inference requires representation; and if a system's states and processes are constituted by the occurrence, transformation, and storage of mental or neural representations, then the system is automatically cut off from direct access to the external world. So, if a system makes use of inference, and inference requires representation, then the system is secluded from its environment, only ever able to access its own projections of it (see, e.g., Chemero 2003; Gallagher 2008; Hohwy 2016).

One response to this argument is to do away with mental representations altogether and maintain that (neuro)cognitive systems need not leverage representational states to behave adaptively and perform their functions. This move has been associated most closely with the so-called embodied, embedded, enactive, extended (4E) approach to cognition. Shapiro (2011, 150) captures this general idea in the following way:

1 Representations are stand-ins for actual objects.
2 An agent is in continuous contact with the objects with which it needs to interact.
3 If an agent is in continuous contact with the objects with which it needs to interact, then it doesn't require stand-ins for these objects.

Therefore, an agent has no need for representational states that stand in for actual objects.

This argument tends to elevate the idea of using the world as its own model above all else (Brooks 1991), such that the strategy should apply in all circumstances at all times. To save bandwidth, the cognitive system will offload the workload by using the world as its own model, thus obviating the need for offline encoding of its properties in the form of representations, and hence for the use of inference. There has been debate within the 4E camp itself regarding how this conception might deal with so-called representation-hungry problems, which require reasoning about "absent, non-existent, or counterfactual states of affairs" (Shapiro 2011, 152). As a result, many theorists have advocated minimizing rather than eliminating the role of representations (and hence inferences) so that they may be employed for a particular set of cognitive tasks but are not required for all tasks at all times.

Even if inferential processes were said to play a relatively minor role in (neuro)cognitive systems' mental capacities and engagements with the world, problems would remain. A prominent image here is that of the skull as a black box sequestering the brain from its environment: all that the brain would have access to are noisy and ambiguous patterns of stimulation impinging upon sensory receptors. Viewed in this way, the brain's job would be to infer the "hidden" (unobservable) causes that could have produced said sensory data. For Hohwy (2016), for example, being in direct touch with the external world is incompatible with the idea that the brain is an inferential engine whose primary task is to minimize the errors in the probabilistic predictions it makes about the environmental causes of its sensory inputs. And so, similarly to brain-in-a-vat style skeptical scenarios, his conception precludes the possibility of any direct mind-world relationship, allowing for something close to Cartesian skepticism. This general argument could be called the black box, solipsism, or seclusion problem (Constant et al. 2020).

But this way of posing the problem is fraught from the beginning, since the metaphor already assumes what it sets out to support, that is, the mind being cut off from its environment (Anderson 2017). Instead, the *explanandum* could be thought of as organisms' adaptive attunement to the world. And so, a possible line of response counters that the notions of inference and representation that matter here are different from the richer, more demanding versions that Hohwy (2013) presents. For example, according to Clark, the relevant notion of inference "is not necessarily defined over internal states that bear richly reconstructive, or symbolic, or propositional contents" (Clark 2017, 734). As a result, the alleged skeptical implications stemming from inference are a "red herring" (Clark 2015, 192–193).

Either way, for the brain in a vat scenario to work properly, it would have to receive fully embodied, action-sensitive, and world-involving sensory flows. But in that case, this allegedly envatted brain would effectively leave the vat and become an embodied agent, structurally and dynamically entangled with its environment. Clark maintains that the picture flowing out of (neuro)cognitive systems minimizing prediction error implies that it is not exclusively the neurocentric skull-bound inner model that is the appropriate locus of inferential processing but the embodied organism as a whole. The general suggestion is that certain representations supporting inference (a) are not propositional or discursive, and they (b) concern themselves only with organism-relevant and organism-relative contents (c) in service of pragmatic success or utility rather than a detached veridical reconstruction of the world out there (Williams 2018).

Relatedly, Piccinini (2022) also argues that 4E approaches do not stand in contrast to computational and representational approaches of this kind. He emphasizes that a viable account of neural representations *requires* embodiment, embeddedness, enaction, and affect, since the representations deployed by (neuro)cognitive systems "emerge diachronically through the dynamical interaction between the nervous system, its body, and its environment in a way that must take into consideration the organism's needs" (Piccinini 2022, 6). From this situated perspective, representation and inference require that mind and world stand in a direct relation. The functional role of neural representations for a system is akin to that of a model instantiating a structural relationship with some target task, which can be exploited by the system to guide its behavior with respect to that target. While such structural representations would not enter the kind of inferential transitions that linguistic symbols may enter, they can nonetheless feature in inferential processes whose outputs are sensitive and attuned to functionally significant mismatches with their targets. As a result, the system can adjust its functionally relevant structure and thus the degree of action-oriented structural accuracy of the representations it deploys. In other words, "without the additional interactions between an active body-brain and the world, the world cannot be made to appear," which is why "the idea of the 'brain in a vat' [is] an empty fantasy" (Ginsburg & Jablonka 2019, 458).

Constant et al. (2020) present the notion of *deontic actions* to capture these types of unmediated adaptive observation-action loops. These are actions that become habitualized and automatized in the form of "social reflexes" to yield typically expected behavior in the relevant context. Regularities in the material and social environment built up collectively over time serve as cues, which allow enculturated agents like us to act fluently and effortlessly. This fluidity obviates the need for representational heavy-lifting, opting for fast, frugal, and dynamic processing strategies geared toward organism-specific needs. Similarly, Sims (2023) argues that this sort of organism-environment coupling enables the framework to make productive use of the classically 4E-related concept of affordances, that is, directly perceived opportunities for action (Gibson 1979), even though it employs representational states. And so, as Clark (2015, 195) put it, the mind-world relation here would be "not-indirect,"[2] "since what we perceive is not itself a hypothesis (or model, or fantasy, or virtual reality). Instead, what we perceive is (when all is going well) the structured external world itself." If this is right, the fact that a (neuro)cognitive system makes use of representations and inferences is not enough in itself to conclude that the system is fundamentally siloed from the surrounding world. As a result, contrary to the skeptical conclusions regularly drawn in this context, we hold that an adequate philosophical account of inference can in fact avoid them.

9.3 Considerations regarding terminological extension

Irrespective of these problematic assumptions found within prominent philosophical formulations, it may be argued that there is no point in extending the scope of the term "inference" much beyond what has been customary in logic and epistemology. After all, this extension is liable to cause some terminological confusion, and it would complicate the distinctive character of inferential transitions commonly used in philosophy. We offer two reasons that speak against this restriction.

First, an expansive notion of inference is already employed in empirical work in cognitive science, going back at least to Helmholtz's theory that perception involves "unconscious inference" in the 19th century. More recently, there has been a range of models, which hold that cognitive systems rely on (approximate) Bayesian inference to perform various cognitive functions (Danks 2014; Tenenbaum et al. 2011; Rescorla 2021; Colombo 2023). There are also some broader theoretical frameworks, which leverage probabilistic inferences with more unificatory ambitions; that is, instead of focusing on explaining some particular mental capacity, they aim to offer more general principles to account for a whole array of capacities. One such umbrella framework is predictive processing, with predictive coding and active inference models falling under it (Parr et al. 2022; Clark 2023; Sprevak & Smith 2023). Another approach used for understanding inferences in (neuro)cognitive systems could be found in deep neural-network modeling. For instance, some architectures have been described as making use of "big activity vectors, big weight matrices and scalar non-linearities to perform the type of fast 'intuitive' inference that underpins effortless commonsense reasoning" (LeCun et al. 2015, 441). In fact, many of the recent advances in artificial intelligence concern the ways in which some deep learning structures can seemingly learn by performing categorical, predictive, and causal inferences quickly and without using explicit declarative knowledge or symbolic rules. All of these are active research programs, in which the concept of inference plays a central role. If one of the desiderata for an account of inference in (neuro)cognitive systems is that it should accommodate the ways in which cognitive scientists use this notion to pursue a variety of epistemic and explanatory goals and needs, then undue restrictiveness or epistemological chauvinism regarding the term "inference" is misguided.

Second, and perhaps more importantly, conscious, deliberate inferential transitions in thought both depend on and are supported by subconscious, subpersonal inferential processes. There are at least two readings of this claim—a diachronic one and a synchronic one. On the one hand, the claim could be concerned with the evolutionary history of more deliberative inferences, pointing to these emerging from more ancient, less elaborate inferential transitions. On the other hand, a possible synchronic reading would submit that (neuro)cognitive systems' ability to engage in explicit conscious inference still relies on their subconscious inferential processes. Both interpretations are compelling possibilities.

Regarding the diachronic evolutionary reading, Ginsburg and Jablonka's (2019) work can serve as a useful reference point again, even though they are mostly interested in the emergence of consciousness as a feature of whole organisms and couch their theory primarily in relation to association. Reber et al. (2023) make the same general claim, expressing it mainly with respect to information processing (see also Reber's (1992) earlier work on the evolutionary history of the "cognitive unconscious"). More generally, there is a sense in which the diachronic point seems almost unavoidable. After all, it is widely accepted that more complex life forms and capacities have evolved from simpler ones—since "unconscious organisms preceded conscious organisms" (Bargh & Hassin 2022, 204), unconscious information processing preceded and enabled conscious information processing to develop. This is not to deny the emergence of novel phenomena over evolutionary time, and yet gradualism and continuity are salient features of the evolutionary process. This observation can also be captured thus: "While some might feel that this link between the senomic[3] information processing of individual cells and that of humans is a bit of a stretch, it really is not," since "evolutionary mechanisms function by building on earlier successful forms and functions" (Reber et al. 2023, 135–136). Crucially, although the use of "consciousness" varies widely as well and could arguably be applied more broadly than is currently done, the relevant notion for our purposes is, in essence, explicit personal-level processing as opposed to implicit subpersonal processing one would not be directly aware of. And so, whatever view one holds as to the emergence of consciousness, the diachronic reading would be hard to reject completely.

When it comes to the synchronic interpretation, various types of neurological damage can provide one fruitful line of evidence—in cases where explicit personal-level processing is seriously impacted, certain implicit unconscious processes are often unaffected, but not vice versa (Bargh & Hassin 2022). More generally, a dominant tenet within much of cognitive neuroscience today is that cognition is structured hierarchically (or heterarchically), for example, higher-level more abstract conscious thought depending on subpersonal, subconscious processing on lower levels that are closer to the sensory channels (Kandel et al. 2021; Piccinini 2020; Trappenberg 2022). In addition to hierarchical structural and functional organization, another important explanatory notion in contemporary cognitive neuroscience is prediction or anticipation (e.g., Luczak et al. 2022), and these are particularly salient among researchers aiming to uncover a common principle of operation within the neocortex. In his critique, Williams (2019) offers two reasons for why researchers in cognitive neuroscience have been drawn to the idea of a common underlying principle of functioning and hence what he calls the "unified inferential hierarchy": the radical plasticity of the neocortex and the relative anatomical uniformity across different regions. It is a widespread view that, when it comes to the functioning of (neuro)cognitive systems, there does not appear to be a strict separation between processes underpinning subpersonal inference and the sort of explicit conscious inference philosophers have tended to focus on. Instead, personal and subpersonal processes increasingly seem to be grounded in similar mechanisms, resulting in the use of similar concepts, too. In light of this, we now turn to formulating an updated philosophical account of inference that is sensitive to the needs of cognitive scientists.

9.4 A minimalist account of inference in (neuro)cognitive systems

Given these considerations and the desiderata we outlined in the introduction, a promising approach is to adapt Susanna Siegel's (2017) account of inference and extend it to (neuro)cognitive systems at multiple scales. According to her *Response Hypothesis*:

> [I]inferring is a distinctive kind of response to an informational state, or to a combination of such states, that produces a conclusion. An

> informational state can take the form of a belief, but it can also take the form of a perceptual experience, a supposition, or an attitude. The distinctively inferential kind of response to information is formed when one reaches a conclusion.
>
> (Siegel 2017, 77)

According to her, this conclusion might consist in a generalization, projection, categorization, posterior probability, causal hypothesis, and so on.

There is much to like about Siegel's Response Hypothesis. First, it is not committed to the idea that inference must always be conscious. She writes that "inference establishes epistemic dependence of a conclusion on inferential inputs" which "need not leave any mark in consciousness" (Siegel 2017, 83). Accordingly, an inferrer need not explicitly "reckon" that the informational states entering its inferences support the conclusion. For Siegel, conscious reckoning is involved in some inferences, but there are other paradigmatic cases that do not depend on it. Second, Siegel's analysis of inference does not require the informational states entering inferential processes to take the form of symbolic, language-like representations. According to Siegel, "inference is not always illuminated by the model of an argument with [discursive] premises and conclusions" (Siegel 2017, 83).

Third, the Response Hypothesis does not entail that inferential processes automatically prevent minds from being in direct contact with the world. Siegel (2017) explains that perceptual experiences arise from inferential processes that recruit both a subject's "prior outlook" and "perceptual input," but the objects of perception for her are neither a prior outlook nor a perceptual input. Perceptual experiences reveal the world itself, since their underlying processes would be aimed at inferring facts about the external world that can explain the subject's perceptual inputs. To the extent that an organism is successful in navigating its environment, meeting its needs, and reaching its goals, it is getting at or at least approximating the actual structure of the world—and, as will be discussed below, it possesses ecological rationality. And so, even though an inferential view of perception generally relies on a system's prior knowledge and informational state, it does not follow that the system lacks direct access to the world. More generally, presuming that inference precludes direct mind-world contact wrongly presupposes that it is possible for the world to come neatly pre-packaged independently from a cognitive system's knowledge, needs, and interests. While cognitive systems do not *control* how reality is by mere thinking, their engagement with the world is mind-*framed*—for us humans it "is the more agent-salient world of tables, chairs, football games, dogs, cats, and (as J. L. Austin famously quipped) 'medium-sized dry goods.' This is a world parsed for human action and choice" (Clark 2013, 492).

Nevertheless, in its current form, Siegel's (2017) formulation is not quite fit to cover the variety of inferences in cognitive science. To extend Siegel's account to inferential transitions across a wide range of (neuro)cognitive systems, some qualifications are needed. As Siegel herself recognizes, "[m]any of the low-level processes called 'inference' in psychology would not qualify [as such]" on her account, mainly because she is interested in "the kind of inferring that redounds well or poorly on the subject's rational standing" (Siegel 2020, 725). Thus, while this picture allows for human perceptual experiences to count as the results of rationally evaluable inferential processes, under a stricter reading of "rationality," this would not hold for capacities possessed by other kinds of cognitive systems—from prokaryotes to plants, to animals with or without nervous systems. However, there is increasing evidence for relevant abilities in such systems; for example, they can also selectively attend to a feature to the exclusion of others, sense and recognize existentially salient features, assign different values to them, combine and weigh these, and select appropriate actions to achieve certain tasks (see Lyon et al. 2021, table 1; Lyon 2006; Levin 2019). Under the Response Hypothesis, processes in such systems would likely not count as rationally evaluable, since it would be implausible that cognitive systems besides humans can be subjects of epistemic blame, credit, and responsibility. And yet, depending on what one considers to be the appropriate reading of "rationality," there may be no theoretical necessity to shut out mental phenomena displayed by non-human cognitive systems from being inferential and rationally appraisable.

As a result, there are two key qualifications that we are going to add to Siegel's Response Hypothesis. One concerns the notion of an informational or representational state; the other relates to the notion of rationality. Both amendments are rooted in biogenic and ecological approaches to cognition.

First, Siegel thinks of inference as "a distinctive way of leveraging information one has already into a new informational state. When it is formed via such a response, the new informational state is the conclusion of the act of inferring" (Siegel 2020, 724–725). Yet she leaves the notion of "information" somewhat underspecified. One empirically oriented and adaptive action-focused way to think about information used within the basal cognition research program is this: "A state of affairs is information for an organism if it triggers a change in physiology or behaviour relative to that state of affairs" (Lyon et al. 2021, 4). On the face of it, this conception may seem to blur the semantic and causal dimensions of information, but we can cash it out in greater detail in terms of *action-oriented, situated representation* (Piccinini 2022; see also Mandik 2005). That is, a system's representational states should present a partial morphism to their targets. Such states are causally activated by their targets and deployed by the system to guide behavior both in the presence and sometimes in the absence of signals coming from the target. They do not require linguistic abilities to play this functional role in action-guidance.

Crucially, information flows from the external milieu of a system as well as its own body can only make sense to the system in relation to its physiological and motivational state, for example, "under-nourished, starving, sated, reproductive, dormant, acting alone, acting with others, experienced, naïve, threatened, secure and so on" (Lyon et al. 2021, 4). What is more, (neuro)cognitive systems are not passive recipients of signals from their environment, they are perennially active—both in terms of regulating their internal functioning to maintain stable conditions and acting on the world to detect or modify incoming signal streams to achieve preferred goals. Conceived of in this way, the range of systems that can possess action-oriented or situated representational states would be broader than commonly presumed.

Second, Siegel (2017, chapter 3) introduces the notion of "epistemic charge." She states that "[b]y virtue of being epistemically charged, a mental state redounds well or badly on the subject. In this way, epistemically charged mental states manifest an epistemic status" and can transmit their charge to other mental states, contributing to the all-things-considered rational standing of the agent (Siegel 2017, 25). But to allow for the rational evaluability of inferential transitions across a diverse variety of (neuro)cognitive systems, the notions of "epistemic charge" and rationality should not only apply, as Siegel (2018, 764) suggests, to the "local," "short-term transitions that determine the epistemic status of the output." Here, we may want to broaden Siegel's notion of epistemic charge and ground it in an ecological view, where epistemic rationality is pragmatically encroached, contextual, and oriented toward some adaptive goal in the world. Following this kind of externalist ecological conception of rationality, the degree to which an inferential transition in a (neuro)cognitive system is rational depends on how it exploits the structure of the information in the environment to achieve a certain task, given its perceptual and evaluative resources, motivational state, preferred goals, and prior informational outlook (Gigerenzer 2000, 57). Besides coherence and accuracy, it is modulated by considerations of adaptivity, frugality, and speed, as well as by its "role in guiding temporally extended cycles of perception and action" (Clark 2018, 747).

There is a distinction to be made between association and inference based on their functional profile, on the type of pre-existing informational or representational states they recruit, on the kinds of epistemic feelings they may elicit, and the tasks in which the outputs of such transitions are deployed. Yet, from the point of view of ecological rationality, both association and inference can in principle be rationally appraised (see also Hurley 2003). This does not overly cheapen the notion of rationality, however.

Ecological rationality would be restricted to cognitive systems, which, in light of basal cognition, comprise a class that is broader than commonly assumed, but nevertheless limited (e.g., on cognitive systems as active matter systems, see Harrison et al. 2022). In cognitive systems, both associative and inferential transitions can follow certain principles, produce intelligent and adaptive behavior, and help a system achieve certain goals. Some transitions between associative states might be brute causal dispositions, but other learned associations can be sensitive to evidential relationships as well as to means-ends relevance, for example, they may be learned and modified via various "rational," normatively respectable reinforcement learning processes (Chater 2009; Niv 2009).

As we explained above, a sharp dichotomy between a-rational association and rational inference does not comport with the explanatory needs and aims in comparative psychology and other fields in cognitive science. In the context of these needs, Dacey (2016) argues that the term "associative" should be understood as a "filler term" for whatever mechanism underpins sequences of representational states displaying certain degrees of freedom and aimed at achieving certain tasks. As a result, the precise content of the contrast between association and inference might vary in different experimental contexts depending on the system and the task at hand. This view is compatible with the claim that there is a continuum between the two processes without dismissing the differences that are helpful to draw between processes at the ends of such a continuum.

Based on the functional profile of a given inferential response, the sort of representational states it recruits, and the kind of task it is aimed at in a given context, one could distinguish between different types of inferences. For instance, in arguing that perception is explained by unconscious inference, Ritchie (2022) draws our attention to the fact that some tasks have features that are characteristic of distinctive species of inference. For the sort of non-deductive inferences that might explain visual perception—inferences that are ampliative and non-monotonic—a task can be considered *inductive* when

> it is *diachronic* in that it concerns how we overtly represent the present in light of the past, and might change what we believe accordingly; . . . it is *empirical* in that it particularly concerns our representation of past experiences or acquired knowledge, which we bring to bear in drawing conclusions about present circumstances . . .; it is *extrapolative* in the sense that when we generalize from past experience to novel circumstances there will typically be various ways in which the past and present circumstances differ.
>
> (Ritchie 2022, 25)

Drawing these distinctions is compatible with the minimalist inference view, so long as they are empirically grounded and fruitful. These will also likely vary, again, based on the system and task in question.

In our view, inference is to be understood as a form of rationally evaluable transition from some inputs and current representations to some subsequent or output representation. We submit that this conception meets the three requirements we set out initially. A minimalist notion of inference tracks the needs and current directions of productive scientific practice across cognitive science, including not only cognitive neuroscience but also areas like comparative psychology and basal cognition. Second, this way of viewing inference nevertheless allows for useful fine-grained distinctions to be drawn across the association-inference continuum, which may be needed in the context of specific experimental settings investigating a given phenomenon. Finally, it has an epistemic upshot and allows for normative evaluation in relation to a broader ecological reading of rationality, which centers the adaptive success of an active organism embedded in its environment.

The conception of inference that we have put forward also opens the door to the possibility of (neuro)cognitive systems being meaningfully in touch with the outside world. There is a real sense in which the image of a brain in a dark box, that is, the skull, may seem compelling. In such a situation, the only way for the brain to tell what may be going on would be to guess—that is, infer—the hidden causes of the various inputs it is receiving from its sensory surfaces. Under these conditions, aspects of the external world could be seen as mere phantasms in the sequestered brain, which does not and cannot really know the actual causes of its sensory stimuli. However, as we have argued, inferences and their accompanying representations do not have to take this isolationist form, nor be understood in the traditional heavy-handed sense, which has given us a picture of the mind as a formal symbol manipulation machine operating on a set of abstract representations. That is, minds need not be shut off from the world—instead, for the most part, temporally extended cycles of adaptive action enable (neuro)cognitive systems to attain and maintain a reliable sensorimotor grip on their environment.

9.5 Conclusions

In this chapter, we sketched a minimalist account of inference in (neuro) cognitive systems. We introduced Siegel's *Response Hypothesis*, whereby "inferring is a distinctive kind of response to an informational state, or to a combination of such states, that produces a conclusion" (Siegel 2017, 77), and we extended this notion by taking a biogenic approach to the concepts of representation (following basal cognition) and rationality (in line with ecological rationality). Specifically, we argued that inferences are rationally evaluable transitions from some inputs or current representations to some subsequent or output representation in the context of organism-relevant information. While our account is minimalistic, it is not uninformative and meets the three desiderata we indicated. Namely, this conception of inference can do justice to how researchers in cognitive science employ the notion to pursue disparate explanatory projects and goals, thus allowing them to see, among certain phenomena, a unity that could otherwise be overlooked. Second, it enables us to make scientifically productive distinctions between different forms of inferential and non-inferential processes in different kinds of (neuro)cognitive systems. That is, depending on the type of system, the sort of task it is facing, and the functional profile of the representational states entering a transition, we can distinguish between different species of inferences, as well as draw empirically fruitful distinctions between association and inference in particular (experimental) contexts. Third, the minimalist view of inference allows for inferential transitions to have normative character. Even if the system did not consciously take the representational states to be premises which support the inferred conclusion, we can still evaluate to what extent and how the inference drawn by the system is epistemically legitimate in virtue of an externalist, pragmatically oriented account of ecological rationality. The upshot of this account of inference is the possibility of a "not-indirect" relationship between mind and world, such that there is no need to posit a deep epistemic chasm between them.

Acknowledgments

We are indebted to a number of people for helpful feedback while developing our account. In particular, we thank Gualtiero Piccinini, Matthew Sims, Aliya Dewey, Pablo Fernandez Velasco, Wiktor Rorot, John Campbell, and an anonymous referee for their comments on earlier versions of this chapter.

Notes

1 The original anecdote concerns the wasp *Sphex*—it tends to inspect the burrow where it will lay its eggs before dragging a prey item inside for its young to feed on, once they hatch. If the prey is moved away while the wasp is inside, it will move it back before inspecting the burrow again. The sequence can be repeated again and again. As a result, it has served as an influential example of apparently inflexible behavior.

2 Clark (2015, 195) uses the term "not-indirect perception" to denote the following: we perceive "the external world itself," but it is a "world parsed according to our organism-specific needs and action repertoire." In other words, a (neuro)

cognitive system is genuinely open to its environment, populated by the processes, structures, and items of relevance to the system.

3 Senomic information is information relevant to "all the sensory experiences of the cognitive cell and its sensing apparatus" (Baluška & Miller 2018, 1).

References

Adler, J. E. (2008). Introduction: Philosophical Foundations. In Adler, J. E., Rips, L. J. (eds.) *Studies of Human Inference and Its Foundations*. Cambridge University Press.

Allen, R., Reber, A. S. (1998). Unconscious Intelligence. In Graham, G., Bechtel, W. (eds.) *A Companion to Cognitive Science*. Blackwell.

Anderson, M. L. (2017). Of Bayes and Bullets: An Embodied, Situated, Targeting-Based Account of Predictive Processing. In Metzinger, T., Wiese, W. (eds.) *Philosophy and Predictive Processing*. Frankfurt am Main: MIND Group.

Baluška, F., Levin, M. (2016). On Having No Head: Cognition Throughout Biological Systems. *Frontiers in Psychology*, 7, 902.

Baluška, F., Miller, W. B. Jr. (2018). Senomic View of the Cell: Senome Versus Genome. *Communicative & Integrative Biology*, 11(3), 1–9.

Bargh, J. A., Hassin, R. R. (2022). Human Unconscious Processes In Situ: The Kind of Awareness that Really Matters. In Reber, A. S., Allen, R. (eds.) *The Cognitive Unconscious: The First Half Century*. Oxford University Press.

Bartolo, R., Averbeck, B. B. (2021). Inference as a Fundamental Process in Behavior. *Current Opinion in Behavioral Sciences*, 38, 8–13.

Birch, J., Ginsburg, S., Jablonka, E. (2020). Unlimited Associative Learning and the Origins of Consciousness: A Primer and Some Predictions. *Biology and Philosophy*, 35, 56.

Boghossian, P. (2014). What is Inference? *Philosophical Studies*, 169(1), 1–18.

Bouton, M. E. (2004). Context and Behavioral Processes in Extinction. *Learning & Memory*, 11(5), 485–494.

Braine, M. D., O'Brien, D. P. (1998). The Theory of Mental-Propositional Logic: Description and illustration. In Braine, M. D., O'Brien, D. P. (eds.) *Mental Logic*. Erlbaum.

Brooks, R. (1991). Intelligence without Representation. *Artificial Intelligence*, 47, 139–159.

Buckner, C. J. (2019). Rational Inference: The Lowest Bounds. *Philosophy and Phenomenological Research*, 98, 1–28.

Buckner, C. J. (2024). *From Deep Learning to Rational Machines: What the History of Philosophy Can Teach Us about the Future of Artificial Intelligence*. Oxford University Press.

Calvo, P., Friston, K. (2017). Predicting Green: Really Radical (Plant) Predictive Processing. *Journal of the Royal Society Interface*, 14(131), 20170096.

Chater, N. (2009). Rational and Mechanistic Perspectives on Reinforcement Learning. *Cognition*, 113(3), 350–364.

Chemero, A. (2003). An Outline of a Theory of Affordances. *Ecological Psychology*, 15(2), 181–195.

Clark, A. (2013). Expecting the World: Perception, Prediction, and the Origins of Human Knowledge. *The Journal of Philosophy*, 110(9), 469–496.

Clark, A. (2015). *Surfing Uncertainty: Prediction, Action, and the Embodied Mind*. Oxford University Press.

Clark, A. (2017). Busting Out: Predictive Brains, Embodied Minds, and the Puzzle of the Evidentiary Veil. *Noûs*, 51(4), 727–753.

Clark, A. (2018). Priors and Prejudices: Comments on Susanna Siegel's The Rationality of Perception. *Res Philosophica*, 95(4), 741–750.
Clark, A. (2023). *The Experience Machine: How Our Minds Predict and Shape Reality*. Pantheon Books.
Colombo, M. (2013). Constitutive Relevance and the Personal/Subpersonal Distinction. *Philosophical Psychology*, 26(4), 547–570.
Colombo, M. (2018). Learning and Reasoning. In Sprevak, M., Colombo, M. (eds.) *The Routledge Handbook of the Computational Mind*. Routledge.
Colombo, M. (2023). Bayesian Cognitive Science. In *Routledge Encyclopaedia of Philosophy*. 10.4324/9780415249126-Q150–1
Constant, A., Clark, A., Friston, K. J. (2020). Representation Wars: Enacting an Armistice through Active Inference. *Frontiers in Psychology*, 11, 598733.
Dacey, M. (2016). Rethinking Associations in Psychology. *Synthese*, 193, 3763–3786.
Danks, D. (2014). *Unifying the Mind: Cognitive Representations as Graphical Models*. MIT Press.
Evans, J. S. B., Stanovich, K. E. (2013). Dual-Process Theories of Higher Cognition Advancing the Debate. *Perspectives on Psychological Science*, 8, 223–241.
Fodor, J. A., Pylyshyn, Z. W. (1988). Connectionism and Cognitive Architecture: A Critical Analysis. *Cognition*, 28(1–2), 3–71.
Gallagher, S. (2008). Direct Perception in the Intersubjective Context. *Consciousness and Cognition*, 17(2), 535–543.
Gibson, J. J. (1979). *The Ecological Approach to Visual Perception*. Houghton Mifflin.
Gigerenzer, G. (2000). *Adaptive Thinking: Rationality in the Real World*. Oxford University Press.
Gigerenzer, G., Brighton, H. (2009). Homo Heuristicus: Why Biased Minds Make Better Inferences. *Topics in Cognitive Science*, 1(1), 107–143.
Gigerenzer, G., Todd, P. M., The ABC Research Group. (1999). *Simple Heuristics that Make Us Smart*. Oxford University Press.
Ginsburg, S., Jablonka, E. (2019). *The Evolution of the Sensitive Soul: Learning and the Origins of Consciousness*. MIT Press.
Gregory, R. L. (1980). Perceptions as Hypotheses. *Philosophical Transactions of the Royal Society of London B, Biological Sciences*, 290(1038), 181–197.
Grosenick, L., Clement, T. S., Fernald, R. D. (2007). Fish Can Infer Social Rank by Observation Alone. *Nature*, 445(7126), 429–432.
Halina, M. (2022). Unlimited Associative Learning as a Null Hypothesis. *Philosophy of Science*, 89(5), 1186–1195.
Harman, G. (1986). *Change in View*. MIT Press.
Harrison, D., Rorot, W., Laukaityte, U. (2022). Mind the Matter: Active Matter, Soft Robotics, and the Making of Bio-Inspired Artificial Intelligence. *Frontiers in Neurorobotics*, 16, 880724.
Hatfield, G. (2002). Perception as Unconscious Inference. In Heyer, D., Mausfeld, R. (eds.) *Perception and the Physical World*. Oxford University Press.
Hohwy, J. (2013). *The Predictive Mind*. Oxford University Press.
Hohwy, J. (2016). The Self-Evidencing Brain. *Noûs*, 50(2), 259–285.
Hurley, S. (2003). Animal Action in the Space of Reasons. *Mind & Language*, 18(3), 231–257.
Hurley, S., Nudds, M. (eds.) (2006). *Rational Animals?* Oxford University Press.
Inhelder, B., Piaget, J. (1958). *The Growth of Logical Thinking*. Basic Books.
Jardri, R., Denève, S. (2013). Circular Inferences in Schizophrenia. *Brain*, 136(11), 3227–3241.

Johnson-Laird, P. N. (1983). *Mental Models*. Cambridge University Press.
Kahneman, D. (2011). *Thinking Fast and Slow*. Farrar, Strauss, Giroux.
Kandel, E. R., Koester, J. D., Mack, S. H., Siegelbaum, S. A. (eds.) (2021). *Principles of Neural Science*, 6th Edition. McGraw Hill.
Keijzer, F. (2013). The Sphex Story: How the Cognitive Sciences Kept Repeating an Old and Questionable Anecdote. *Philosophical Psychology*, 26(4), 502–519.
Kihlstrom, J. F. (1987). The Cognitive Unconscious. *Science*, 237(4821), 1445–1452.
LeCun, Y., Bengio, Y., Hinton, G. (2015). Deep Learning. *Nature*, 521(7553), 436–444.
Levin, M. (2019). The Computational Boundary of a "Self": Developmental Bioelectricity Drives Multicellularity and Scale-Free Cognition. *Frontiers in Psychology*, 10, 2688.
Levin, M. (2023). Bioelectric Networks: The Cognitive Glue Enabling Evolutionary Scaling from Physiology to Mind. *Animal Cognition*, 26(6), 1865–1891.
Luczak, A., McNaughton, B. L., Kubo, Y. (2022). Neurons Learn by Predicting Future Activity. *Nature Machine Intelligence*, 4, 62–72.
Lupyan, G. (2023). Is Language-of-Thought the Best Game in the Town We Live? *Behavioral and Brain Sciences*, 46, e281.
Lyon, P. (2006). The Biogenic Approach to Cognition. *Cognitive Processing*, 7(1), 11–29.
Lyon, P. (2015). The Cognitive Cell: Bacterial Behavior Reconsidered. *Frontiers in Microbiology*, 6, 264.
Lyon, P., Cheng, K. (2023). Basal Cognition: Shifting the Center of Gravity (Again). *Animal Cognition*, 26, 1743–1750.
Lyon, P., Keijzer, F., Arendt, D., Levin, M. (2021). Reframing Cognition: Getting Down to Biological Basics. *Philosophical Transactions of the Royal Society B*, 376, 20190750.
Maley, C. J. (2020). Continuous Neural Spikes and Information Theory. *Review of Philosophy and Psychology*, 11(3), 647–667.
Mandelbaum, E., Quilty-Dunn, J. (2020). Non-Inferential Transitions: Imagery and Association. In Nes, A., Chan, T. (eds.) *Inference and Consciousness*. Routledge.
Mandik, P. (2005). Action-Oriented Representation. In Brook, A., Akins, K. (eds.) *Cognition and the Brain: The Philosophy and Neuroscience Movement*. Cambridge University Press.
Mollo, D. C., Vernazzani, A. (2024). The Formats of Cognitive Representation: A Computational Account. *Philosophy of Science*, 91(3), 682–701.
Money, N. P. (2021). Hyphal and Mycelial Consciousness: The Concept of the Fungal Mind. *Fungal Biology*, 125(4), 257–259.
Nes, A., Chan, T. H. W. (eds.) (2020). *Inference and Consciousness*. Routledge.
Niv, Y. (2009). Reinforcement Learning in the Brain. *Journal of Mathematical Psychology*, 53(3), 139–154.
Oaksford, M., Chater, N. (2007). *Bayesian Rationality: The Probabilistic Approach to Human Reasoning*. Oxford University Press.
Papineau, D., Heyes, C. (2006). Rational or Associative? Imitation in Japanese Quail. In Hurley, S., Nudds, M. (eds.) *Rational Animals*. Oxford University Press.
Parr, T., Pezzulo, G., Friston, K. J. (2022). *Active Inference: The Free Energy Principle in Mind, Brain, and Behavior*. MIT Press.
Pettit, P. (2023). Five Elements of Group Agency. *Inquiry*, 1–21.
Piccinini, G. (2020). *Neurocognitive Mechanisms: Explaining Biological Cognition*. Oxford University Press.

Piccinini, G. (2022). Situated Neural Representations: Solving the Problems of Content. *Frontiers in Neurorobotics*, 16, 1–13.
Piccinini, G., Bahar, S. (2013). Neural Computation and the Computational Theory of Cognition. *Cognitive Science*, 34, 453–488.
Quilty-Dunn, J., Mandelbaum, E. (2018). Inferential Transitions. *Australasian Journal of Philosophy*, 96(3), 532–547.
Reber, A. S. (1992). The Cognitive Unconscious: An Evolutionary Perspective. *Consciousness and Cognition: An International Journal*, 1(2), 93–133.
Reber, A. S., Baluška, F., Miller, W. B. (2023). *The Sentient Cell: The Cellular Foundations of Consciousness*. Oxford University Press.
Redish, A. D., Jensen, S., Johnson, A., Kurth-Nelson, Z. (2007). Reconciling Reinforcement Learning Models with Behavioral Extinction and Renewal: Implications for Addiction, Relapse, and Problem Gambling. *Psychological Review*, 114(3), 784–805.
Rescorla, M. (2021). Bayesian Modeling of the Mind: From Norms to Neurons. *WIREs Cognitive Science*, 12, e1540.
Rips, L. J. (1994). *The Psychology of Proof*. MIT Press.
Ritchie, B. J. (2022). Recognizing Why Vision is Inferential. *Synthese*, 200, 25.
Roskies, A. L., Allen, C. (2023). Language-of-Thought Hypothesis: Wrong, But Sometimes Useful? *Behavioral and Brain Sciences*, 46, e288.
Schöner, G. (2020). The Dynamics of Neural Populations Capture the Laws of the Mind. *Topics in Cognitive Science*, 12(4), 1257–1271.
Shapiro, L. (2011). *Embodied Cognition*. Routledge.
Siegel, S. (2017). *The Rationality of Perception*. Oxford University Press.
Siegel, S. (2018). Perception as Guessing vs. Perception as Knowing: Replies to Clark and Peacocke. *Res Philosophica*, 95, 761–784.
Siegel, S. (2020). Precis of The Rationality of Perception. *Philosophy and Phenomenological Research*, 101, 724–726.
Sims, M. (2023). Ecological in Spirit? Predictive Processing and the Limits of Direct Perception. In Curado, M., Gouveia, S. S. (eds.) *Predictive Minds: Old Problems and New Challenges*. Vernon Press.
Sprevak, M., Smith, R. (2023). An Introduction to Predictive Processing Models of Perception and Decision-Making. *Topics in Cognitive Science*. https://doi.org/10.1111/tops.12704
Stevens, J. R., King, A. J. (2013). The Lives of Others: Social Rationality in Animals. In Hertwig, R., Hoffrage, U., The ABC Research Group (eds.) *Simple Heuristics in a Social World*. Oxford University Press.
Tenenbaum, J. B., Kemp, C., Griffiths, T. L., Goodman, N. D. (2011). How to Grow a Mind: Statistics, Structure, and Abstraction. *Science*, 331(6022), 1279–1285.
Trappenberg, T. P. (2022). *Fundamentals of Computational Neuroscience*, 3rd Edition. Oxford University Press.
Völter, C. J., Call, J. (2017). Causal and Inferential Reasoning in Animals. In *APA Handbook of Comparative Psychology: Perception, Learning, and Cognition*, Vol. 2, pp. 643–671. American Psychological Association.
Williams, D. (2018). Predictive Processing and the Representation Wars. *Minds & Machines*, 28, 141–172.
Williams, D. (2019). Hierarchical Minds and the Perception/Cognition Distinction. *Inquiry*, 66(2), 275–297.

10 Interventionist methods for interpreting deep neural networks

Raphaël Millière and Cameron Buckner

10.1 Introduction

The maturation of connectionist models into modern deep neural networks has sparked renewed interest in the computational foundations of cognition (Buckner 2019). This development is particularly evident in the progress of large language models (LLMs) like GPT-4 (OpenAI 2023). While the simpler artificial neural networks (ANNs) that preceded the advent of deep learning were initially inspired by abstract features of brain structure, recent architectural innovations in DNNs are not biologically motivated. Nonetheless, DNNs' impressive performance on complex tasks, when rigorously evaluated under controlled conditions, can inform comparisons with human cognitive abilities (Millière 2024). This parallel has led to a growing trend in cognitive neuroscience of using DNNs as tools for probing neural computation and representation (Doerig et al. 2023). However, the utility of such comparisons arguably hinges on our ability to interpret the internal workings of DNNs, to address ongoing concerns about their status as neurocognitive models (Räz & Beisbart 2024). Just as the brain remains a "black box" in many respects, information processing within DNNs is typically opaque, despite our having full access to their architecture and weights. This opacity presents both a challenge and an opportunity: by developing methods to uncover the computational and representational structures within DNNs, we may not only enhance our understanding of these artificial systems but also refine our approaches to investigating biological neural networks. In this respect, the development of rigorous interpretability techniques for ANNs may inform future methodological approaches in computational cognitive neuroscience. In this chapter, we focus on interventionist methods for interpreting DNNs, with a particular emphasis on LLMs. First, we review concerns about many common methods used to evaluate and understand LLM performance, such as benchmarking and probing (Section 10.2). Second, we argue that more recent methods based upon increasingly sophisticated and hypothesis-driven

DOI: 10.4324/9781003458531-10

interventions on the internal workings of LLMs can address some of these concerns, allowing us to gain a more robust understanding of the internal organization of LLMs (Section 10.3).

10.2 Challenges to previous interpretation methods

LLMs have made rapid progress on producing human-like linguistic behavior in many different scenarios that were challenging for previous methods in artificial intelligence. They can readily produce grammatically correct and generally semantically coherent text. They can respond flexibly to a wide range of questions, including various aptitude and proficiency tests that prove challenging even for educated adults (OpenAI 2023). Moreover, LLMs can produce functional blocks of computer code in multiple programming languages, as well as code in visual markup languages that create coherent images based on textual descriptions (Bubeck et al. 2023). More broadly, their performance on a diverse range of behavioral benchmarking tasks designed to assess specific abilities is converging toward matching or exceeding human baselines. However, there remain significant concerns that these achievements may have less profound explanations than they first appear to merit.

One straightforward concern about all these performance achievements is that any apparently intelligent behavior produced by a DNN-based system could be due to mere memorization of training data. We have elsewhere argued that memorization serves as a useful "null hypothesis" for evaluating the capacities of AI systems (Millière & Buckner 2024). For example, many of GPT-4's feats could in principle be produced by an inefficient and inflexible memory retrieval operation. GPT-4's training set likely encompasses trillions of tokens[1] in millions of textual documents, a significant subset of the whole internet. These training sets include dialogues generated by hundreds of millions of individual humans and hundreds of thousands of academic publications covering potential question-answer pairs. Empirical studies have discovered that the many-layered architecture of DNNs grants them an astounding capacity to efficiently encode their training data, which can allow them to retrieve the right answers to millions of randomly labeled data points in artificially constructed datasets where we know a priori there are no abstract principles governing the correct answers (Zhang et al. 2021). This suggests that GPT-4's responses could be generated by approximately—and, in some cases, exactly—reproducing samples from its training data (McCoy et al. 2023).[2] Compare this to a human student who had found a test's answer key on the Internet and reproduced its answers without any deeper understanding; such regurgitation would not be good evidence that the student was intelligent or understood the subject material. For these reasons, "data contamination"—when the training set contains the very question on which the LLM's abilities are assessed—is considered a serious concern in any report of an LLM's performance, and many think it must be ruled out by default when comparing human and LLM performance (Aiyappa et al. 2023).

While we have argued that skeptical interpretations appealing to mere memorization cannot account for the full range of behaviors exhibited by these models (Millière & Buckner 2024), our previous analysis did not provide a positive account of the mechanisms that might enable LLMs to achieve such breakthrough performance. Providing such an account is challenging due to familiar concerns about the opacity of neural networks, which is exacerbated by LLMs' relatively novel architecture, their enormous number of adjustable parameters, and the sheer magnitude of their training data. To address these challenges, the research community has developed new methods for evaluating the performance of these systems and probing their internal organization, which we explore in this section.

10.2.1 Mechanistic explanation

In science, mechanistic explanations aim to reveal the causal structure underlying a phenomenon of interest by describing the organized entities and activities that are responsible for producing or maintaining that phenomenon (Machamer et al. 2000). More precisely, a mechanistic explanation identifies the component parts of a mechanism, characterized by their properties and capacities, the causal interactions between these component parts, and the organization of the parts and activities such that they give rise to the phenomenon. Such explanations stand in contrast to purely descriptive or phenomenological models that simply re-describe the phenomenon itself, as well as covering-law explanations that explain by subsuming the phenomenon under empirically discovered regularities or governing laws. Mechanisms explain by revealing how the phenomenon arises from the causal structure of the system, beyond merely capturing the empirical regularities in which it partakes. As such, mechanistic explanations reveal opportunities for manipulation and control over the phenomenon in a way that descriptive or law-based explanations do not.

Understanding the behavior of a simple system like a mechanical clock is easy enough—one can simply open it up and observe the mechanism at work. This is not so straightforward with more complex systems, like the weather, the brain, or ANNs. Neural networks are often described as "black boxes" precisely because the causal mechanisms that explain their behavior seem opaque to scrutiny. As we emphasized elsewhere (Millière & Buckner 2024), knowing the learning objective, architecture, or size of LLMs is insufficient to explain their remarkable performance on challenging tasks, and to determine what functional capacities can be meaningfully ascribed to them. In principle, one could even provide a complete mathematical description of an LLM as a giant composite function, consisting in an absurdly complex sequence of linear and nonlinear transformations across many layers, but such a description, on its own, would be useless to provide a genuine explanation of the network's behavior in specific contexts.[3] The "black box" metaphor underscores this chasm: it highlights the difficulty to trace precise causal pathways in the network through which specific inputs are transformed into specific outputs. This is why merely re-describing what an LLM does in terms of next-token prediction or matrix multiplication—which we have elsewhere called the re-description fallacy—cannot possibly settle philosophical debates that are fundamentally about causal organization.

The search for causal mechanisms has become central across the life sciences and cognitive science. Like their artificial counterparts, biological neural networks are "black boxes"; yet neuroscientists are engaged in the project of uncovering multilevel mechanisms underlying psychological capacities and nervous system functions (Craver 2007; Piccinini 2020). The notion of intervention is central to this explanatory project. In the philosophy of science, interventionism holds that causal relationships are best understood in terms of what would change under interventions or manipulations to parts of the system (Woodward 2005). More specifically, X is considered a direct cause of Y if and only if there exists a possible intervention on X that will change Y (or the probability distribution of Y) when all other variables in the system are held fixed. This relationship is asymmetric—intervening on Y does not change X if the causal arrow truly points from X to Y. Interventionism eschews regularity or correlational notions of causation, recognizing that a system's behavior depends on more than merely observing regular successions of events. In the context of mechanistic explanation, interventions involve altering specific parts of a posited mechanism piecemeal in order to learn about their causal contribution to the target phenomenon. For example, pharmacology targets specific receptor mechanisms or cell signaling pathways; brain stimulation techniques activate or inhibit activity in restricted brain areas; knockout models remove genes hypothesized to be critical for mechanisms; and optogenetics uses light to activate neurons genetically modified to express light-sensitive channels.

A similar interventionist approach can be applied to unravel causal mechanisms in ANNs like LLMs. Of course, there are major disanalogies between biological nervous systems evolved over millions of years and ANNs designed by human engineers. Nevertheless, the motivation of interventionist research is similar: to achieve explanatory understanding by revealing multilevel mechanisms, not merely observing input–output patterns. Like neuroscientists, computer scientists can aim for explanatory understanding linking particular components of neural networks and patterns of internal activations to specific capacities, such as translating between languages or answering arithmetic questions. This explanatory project is often described as mechanistic interpretability. In a broad sense, mechanistic interpretability could be characterized as the search for mechanistic explanations of the behavior of deep neural networks, including LLMs. As we shall see, however, the phrase is often used in a more restrictive sense, to denote a particular set of theoretical assumptions and intervention methods used to achieve mechanistic explanations in machine learning.[4]

10.2.2 Opening up the black box

There are three main methodological approaches used to investigate the inner structure of neural networks: probing, attribution, and causal intervention. Probing involves training a separate supervised classifier, also known as a diagnostic probe, to predict certain features of the input (e.g., part-of-speech tags, dependency relations) from the model's internal activations (Alain & Bengio 2018; Hupkes et al. 2018). High accuracy in decoding a particular linguistic feature F from a probe tuned to a pattern of activation A in a subset of the network can be thought to provide evidence that A is sensitive to F and provide information about the presence of F—information for downstream processing. However, a probe's successful prediction of a certain linguistic feature from a model's activations does not necessarily mean that the feature plays a causal role in the model's behavior (Belinkov 2022). By way of analogy, suppose you throw block letters spelling a word into a pond. With the right apparatus, you might be able to decode the word's identity from the ripples formed by letters over the pond's surface, but this does not provide evidence that the pond represents (let alone "understands") the word in any meaningful sense. It merely shows that the block letters created surface patterns whose origin could be recovered by a sufficiently powerful decoding method. In addition, probes can pick up on spurious correlations, thereby failing to distinguish between genuine representation and incidental associations (Hewitt & Liang 2019).

Methods that do not involve training a separate classifier—also known as nonparametric methods—have been explored as an alternative to probing (Yousefi et al. 2024). For example, one can directly investigate the weights (or attention scores) that attention heads[5] place on different parts of the input sequence. A high attention score on a particular word might be thought to provide evidence that the head is playing a role in processing the semantic or syntactic role of that word in the rest of the sentence. This approach to interpretability falls within the broader and somewhat loose umbrella of "attribution methods" in deep learning. Attribution methods assign importance scores to input features to explain individual model predictions; in other words, they are meant to identify parts of the input that are most influential for the output. While attention patterns in Transformers—the architecture of nearly all recent LLMs like BERT, Claude, Llama, and the GPT-series—intuitively provide clues about the relevance of each input token to model predictions, simply analyzing these patterns in a given layer of the network only offers limited explanatory power (Chefer et al. 2021). In particular, visualizing attention patterns is thoroughly insufficient to draw strong conclusions about the flow of information through the network. This fact is further illustrated by the empirical observation that many attention heads in trained LLMs appear to be redundant or superfluous, in that they can be deactivated without much observable effect on important kinds of performance in the network (Voita et al. 2019; Bian et al. 2021; He et al. 2024; Bansal et al. 2023; Gromov et al. 2024). Ultimately, a common question about probing and attribution methods is whether they provide evidence that is strong enough to disconfirm false hypotheses about the representational organization of these networks. If we want to understand what information the model represents in the input and how that information is processed through step-by-step computations across layers to drive output predictions, we need to intervene directly on the network to reveal its causal structure.

10.3 Interventionist methods for understanding LLMs

To assert that a certain activation pattern in a model genuinely represents a given feature, mainstream philosophical theories of representation require that three criteria be satisfied: (a) the activation pattern should carry information about the feature; (b) the activation pattern should influence the model's behavior in a task-relevant way, and (c) the model should be capable of misrepresenting the feature (Harding 2023). Standard probing methods only provide evidence for the first criterion. Showing that some information about a feature is actually used by the LLM to generate its outputs requires suitable interventions on patterns of activation encoding that information. Changes in model behavior caused by such interventions should be consistent with the hypothesis that the model represents the target feature, but should not be explainable by mere perturbations from the training set. Accordingly, an increasingly large body of work uses targeted intervention methods to establish causal relationships between language models' internal representations and their behavior.

10.3.1 From ablation to causal probing

The simplest kind of intervention on neural networks—both artificial and biological—is an ablation. In the context of ANNs, ablation involves disabling or eliminating individual neurons or groups of neurons within a trained model to observe the resulting changes in behavior. By ablating neurons and observing the resulting change in performance metrics such as classification accuracy or reconstruction error, researchers can determine how much each neuron or group of neurons contributes to the network's overall function. Neurons that are critical to network performance will result in a substantial decline in metrics when ablated.

Indiscriminate ablations, common in the early days of connectionist research, involved disabling nodes at random. This approach was aimed at demonstrating general properties of neural networks. One key finding from such studies was the concept of graceful degradation: like biological brains, the performance of neural networks tends to decline in a gradual, rather than abrupt, manner when parts of the network are damaged or disabled (Sejnowski & Rosenberg 1987; Smolensky 1988).

Targeted ablations, on the other hand, involve disabling specific nodes or modules believed to serve distinct representational or functional roles within the network. By analyzing the network's altered behavior, researchers could infer the impact of the disabled nodes, thereby gaining insights into their hypothesized functions (Meyes et al. 2019). This method mirrors a common approach in neuroscience where the study of natural or induced brain lesions helps to unravel the functions of specific brain areas.

However, both indiscriminate and targeted ablation studies are somewhat limited in their ability to fully uncover representational or functional roles in neural networks. This limitation is partly due to the nature of how information is represented within these networks. Traditional views centered around localized representations, positing that specific neurons or small groups of neurons could be responsible for representing distinct, complex stimuli or properties. The classic example in neuroscience is the hypothetical "grandmother cell"—a neuron that would activate exclusively in response to the mental image or property of one's grandmother. While intuitively appealing, this model of localized representation has largely been abandoned in favor of a model on which information about meaningful features is distributed across multiple network components (Plaut & McClelland 2010; Barwich 2019).

In practice, individual neurons in neural networks often encode information about multiple properties or concepts, a phenomenon known as "polysemanticity." This concept has been studied since the early days of connectionist research under the name of distributed representations (Smolensky 1986; Rumelhart et al. 1987). When features are represented as directions in the network's activation space, they may not align with individual neurons (the standard basis of the vector space). In such cases, each neuron captures a mixture of underlying features, resulting in polysemantic neurons. This is characteristic of distributed representations, where features are encoded by the activations of multiple neurons.

More recently, researchers have focused on a specific phenomenon called "superposition," as described by Elhage et al. (2022). Superposition occurs when neural networks represent more features than they have neurons or dimensions. In this scenario, the network is forced to encode multiple features in overlapping directions, leading to interference between features. This inherently creates polysemantic neurons, as each neuron must encode information about multiple features to accommodate the excess of features relative to dimensions. Superposition allows neural networks to efficiently encode a large number of sparse features, but it also introduces challenges for interpreting individual neurons and understanding the network's internal representations.

The distributed model of representations in neural networks reflects a more integrated and holistic approach to neural processing, where information is not stored in isolation but as part of a dynamic, interconnected system. The shift from localized to distributed representation models has significant implications for interpreting the results of ablation studies. In a distributed system, disabling a node or a set of nodes can impact the network-wide interplay of neural activities. Therefore, determining the specific effects of localized damage on the overall behavior of the network becomes increasingly complex (Jonas & Kording 2017).

Fortunately, the level of control we have over ANNs allows for far more sophisticated causal interventions that can allow us to "edit" distributed representations in ways consistent with representational or functional hypotheses about their meanings, and then verify corresponding changes in behavior. For example, a more sophisticated approach called "iterative nullspace projection" was developed by Ravfogel et al. (2020). Iterative nullspace projection can determine whether some particular information is causally involved in a neural network's predictions by identifying and removing that information from distributed neural representations and then assessing the consequence on model behavior. The approach involves iteratively projecting neural representations onto the "nullspace" of a probe—that is, the subspace of the model's activations for which the probe makes the same constant prediction with respect to that feature—to remove detectable information about user-defined target properties.

More specifically, the first step is to train linear probes on internal neural representations to predict values of the property of interest. For example, probes could be trained to classify grammatical number (e.g., whether a noun is plural or singular in Romance languages) from the hidden state of a language model. Projecting representations onto these nullspaces removes information that linearly correlates with property values while preserving unrelated information. After projecting onto the first probe's nullspace, a second probe is trained on the transformed representations to predict the target property. This process is repeated iteratively, with projections onto each new probe's nullspace removing additional detectable information about the target property. Finally, the network makes predictions using the fully projected representations where linear information about the target property has putatively been eliminated. If performance degrades, the target property can be inferred to be causally significant for model performance, suggesting that the network exploits it for its original computations.

Suppose we want to remove grammatical number information from the embedding space of a word embedding model. We would first train a linear probe to predict whether a word is singular or plural from its embedding. This probe could be taken to identify a "grammatical number" direction in the geometry of the LLM's encoding space. We then project all word embeddings onto the nullspace of this probe, removing the number information. Next, we train a second probe to again predict grammatical number on the projected embeddings and project onto its nullspace, and so on iteratively. The key advantage of this iterative approach is that multiple orthogonal directions may encode relevant information about the property of interest. By repeating the process, we can identify and strip out multiple directions rather than just the most prominent ones. Once the dimensions are identified, vectors could also be moved in the opposite direction to enhance the information contained about the relevant property, rather than remove it.

For example, Ravfogel et al. (2021) investigated whether information about relative clause boundaries is used by language models like BERT to predict subject-verb agreement across a relative clause. They trained a set of linear probes to predict whether a word is inside a relative clause, based on the model's contextual word embeddings. Each probe defines a direction in the representation space that separates words inside relative clauses from words outside relative clauses. Then, iterative nullspace projection is used to generate "counterfactual representations" for the masked verb token. The representation is projected into the relative clause feature subspace and then flipped to the opposite side of the separating hyperplane, either toward the side containing relative words ("positive counterfactual") or away from that side ("negative counterfactual"). This process minimally modifies the representation to incorrectly encode that the word is inside or outside a relative clause. Finally, the effect of the counterfactual representations on the model's number agreement predictions is measured. If swapping in the positive counterfactual increases error rate and the negative counterfactual decreases error rate, this alignment with predictions from linguistic theory suggests the model uses relative clause boundary information appropriately for agreement, confirming the causal effect (Figure 10.1).

This method is directly inspired by counterfactual approaches to causal explanation in philosophy of science mentioned above (Woodward 2005).

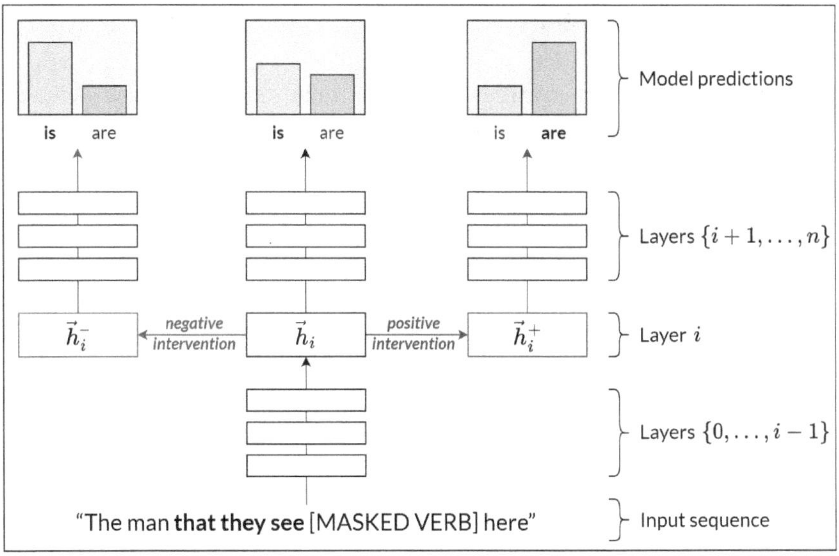

Figure 10.1 Iterative nullspace projection. Given the representation $\vec{h}_i$ of a masked word in layer i of a Transformer model, a probe is trained to predict relative clause boundaries. The probe's nullspace, which encodes information not relevant to relative clause boundaries, is identified. Two counterfactual representations, $\vec{h}_i^-$ and $\vec{h}_i^+$, are derived by projecting $\vec{h}_i$ onto the nullspace and then performing negative and positive interventions, respectively, along the probe's decision boundary. $\vec{h}_i^-$ encodes that the word is outside a relative clause, while $\vec{h}_i^+$ encodes that it is inside a relative clause, with other information preserved. The model's predictions using these counterfactual representations are compared to its original prediction to assess the causal effect of the relative clause boundary information on the model's behavior in number agreement.

Such approaches aim to isolate the causal contribution of some factor X to an outcome Y by minimally altering the value or presence of information about X in the system, while holding all other factors fixed. This allows assessment of whether and how much the factor X is exploited causally by the system to produce Y. The removal of detectable information about a target variable through iterative nullspace projection is analogous to a hypothetical intervention that breaks the links between that variable and the system while leaving other causal mechanisms intact (i.e., the vectors' other meaningful dimensions). The comparison between original model outputs and outputs based on projected representations missing information about X mirrors the evaluation of interventionist counterfactuals, which aims to answer questions such as "What would happen to Y if an intervention prevented information about X from influencing the mechanism?".

10.3.2 Features and circuits

Intervention methods such as iterative nullspace projection can help us determine whether some target property, such as a particular syntactic feature, is represented by a language model and causally efficacious in its behavior. But this is insufficient to understand language models and other neural networks the way we understand classical computer programs. Indeed, the guiding ideal of interpretability research is to model the internal causal structure of neural networks, such that we can explain some behavior of interest in terms of a series of computational steps applied on the input. Mechanistic interpretability refers to this concerted effort to reverse engineer the internal computations performed by ANNs. Rather than focusing solely on interpreting individual model predictions, it aims to provide a detailed and systematic understanding of how the model transforms inputs to outputs (Elhage et al. 2021). The overarching goal of mechanistic interpretability is to open the "black box" of neural networks by providing human-intelligible descriptions of the functional modules that drive the emergence of model behaviors.

Like systems neuroscience, mechanistic interpretability specifically targets the algorithmic level of analysis (Marr 1982; Lindsay & Bau 2023). An algorithmic explanation elucidates the flow of information through a network and the sequence of operations performed upon it. Crucially, it abstracts away from engineering (or biological) details, instead focusing on the computations performed by the system. Algorithmic explanations support counterfactual inferences about how changing inputs or network components would affect computations and outputs. This involves identifying structural elements like motifs and circuits that are reusable, such that manipulating them produces systematic effects across many inputs and conditions.

Mechanistic interpretability specifically seeks to reverse engineer neural networks in terms of learned features and reusable circuits that operate on those features. Features refer to human-interpretable properties of the input data that the model encodes internally. For computer vision models, such features might be edges, textures, or shapes. For language models, features may correspond to part-of-speech tags, named entities, or semantic relationships. By studying a network's internal activations, researchers aim to determine which features are encoded where.

Circuits, in turn, are chains of operations that detect certain input features and transform them into output features. The goal is to decompose an entire neural network into a hierarchy of understandable features and circuits that process information step by step. This modular view would explain how trained models transform inputs to outputs via learned features and program-like circuits, providing a form of algorithmic understanding.[6]

More formally, a given neural network can be modeled as a graph (sometimes called a computational graph) whose nodes correspond to components of the network at some level of granularity—such as attention heads or individual neurons. The directed edges between nodes represent the flow of information and computations in the neural network. The connectivity defined in the computational graph needs to faithfully represent the actual computation flow in the neural network. In this model, a circuit is a subgraph within the overall computational graph that implements some specific behavior or functionality of interest. The goal of representing the neural network as a computational graph is to systematically analyze it and localize particular circuits that are causally responsible for certain model behaviors.

Once candidate circuits are found, their functionality must be validated through causal interventions. This often involves surgically replacing components of suspected circuits and observing the effect on model outputs. Validated circuits can then be composed into a hierarchical understanding of the succession of transformations enacted by the model. The resulting mechanistic explanation should provide significant behavioral control, for instance allowing targeted editing of model computations.

The picture that emerges from mechanistic interpretability research is that Transformer models can be viewed as containing parallel processing streams (also known as "residual streams"), one at each input token position (Figure 10.2). While each residual stream encodes information in a very high-dimensional vector space, attention heads at each layer operate over much smaller (low-dimensional) subspaces of the stream that may not overlap with one another. Making use of these disjoint subspaces allows Transformers to route information about tokens and their dependencies dynamically across layers and positions. Specifically, each stream is functionally analogous to an addressable memory, in which attention heads can write to and read from subspaces of the main embedding space. Such information may include, for example, syntactic dependencies between tokens in the input sequence.

A common intervention method used in mechanistic interpretability research is activation patching (Zhang & Nanda 2023), also known as causal tracing (Meng et al. 2022) and interchange intervention (Geiger et al. 2021) (Figure 10.3).[7] The basic method involves three steps. First, the neural network must be run on an original input that relates to some behavior of interest (e.g., answering factual questions). For example, the original input might be "The capital of France is . . .," with the expected output being "Paris." Importantly, the activations of the network during the forward pass on this original input are cached for later use. The second step is to run the model again with an alternative input that introduces a key variation on the original input that changes the behavior (output). For

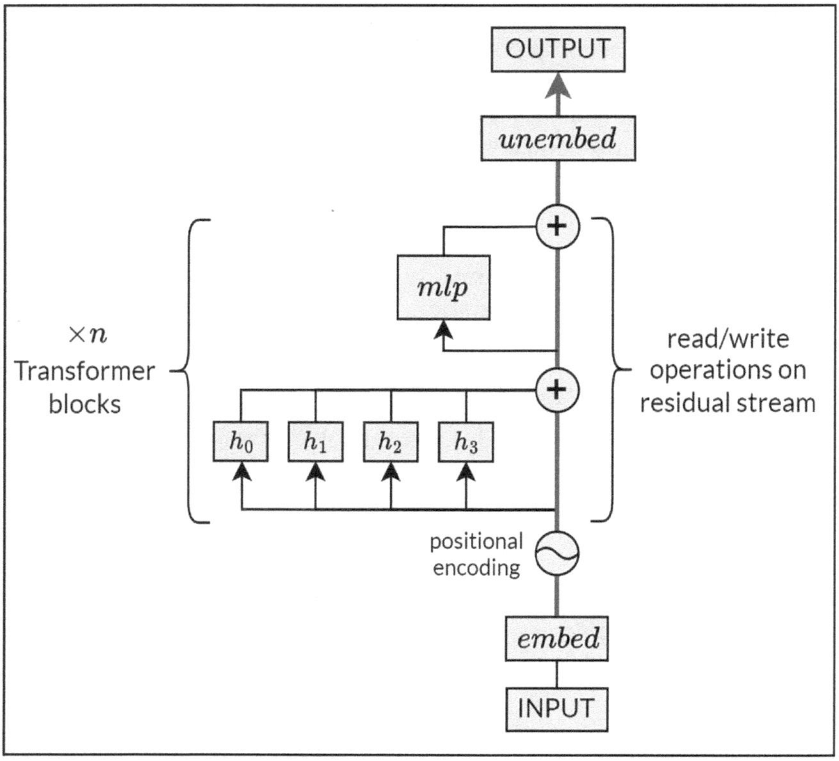

Figure 10.2 The residual stream view of the Transformer. Each input token is first embedded into a dense vector representation and combined with a positional encoding injecting information about their position in the input sequence. This forms the initial state of the "residual stream" (depicted by the red arrow), which flows through the entire network. Each Transformer block, consisting of multi-head self-attention and a multilayer perceptron (MLP), reads from the residual stream, transforms the representation, and writes the result back into the stream via residual connections. This process is repeated across multiple Transformer blocks. Finally, the output of the residual stream is "unembedded" to map the transformed representation back to the original token space. In this view, the Transformer's components are seen as operators that successively refine the residual representation.

example, an alternative input might be "The capital of Germany is . . .," with the expected output being "Berlin." Finally, the alternative input is run again but a specific component of the network's activation is swapped for its cached value from the original forward pass—an intervention known as "patching," because it patches the alternative forward pass with

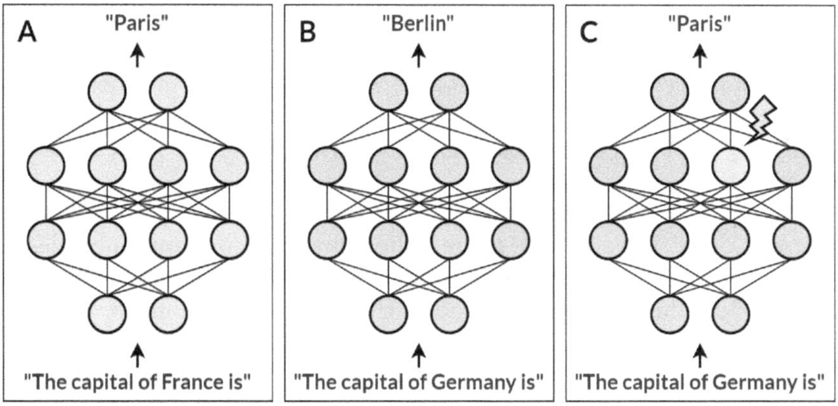

Figure 10.3 Activation patching. (A) In the original forward pass, the model takes as input the prompt "The capital of France is" and outputs the correct answer "Paris." The model activations from this forward pass are cached. (B) In the alternative forward pass, the prompt is changed to "The capital of Germany is." The model now outputs "Berlin" as the answer. (C) Activation patching is applied in a third forward pass. The model is given the alternative prompt "The capital of Germany is" once again, but a specific component of the model has its activations replaced (patched) with those from the original forward pass on the France prompt. This causes the model to output "Paris" instead of "Berlin," despite being given the Germany prompt. The restoration of the original output through patching the activations of a particular model component provides evidence that this component encodes information that is causally implicated in the target behavior. (Note that nodes in these diagrams will typically not correspond to individual neurons in a network, but rather more complex patterns or directions in activation space.)

a component of the original forward pass. The effect of patching a component's activation is then evaluated by comparing the model's performance in the regular forward pass on the alternative input versus the patched forward pass on the same input.

For example, the metrics used could be the probability the model assigns to the original, correct token ("Paris"), or the difference in logits[8] between the correct and incorrect tokens ("Paris" vs. "Berlin"). Intuitively, changing the input hurts model performance on the expected behavior, while patching activations from the original run helps restore it. So if patching a particular component leads to a significant restoration of performance (e.g., an increase in the probability of "Paris" being the output), it suggests that component is important for the model's behavior on the task (i.e., the component not only contains information on the target feature, but that information is causally implicated in producing the desired result). By iterating this procedure over many components of the computational graph, such as attention heads in a Transformer model, activation patching aims to identify the key circuits that enable behaviors like factual recall or logical reasoning.

The mechanistic interpretability framework has been applied to many different problems and has started shedding light on some abilities of Transformers. It is worth noting that much of this research is painstaking work conducted with toy models that are easier to interpret. While these toy models are based on the Transformer architecture, they may differ from LLMs in terms of architectural details, learning objective, dataset, and of course size (parameter count). Nonetheless, efforts are underway to automate and scale mechanistic interpretability techniques to bona fide LLMs, with promising initial results (Wu et al. 2023; Conmy et al. 2023; Syed et al. 2023). In what follows, we will briefly illustrate the fruitfulness of the mechanistic interpretability program through three case studies.

10.3.2.1 Case study 1: Induction heads

A canonical example of circuit discovery in language models through the tools of mechanistic interpretability is that of so-called induction heads (Olsson et al. 2022). These are specialized attention heads that emerge through training even in very small Transformer models and implement a form of on-the-fly pattern completion. In other words, they allow models to repeat or generalize sequences in novel patterns that appear within prompts—potentially explaining a large amount of the so-called in-context learning exhibited by LLMs, where they can exhibit impressive few-shot learning of patterns not present in their training sets (Mirchandani et al. 2023).

Specifically, induction heads use a "prefix matching" attention pattern to look back over previous tokens in the current context window and detect if any match the current token. Rather than relying merely on memorized statistics about which tokens tend to follow others, they flexibly attend to whichever prior token is most similar to the current one based on learned representations. If a previous token is sufficiently similar, the induction head will attend to the next token after it. The head then increases the probability of that attended next token, effectively predicting that the current sequence will continue like the previous matched sequence. For example, if the input contains the sequence "the cat sat on the mat. The cat," the induction head matches the second "cat" token to the first one, attends to "sat" from the first sequence, and increases the likelihood of outputting "sat" again. This allows Transformer models to repeat sequences and generalize patterns.

Importantly, the computations performed by induction head circuits do not rely on bigram statistics memorized from the training data; rather, they operate over abstract patterns in the input sequence (prompt), even if the latter does not contain familiar strings. When the network is processing a sequence [A][B] . . . [A], the induction head circuit can be characterized algorithmically as storing the value of the first [A] token in a specific subspace of the residual stream—which we may call the previous_token subspace—at the position of the [B] token. Importantly, this operation occurs whatever that specific value (content) of [A] might be. Functionally, the previous_token subspace in which this information is stored operates rather like a variable in a classical symbolic program. Its value is accessed at the next layer by the second part of the induction head circuitry, which reads the content of the previous_token subspace in the residual stream at the position of the [B] (Figure 10.4).[9]

10.3.2.2 Case study 2: Modular addition

Mechanistic interpretability is helpful to investigate not only how trained neural networks process information through algorithmic circuits but also how they learn such algorithms during training. Studying the training dynamics of neural networks is important, because it can provide insights into learning transition phases that are highly relevant to ongoing debates about the capacities of LLMs. Thus, Nanda et al. (2022) investigated the puzzling phenomenon of grokking, where neural networks trained on algorithmic tasks with regularization initially overfit to the training data but later suddenly generalize after many training steps. As a case study, the authors trained small Transformer models on modular addition tasks. They find these networks exhibit "grokking," initially overfitting but later learning to generalize.

To understand this phenomenon, they reverse engineered the mechanisms learned by these networks using techniques from mechanistic interpretability. That found that the network learns to perform modular addition tasks by mapping inputs onto rotations in the plane and composing those rotations using trigonometric identities (Figure 10.5). This clever algorithm, dubbed "Fourier multiplication," allows the network to perform addition given a particular prime number chosen as a modulus (where the modulus determines the size of the "circle" around which the addition operation travels—see Figure 10.5).

Using this understanding, Nanda et al. defined new progress metrics allowing them to study the training dynamics of the models as they learn

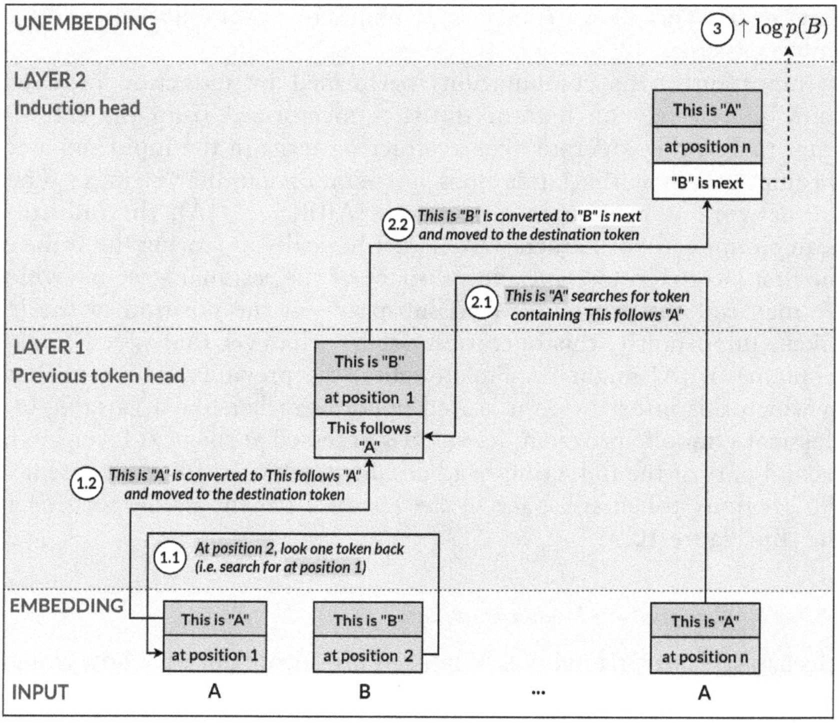

Figure 10.4 A schematic illustration of the induction head circuit in a two-layer Transformer model. At the embedding stage, each token from the input sequence is encoded as a vector, together with information about its position in the sequence. The first layer contains an attention head—known as the previous token head—that acquired a specialized function during training. When processing token [B] at position 2, the previous token head does the following: (1.1) it attends to the previous token at position 1; (1.2) it writes the identity of this preceding token to a dedicated subspace of the residual stream at the current position (position 2), effectively storing the information "the token before me is [A]." Layer 2 contains another specialized attention head known as the induction head. When processing the second instance of [A] at position n, the induction head does the following: (2.1) it queries the residual stream for information in the "previous token" subspace matching the current token's identity; (2.2) having located this previous token information in the residual stream at position 2, it retrieves the identity of the token at that position ([B]), then writes this identity to a dedicated subspace of the residual stream at the current position (position n), effectively storing the information "predict that the next token will be [B]." (3) The unembedding layer maps information in the "next token" subspace at position n to an increased logit for [B] at position $n + 1$, which translates to an increased log likelihood of [B] being predicted as the next token.

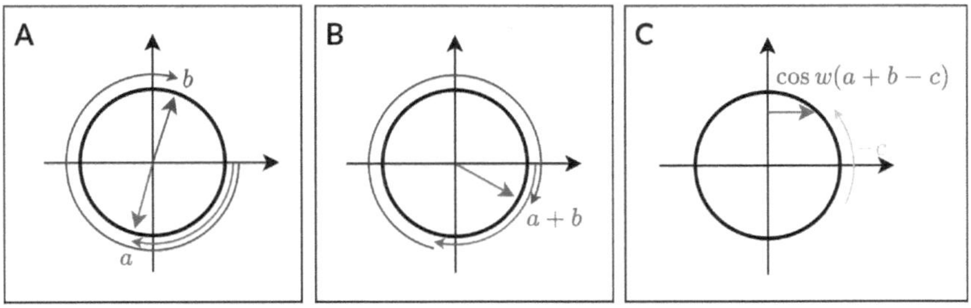

Figure 10.5 A learned algorithm for modular addition. (A) Embedding projection. Given two input numbers a and b in the modular addition $a + b \equiv c$ mod P, the model uses its embedding matrix to project each number onto a corresponding rotation around the unit circle. The embedding matrix essentially memorizes a mapping between each possible input number and a specific rotation amount, converting the numbers into geometric representations. (B) Rotation composition. The model composes the two rotations generated for a and b. This step effectively adds the two rotation amounts together, resulting in a new, single rotation that represents the sum $a + b$ in modular arithmetic. In modular arithmetic, numbers "wrap around" after exceeding the modulus P, so if $a + b$ is greater than P, the resulting rotation will correspond to $a + b$ mod P, which is the remainder when $a + b$ is divided by P. (C) Output decoding. To produce the output logits (raw scores used for next-token prediction), the model considers each possible result c (ranging from 0 to $P - 1$) and performs a reverse rotation by $-c$. This step essentially checks, for each c, whether undoing the rotation by c results in a rotation that matches the one representing $a + b$ mod P. The output c that produces the rotation most closely matching the $a + b$ mod P rotation is assigned the highest logit. This works because the trained model ensures that the correct c satisfying $a + b \equiv c$ mod P will undo the rotation by exactly the right amount to point back to the $a + b$ mod P rotation. The trigonometric functions cosine and sine are used to implement these rotations and achieve the desired result mathematically using angle addition identities, but conceptually, the algorithm is based on representing numbers as rotations and composing these rotations together.

Source: Adapted from Nanda et al. (2022)

to perform modular addition algorithmically. They identified three distinct learning phases, each marked by continuous progress on certain metrics:

1. *Memorization phase*: In the early part of training, the models fit to the training data by simply memorizing input–output pairs. Performance on the test set remains low, while performance on the training set increases rapidly, indicating the models are overfitting.
2. *Circuit formation phase*: After memorization, there is a transition period where the models internalize the algorithm for modular addition using

trigonometric identities and rotations (the "Fourier multiplication" circuit). However, performance on the test set remains low, implying memorization components still persist.
3. *Cleanup phase*: Finally, weight decay drives the removal of the initial memorization components. Performance on the test set abruptly improves to match performance on the training set at the end of this phase, corresponding to the "grokking" transition in generalization capability.

The entire phenomenon is thus not a sudden onset of generalization ability, but rather a gradual amplification of structured mechanisms encoded in the weights, followed by pruning of unnecessary components. Importantly, these findings speak against the skeptical view of LLMs discussed in Section 10.1, according to which their abilities can be explained by mere memorization. While memorization dominates in the initial phase of training, subsequent phase transitions show that Transformer models can learn general, rule-like algorithms to solve tasks, including tasks as rigid and well-defined as arithmetic problems.

10.3.2.3 Case study 3: World models

There is an ongoing debate about whether LLMs may acquire "world models," characterized as sets of structure-preserving internal representations that encode abstract information about real-world or domain-specific entities, relations, and processes—rather than merely intralinguistic features (Yildirim & Paul 2023; Mollo & Millière 2023). Inducing such representations could allow LLMs to generate outputs consistent with real-world constraints, reason about worldly phenomena, and potentially misrepresent aspects of reality, despite lacking direct sensorimotor experience. While behavioral evaluations are generally insufficient to settle this debate, mechanistic interpretability shows promise to uncover what the LLMs and related Transformer models represent internally.

To investigate whether Transformer models trained to predict sequences can learn interpretable world representations (rather than mere surface statistics), Li et al. (2023) focus on the board game Othello as a simplified yet nontrivial domain. They trained a GPT variant (Othello-GPT) to sequentially predict tokens representing Othello board positions. The only inputs to the model are move sequences derived from game transcript; no explicit game rules or board structure are provided. After training on championship games and synthetic games, Othello-GPT recommends legal moves with high accuracy, suggesting it has learned more than surface statistics. Furthermore, nonlinear probes reliably predict board states from model activations, implying that a nonlinear representation of the board state had emerged during training. To validate the probes' accuracy, the authors performed intervention experiments that modify Othello-GPT's internal activations to reflect altered board states. The model's subsequent move predictions change accordingly, confirming the causal role of these latent board state representations.

In a follow-up study, Nanda et al. (2023) discovered that rather than representing the board state (e.g., representing each board tile as black, white, or empty), Othello-GPT actually encodes tiles relative to the current player (as player, opponent, or empty). By re-orienting probes to classify this player-centric representation, Nanda et al. demonstrated that the board state is in fact linearly encoded with high accuracy in the network, contrary to Li et al.'s (2023) claim that the board state is only encoded non-linearly (Figure 10.6). They further demonstrated behavioral control by conducting simple vector arithmetic interventions to alter the model's encoding of board states and change predictions accordingly. Hazineh et al. (2023) found similar evidence that information about board state is encoded in a simple, linear way in the deeper layers of Othello-GPT models. Like Nanda et al. (2023), they decoded a representation corresponding to tiles marked player, opponent, or empty, which aligns well with the role of the model in alternating between playing as white or black. To test whether these internal representations play a causal role in the model's predictions, they also intervened by manipulating activations to trick the model about the state of the board. Through visualizing effects on predicted logits and comparing distributional similarity of logit outputs, they demonstrated layers in which this internal representation steers next-move predictions. The internal representation appears fully developed and utilized in middle layers of deeper models, while shallow models fail to use the representation causally.

Studies on Othello-GPT provide tentative evidence that Transformer models can acquire world models at least in toy domains.[10] While Othello-GPT is not a language model properly speaking, it generates legal game moves in written form, and interventionist experiments reveal that it generates such moves on the basis of linearly decodable representations of the board state. To what extent can we generalize these findings to actual LLMs? This is a challenging question. A reasonable assumption is that Othello-GPT acquires a world model—an internal representation of the game world—because this is useful to improve its predictions of legal moves past a certain threshold. However, neural networks are notorious in their ability to achieve high accuracy scores on problems by mastering surface statistics; they tend to learn shortcuts (e.g., shallow heuristics) to achieve good performance on their learning objective until they hit a bottleneck. Furthermore, acquiring world models about the real world, rather than an extremely simple game world, is presumably costly in terms of training dynamics; it would require a substantial reorganization of the model's internal structure, akin to the phase transition undergone by Nanda et al.'s (2022) model to "grok" modular addition, albeit on a much broader scale.

It may well be that LLMs do undergo such phase transitions during training, and acquire representations akin to world models at least in some limited domains. This might be required to unlock certain abilities such

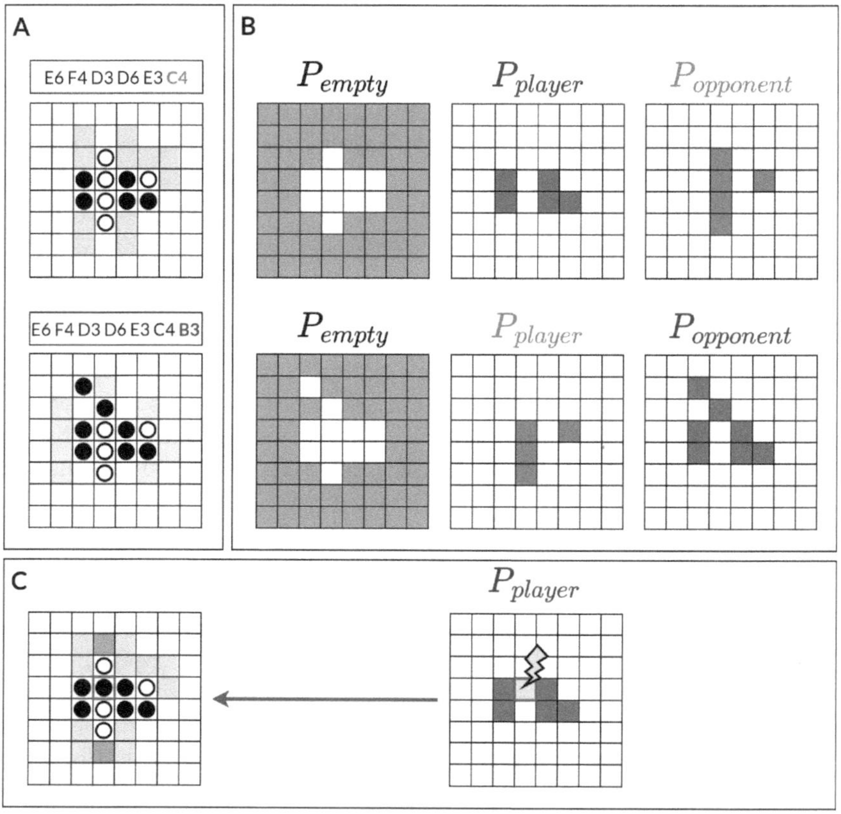

Figure 10.6 Emergent representations of the board state in Othello-GPT. The model represents board states relative to the current player. (A) The ground truth board states at two consecutive time steps. Colored tiles show legal moves for the current player (light blue for black and light orange for white). (B) The board states at the same two time steps decoded by linear probes. The probes are trained to classify the board relative to the current player: empty for empty tiles, player for tiles occupied by the current player, and opponent for tiles occupied by the current player's opponent. Note that the player and opponent colors are flipped between the two time steps, as the current player changes. (C) We can intervene on the model's internal board state representation by pushing an empty tile's representation in the direction of the player vector, to make the model represent that tile as occupied by the current player. This simple linear intervention is sufficient to alter Othello-GPT's move predictions (red tiles on the left), demonstrating that the linearly represented board state causally determines the model's outputs.

Source: Adapted from Nanda et al. (2023)

as commonsense reasoning about intuitive physics, which in turn would further reduce the loss (i.e., improve next-token prediction performance) in certain contexts. For this to happen, two conditions must presumably be satisfied: (a) the text-based training data should provide enough information to induce the relevant components of world models; (b) the learning pressure to reduce loss on the next-token prediction objective should be sufficient to push the model to acquire such representations given long enough training trajectories. Neither condition is antecedently guaranteed, and there is currently limited evidence that they apply to most existing LLMs. In particular, interventionist evidence regarding the most capable LLMs such as GPT-4, whose behavior on certain tasks is most consistent with the acquisition of world models, is still lacking. As far as behavioral evidence is concerned, lingering failure modes on out-of-distribution tasks (those with an informational structure different from the training set, e.g., involving the same physical principles but applied to different objects or in different contexts) cast doubt on the hypothesis that current LLMs acquire world models that are as robust as those discovered by humans and animals (McCoy et al. 2023; Yildirim & Paul 2023).

There are nonetheless ongoing efforts to assess the existence of world models in LLMs using the tools of mechanistic interpretability. For example, Li et al. (2021) used probing and intervention methods on models fine-tuned on the Alchemy and TextWorld datasets. Alchemy contains sequences of instructions for manipulating colored liquids in beakers (i.e., instructions for performing fictional chemical experiments and their outcomes), while TextWorld consists of textual transcripts of navigation in simulated worlds. Li et al. designed probes to test if the models' contextual token embeddings encode and track the state of entities mentioned in the discourse. For example, in Alchemy the probe tries to determine if a representation encodes that a beaker is empty after its contents are drained. Li et al. used these probes to test whether intervening on the decoded entity representations would change model behavior. Specifically, they constructed two discourses $x1$ and $x2$ that describe draining liquid from two different beakers, $b1$ and $b2$, resulting in one empty beaker per discourse. After encoding each discourse, they created a synthetic representation $Cmix$ by taking the encoding $C1$ of discourse $x1$ and replacing the vector representations corresponding to the initial description of beaker $b2$ with those from $C2$. Although $Cmix$ does not correspond to any real textual input, it implicitly represents a situation in which both $b1$ and $b2$ are empty. When generating text conditioned on $Cmix$, they found that the model generates instructions that are more often consistent with both beakers being empty compared to generating from $C1$ or $C2$ alone. However, the generated instructions from $Cmix$ are still not always fully consistent with the implicit state, suggesting the induced representation is approximate rather than perfect. By editing the entity representations to model a new state not seen in the actual training data or prompt, and observing changes in the model's outputs that are often consistent with this new state, Li et al. provided tentative evidence that the model can induce an approximate implicit representation of the state of discourse entities purely from text. Note that producing text that is consistent with the hypothesis-driven manipulation of the LLMs activation space is not something that the system was trained to do. Furthermore, the method was found to be robust against "sanity checks" to confirm that the intervention itself is not wholly responsible for injecting appropriate information into the system, such as finding negative results when attempting the intervention method on similarly complex Transformer architecture with randomized weights. However, the generality of these findings remains uncertain, as the experiments focused on narrow domains with simple objects and dynamics. More systematic testing would be needed to determine how well they generalize to more complex and open-ended settings.

10.3.3 Dictionary learning with sparse autoencoders

While early research in mechanistic interpretability has largely focused on toy models, where probing experiments and causal interventions are more tractable, recent work has explored a new method called dictionary learning which aims to automatically identify interpretable features in LLMs. This approach seeks to address the challenge of superposition, whereby neural networks represent more features than they have neurons by encoding each feature as a combination of multiple neurons' activations.

Dictionary learning uses sparse autoencoders (SAEs) to decompose the dense activation patterns of LLMs into sparse combinations of interpretable features. SAEs are simple unsupervised neural networks that learn to reconstruct their input data using a large hidden layer with many more neurons than the input, while enforcing sparsity constraints to ensure that only a small number of hidden neurons activate for any given input. The process of dictionary learning involves training an SAE on activation vectors from a specific layer of the LLM. The autoencoder learns to reconstruct these activations using a larger set of sparse features. Sparsity constraints are applied during training to encourage each feature to represent a single, interpretable concept. The result is a "dictionary" of features that ideally correspond to human-understandable concepts.

Bricken et al. (2023) established the effectiveness of this approach by showing that SAEs can extract features representing specific contexts, such as Arabic script and DNA sequences, as well as abstract concepts like code errors and mathematical operations from language models. Building on this work, Templeton et al. (2024) and Gao et al. (2024) further demonstrated that this approach scales to state-of-the-art LLMs like Claude and GPT-4, successfully extracting millions of sparse features spanning a wide range of concepts. For example, a sparse "Golden Gate Bridge" feature in Claude activates strongly on both images and text inputs—across multiple languages—related to the Golden Gate Bridge (Templeton et al. 2024).

This approach offers several advantages over older methods like probing. It is highly scalable, efficiently producing millions of reusable features. It can be used to automate the discovery of unanticipated features without relying on hand-crafted probes, which potentially allows for a more comprehensive exploration of the model's internal representations. The extracted sparse features often seem to map onto human-understandable concepts, much more reliably than the activations of individual neurons in the original model. Once extracted, these interpretable sparse features can be reused for various analyses and interventions.

Despite the excitement surrounding dictionary learning, however, it still lacks robust theoretical grounding compared to more established intervention techniques in mechanistic interpretability. There is a notable absence of well-established metrics for evaluating the quality of SAE-extracted features and the overall performance of the SAE. Without a consensus on such metrics, it is difficult to objectively assess whether a given dictionary learning experiment has succeeded in extracting sparse features that are actually involved in the LLMs' computations. The allure of "autointerpretability"—using LLMs themselves to automatically label SAE-extracted features—introduces additional methodological concerns. This approach may create an illusion of understanding by latching onto spurious correlations rather than identifying genuine causal relationships. LLM-generated labels may seem to make sense at first glance based on a subset of highly activating samples for a given sparse feature but can be misleading on closer inspection. Addressing these limitations on firm theoretical and methodological grounding will be important to ensure that dictionary learning actually provides reliable, meaningful insights into the inner workings of LLMs.

10.3.4 Interpretability and causal abstraction

The general project of mechanistic interpretability can be seen through the lens of causal abstraction developed by Geiger et al. (2021). Causal abstraction is a theoretical framework that aims to provide an interpretable high-level causal explanation of the behavior of a complex system, such as a neural network, that is consistent with the low-level causal structure of that system.

The core intuition is that the variables of a large, complex causal model can be clustered into sets and aligned with the variables of a sparse, low-dimensional model with fewer, high-level variables. The high-level model provides an abstract characterization of the low-level model if aligned high-level and low-level variables have equivalent causal roles and information content. This equivalence is experimentally verified with interventions on both levels that produce the same counterfactual behavior. More formally, an alignment between a low-level model L and high-level model H is a partition of the low-level variables into sets, with each set aligned to a high-level variable. A translation function maps low-level variable values to high-level values (Geiger et al. 2025). The alignment is causally consistent if, for any low-level intervention that has a corresponding high-level intervention under the translation function, intervening on both models produces equivalent results after translation. If an alignment is causally consistent, then H is said to be a constructive causal abstraction of L.

Any program or algorithm can be represented as a causal model (Icard 2017). In the context of interpretability research in deep learning, the low-level causal model is a neural network, containing many interconnected nodes and weights that determine its function. The high-level causal model proposed by mechanistic interpretability researchers offer hypotheses about the abstract computations and algorithms implemented by the network. Aligning groups of neurons and weights to single variables in the high-level model allows interpreting their collective causal role. The causal consistency condition ensures the high-level causal model faithfully captures the causal mechanisms embodied by the low-level neural-network model. Causal abstraction thus enables the development of human-interpretable high-level causal models that accurately explain the reasoning and computations inside opaque neural networks.

A key method for assessing causal abstraction involves "interchange interventions," where neural representations created for one input are swapped into the model when it is processing another input. If the low-level neural model and high-level algorithm have the same counterfactual behavior under aligned interchange interventions, that provides evidence that the alignment represents and exploits a causal abstraction. Many of the methods described above, such as activation patching, can be used as interchange interventions. In fact, iterative nullspace projection can also be formalized within a causal abstraction framework (Geiger et al. 2025). Thus, causal abstraction is a useful framework to unify interventionist approaches to interpretability.

In practice, the project of mechanistic interpretability with large neural networks like LLMs can be seen as aiming for approximate causal abstraction, which relaxes the criteria for alignment (Beckers et al. 2020). The degree of abstraction can be quantified as the proportion of aligned interchange interventions that have equivalent effects in the high-level causal model and low-level causal model or target neural network (Geiger et al. 2025). When interchange intervention accuracy is 100%, the high-level model exactly abstracts the neural network. Otherwise, the high-level model abstracts the neural network approximately, to the degree quantified by interchange intervention accuracy. Approximate abstraction can provide an interpretable high-level explanation that still reflects the causal structure of the target neural network with a degree of faithfulness suitable for genuine explanation.

10.4 Conclusion

In this chapter, we motivated and reviewed recent interventionist approaches to interpreting the inner workings of deep neural networks, with a particular focus on Transformer-based language models. We emphasized that purely behavioral methods have significant limitations and argued that mechanistic approaches to interpretability, which aim to elucidate the causal structure of a model's internal organization, offer a more promising avenue from a philosophy of science perspective. To illustrate these points, we presented a series of case studies showcasing recent interpretability techniques. These examples demonstrate the potential of interventionist methods to provide insights into the complex operations of neural networks that were previously considered opaque. We concluded by reviewing recent attempts to develop a general theory of causal abstraction for neural networks; this emerging theoretical framework has the potential to enable more ambitious and systematic applications of interventionist methods. Our analysis challenges the prevalent view that deep neural networks are inscrutable black boxes whose inner workings defy comprehension. Notably, techniques such as activation patching or interchange interventions offer a degree of control over the underlying system that neuroscientists cannot currently aspire to achieve with biological neural networks.

It may be helpful to reiterate some limitations of these methods in closing. For one, while these methods offer potential insights into the causal structure of complex large language models, they typically require full access to their internal weights. Given the proprietary nature of state-of-the-art LLMs like GPT-4 and Claude, academics and regulatory agencies without internal access may find limited practical applications for these methods in addressing scientific questions or trust and safety concerns. Additionally, many of these methods are computationally intensive. For instance, researchers at Anthropic have noted that systematic dictionary learning experiments with Claude would require more computational power than the initial training of the model itself (Templeton et al. 2024). Lastly, as with all emerging scientific methods, these techniques will likely present their own unique methodological challenges, including interpretability illusions resulting from spurious features or causal dependencies. We review these methods here in the hope that it will facilitate future research to address the ethical, practical, and scientific concerns associated with these novel approaches.

Notes

1 "Tokens" are the basic lexical units—whole words or parts of words—used to encode linguistic sequences; for example "camembert" might be tokenized as "cam" + "emb" + "ert." Details about the training data and model size of proprietary LLMs like GPT-4 are not publicly available, although we can turn to other LLMs for clues. For example, PaLM 2 has 340 billion parameters and

was trained on 3.6 trillion tokens (Anil et al. 2023), while the largest version of Llama 2 has 70 billion parameters and was trained on two trillion tokens (Touvron et al. 2023). GPT-4 is rumored to have well over a trillion parameters (Karhade 2023).

2 This concern is highlighted by lawsuits against OpenAI, notably from the New York Times (Grynbaum & Mac 2023). These cases document instances where LLMs like GPT-4 have been shown to reproduce substantial portions of copyrighted text verbatim, raising questions about the originality of their outputs.

3 It is very common to refer to "linear" and "nonlinear" operations in machine learning. Briefly, a linear operation consists of an additive combination of a discrete set of features or inputs where each increase of magnitude in a feature increases the composite function in a constant way, and a nonlinear operation involves non-additive or variable influences of features. For example, multiple linear regression functions are linear, whereas functions which involve thresholding (the effect of a feature changes above a certain threshold value) or fitting exponential curves to data are nonlinear. Linear operations are simpler, more predictable, and require less training data to fit, whereas nonlinear operations are more complex, unpredictable, and require more training data, but the latter can also predict more complex patterns in data. Most currently successful machine learning architectures involve complex combinations of linear and nonlinear operations.

4 The lack of a common lexicon in interventionist research on neural networks is rather unfortunate—many different labels exist for similar ideas and methods, with "interpretability" being a particularly confusing term (Lipton 2018). In what follows, we will attempt to give a cohesive overview of this fragmented literature.

5 Self-attention is a key mechanism that distinguishes Transformer-based language models from previous architectures. In essence, self-attention allows each token's representation to be dynamically influenced by other tokens in the input sequence. The process works as follows: (1) for each token, three vectors are computed—a "query," "key," and "value" vector; (2) the attention weight for each pair of tokens is calculated as the dot product of the query vector of one token with the key vector of another, scaled and normalized (via softmax); and (3) the output for each token is a weighted sum of all tokens' value vectors, using these attention weights. Multiple attention heads operate in parallel, each potentially focusing on different aspects of the relationships between tokens. These heads are stacked in multiple layers, allowing the model to capture complex hierarchical relationships. The effectiveness of self-attention stems from its ability to model long-range dependencies and its computational efficiency. It has demonstrated success in capturing various syntactic and semantic relationships, contributing significantly to the performance improvements seen in Transformer models. Recent research has uncovered specialized behaviors in attention heads, such as specific syntax-related functions, which together enable the model to perform sophisticated language generation tasks.

6 In fact, it is even possible to program a Transformer from scratch using a specialized programming language like RASP (Restricted Access Sequence Processing Language). RASP allows mapping the basic components of a Transformer, like attention and feedforward layers, into simple primitives that can be composed into programs. These RASP programs can then be compiled into the weights of a Transformer network that implements the specified computation (Weiss et al. 2021). More recently, methods have been developed to train Transformers whose weights are constrained to implement human-interpretable RASP-like programs. These Transformer Programs can be learned end-to-end from data

and decompiled back into discrete, readable code (Friedman et al. 2023). Such approaches provide a more direct way to understand the computations of a Transformer in terms of modular algorithmic components.

7 Note that some of these labels have been used in slightly different ways by different researchers. The most neutral term is probably "interchange intervention," as it arguably subsumes the others (Geiger et al. 2025).

8 Logits are the non-normalized model outputs taken as estimates of the probability of a particular token being the likeliest next output.

9 For a discussion of induction heads circuitry as implementing a form of variable binding, and the implications of this view for the debate about compositionality in connectionist models, see Millière (2024).

10 For example, similar results have been found by studying Chess-GPT, a Transformer model trained on chess moves (Karvonen 2024).

References

Aiyappa, R., An, J., Kwak, H. & Ahn, Y.-Y. (2023), 'Can We Trust the Evaluation on ChatGPT?'. *Proceedings of the 3rd Workshop on Trustworthy Natural Language Processing* (TrustNLP 2023) (July 2023) 47–54.

Alain, G. & Bengio, Y. (2018), 'Understanding Intermediate Layers Using Linear Classifier Probes'. https://doi.org/10.48550/arXiv.1610.01644

Anil, R., Dai, A. M., Firat, O., Johnson, M., Lepikhin, D., Passos, A., Shakeri, S., Taropa, E., Bailey, P., Chen, Z., Chu, E., Clark, J. H., Shafey, L. E., Huang, Y., Meier-Hellstern, K., Mishra, G., Moreira, E., Omernick, M., Robinson, K., Ruder, S., Tay, Y., Xiao, K., Xu, Y., Zhang, Y., Abrego, G. H., Ahn, J., Austin, J., Barham, P., Botha, J., Bradbury, J., Brahma, S., Brooks, K., Catasta, M., Cheng, Y., Cherry, C., Choquette-Choo, C. A., Chowdhery, A., Crepy, C., Dave, S., Dehghani, M., Dev, S., Devlin, J., Díaz, M., Du, N., Dyer, E., Feinberg, V., Feng, F., Fienber, V., Freitag, M., Garcia, X., Gehrmann, S., Gonzalez, L., Gur-Ari, G., Hand, S., Hashemi, H., Hou, L., Howland, J., Hu, A., Hui, J., Hurwitz, J., Isard, M., Ittycheriah, A., Jagielski, M., Jia, W., Kenealy, K., Krikun, M., Kudugunta, S., Lan, C., Lee, K., Lee, B., Li, E., Li, M., Li, W., Li, Y., Li, J., Lim, H., Lin, H., Liu, Z., Liu, F., Maggioni, M., Mahendru, A., Maynez, J., Misra, V., Moussalem, M., Nado, Z., Nham, J., Ni, E., Nystrom, A., Parrish, A., Pellat, M., Polacek, M., Polozov, A., Pope, R., Qiao, S., Reif, E., Richter, B., Riley, P., Ros, A. C., Roy, A., Saeta, B., Samuel, R., Shelby, R., Slone, A., Smilkov, D., So, D. R., Sohn, D., Tokumine, S., Valter, D., Vasudevan, V., Vodrahalli, K., Wang, X., Wang, P., Wang, Z., Wang, T., Wieting, J., Wu, Y., Xu, K., Xu, Y., Xue, L., Yin, P., Yu, J., Zhang, Q., Zheng, S., Zheng, C., Zhou, W., Zhou, D., Petrov, S. & Wu, Y. (2023), 'PaLM 2 Technical Report'.

Bansal, H., Gopalakrishnan, K., Dingliwal, S., Bodapati, S., Kirchhoff, K. & Roth, D. (2023), 'Rethinking the Role of Scale for in-Context Learning: An Interpretability-based Case Study at 66 Billion Scale', in A. Rogers, J. Boyd-Graber & N. Okazaki, eds, *Proceedings of the 61st Annual Meeting of the Association for Computational Linguistics (Volume 1: Long Papers)*, Association for Computational Linguistics, Toronto, Canada, pp. 11833–11856.

Barwich, A.-S. (2019), 'The Value of Failure in Science: The Story of Grandmother Cells in Neuro-science', *Frontiers in Neuroscience* 13.

Beckers, S., Eberhardt, F. & Halpern, J. Y. (2020), 'Approximate Causal Abstractions', in *Proceedings of the 35th Uncertainty in Artificial Intelligence Conference*, PMLR, pp. 606–615.

Belinkov, Y. (2022), 'Probing Classifiers: Promises, Shortcomings, and Advances', *Computational Linguistics* 48(1), 207–219.

Bian, Y., Huang, J., Cai, X., Yuan, J. & Church, K. (2021), 'On Attention Redundancy: A Comprehensive Study', in K. Toutanova, A. Rumshisky, L. Zettlemoyer, D. Hakkani-Tur, I. Beltagy, S. Bethard, R. Cotterell, T. Chakraborty & Y. Zhou, eds, *Proceedings of the 2021 Conference of the North American Chapter of the Association for Computational Linguistics: Human Language Technologies*, Association for Computational Linguistics, Online, pp. 930–945.

Bricken, T., Templeton, A., Batson, J., Chen, B., Jermyn, A., Conerly, T., Turner, N., Anil, C., Denison, C., Askell, A., Lasenby, R., Wu, Y., Kravec, S., Schiefer, N., Maxwell, T., Joseph, N., Hatfield-Dodds, Z., Tamkin, A., Nguyen, K., McLean, B., Burke, J. E., Hume, T., Carter, S., Henighan, T. & Olah, C. (2023), 'Towards Monosemanticity: Decomposing Language Models with Dictionary Learning', *Transformer Circuits Thread* 2. https://transformer-circuits.pub/2023/monosemantic-features

Bubeck, S., Chandrasekaran, V., Eldan, R., Gehrke, J., Horvitz, E., Kamar, E., Lee, P., Lee, Y. T., Li, Y., Lundberg, S., Nori, H., Palangi, H., Ribeiro, M. T. & Zhang, Y. (2023), 'Sparks of Artificial General Intelligence: Early Experiments with GPT-4'. https://doi.org/10.48550/arXiv.2303.12712

Buckner, C. (2019), 'Deep Learning: A Philosophical Introduction', *Philosophy Compass* 14(10), e12625.

Chefer, H., Gur, S. & Wolf, L. (2021), 'Transformer Interpretability Beyond Attention Visualization', in *Proceedings of the IEEE/CVF Conference on Computer Vision and Pattern Recognition*, pp. 782–791. IEEE Computer Society.

Conmy, A., Mavor-Parker, A. N., Lynch, A., Heimersheim, S. & Garriga-Alonso, A. (2023), 'Towards Automated Circuit Discovery for Mechanistic Interpretability', *Advances in Neural Information Processing Systems* 36, 16318–16352.

Craver, C. F. (2007), *Explaining the Brain: Mechanisms and the Mosaic Unity of Neuroscience*, Oxford University Press, Clarendon Press, New York.

Doerig, A., Sommers, R. P., Seeliger, K., Richards, B., Ismael, J., Lindsay, G. W., Kording, K. P., Konkle, T., van Gerven, M. A. J., Kriegeskorte, N. & Kietzmann, T. C. (2023), 'The Neuroconnectionist Research Programme', *Nature Reviews Neuroscience* 24(7), 431–450.

Elhage, N., Hume, T., Olsson, C., Schiefer, N., Henighan, T., Kravec, S., Hatfield-Dodds, Z., Lasenby, R., Drain, D., Chen, C., Grosse, R., McCandlish, S., Kaplan, J., Amodei, D., Wattenberg, M. & Olah, C. (2022), 'Toy Models of Superposition', *Transformer Circuits Thread*. https://transformer-circuits.pub/2022/toy_model/index.html

Elhage, N., Nanda, N., Olsson, C., Henighan, T., Joseph, N., Mann, B., Askell, A., Bai, Y., Chen, A., Conerly, T., DasSarma, N., Drain, D., Ganguli, D., Hatfield-Dodds, Z., Hernandez, D., Jones, A., Kernion, J., Lovitt, L., Ndousse, K., Amodei, D., Brown, T., Clark, J., Kaplan, J., McCandlish, S. & Olah, C. (2021), 'A Mathematical Framework for Transformer Circuits', *Transformer Circuits Thread*. https://transformer-circuits.pub/2021/framework/index.html

Friedman, D., Wettig, A. & Chen, D. (2023), 'Learning Transformer Programs', *Advances in Neural Information Processing Systems* 36, 49044–49067.

Gao, L., la Tour, T. D., Tillman, H., Goh, G., Troll, R., Radford, A., Sutskever, I., Leike, J. & Wu, J. (2024), 'Scaling and Evaluating Sparse Autoencoders'. https://doi.org/10.48550/arXiv.2406.04093

Geiger, A., Ibeling, D., Zur, A., Chaudhary, M., Chauhan, S., Huang, J., Arora, A., Wu, Z., Goodman, N., Potts, C. & Icard, T. (2025), 'Causal abstraction:

A theoretical foundation for mechanistic interpretability'. *Journal of Machine Learning Research* 26(83), 1–64.
Geiger, A., Lu, H., Icard, T. & Potts, C. (2021), 'Causal Abstractions of Neural Networks', in *Advances in Neural Information Processing Systems*, Vol. 34, Curran Associates, Inc., pp. 9574–9586.
Gromov, A., Tirumala, K., Shapourian, H., Glorioso, P. & Roberts, D. A. (2024), 'The Unreasonable Ineffectiveness of the Deeper Layers'. In *Proceedings of NeurIPS 2024 Workshop on Scientific Methods for Understanding Deep Learning*. https://openreview.net/forum?id=jwhPErqvdS
Grynbaum, M. M. & Mac, R. (2023), 'The Times Sues OpenAI and Microsoft Over A.I. Use of Copyrighted Work', *The New York Times*.
Harding, J. (2023), 'Operationalising Representation in Natural Language Processing'. *British Journal for the Philosophy of Science, Just accepted*. https://www.journals.uchicago.edu/doi/abs/10.1086/728685
Hazineh, D. S., Zhang, Z. & Chiu, J. (2023), 'Linear Latent World Models in Simple Transformers: A Case Study on Othello-GPT'. In *Proceedings of Socially Responsible Language Modelling Research 2023*. https://openreview.net/forum?id=6mreYNKLKv}
He, S., Sun, G., Shen, Z. & Li, A. (2024), 'What Matters in Transformers? Not All Attention is Needed'. https://doi.org/10.48550/arXiv.2406.15786
Hewitt, J. & Liang, P. (2019), 'Designing and Interpreting Probes with Control Tasks', in *Proceedings of the 2019 Conference on Empirical Methods in Natural Language Processing and the 9th International Joint Conference on Natural Language Processing (EMNLP-IJCNLP)*, Association for Computational Linguistics, Hong Kong, China, pp. 2733–2743.
Hupkes, D., Veldhoen, S. & Zuidema, W. (2018), 'Visualisation and 'Diagnostic Classifiers' Reveal How Recurrent and Recursive Neural Networks Process Hierarchical Structure', *Journal of Artificial Intelligence Research* 61, 907–926.
Icard, T. F. (2017), 'From Programs to Causal Models', in *Proceedings of the 21st Amsterdam Colloquium*, ILLC, Amsterdam, pp. 35–44.
Jonas, E. & Kording, K. P. (2017), 'Could a Neuroscientist Understand a Microprocessor?', *PLOS Computational Biology* 13(1), e1005268.
Karhade, M. (2023), 'GPT-4: 8 Models in One; The Secret is Out'. https://www.kdnuggets.com/2023/08/gpt4-8-models-one-secret.html
Karvonen, A. (2024), 'Chess-GPT's Internal World Model', https://adamkarvonen.github.io/machine_learning/2024 world-models.html.
Li, B. Z., Nye, M. & Andreas, J. (2021), Implicit Representations of Meaning in Neural Language Models, in 'Proceedings of the 59th Annual Meeting of the Association for Computational Linguistics and the 11th International Joint Conference on Natural Language Processing (Volume 1: Long Papers)', Association for Computational Linguistics, Online, pp. 1813–1827.
Li, K., Hopkins, A. K., Bau, D., Viégas, F., Pfister, H. & Wattenberg, M. (2023), 'Emergent World Representations: Exploring a Sequence Model Trained on a Synthetic Task'. In *Proceedings of ICLR 2023*.
Lindsay, G. W. & Bau, D. (2023), 'Testing methods of neural systems understanding', *Cognitive Systems Research* 82, 101156.
Lipton, Z. C. (2018), 'The Mythos of Model Interpretability', *Communications of the ACM* 61(10), 36– 43.
Machamer, P., Darden, L. & Craver, C. F. (2000), 'Thinking about Mechanisms', *Philosophy of Science* 67(1), 1–25.
Marr, D. (1982), *Vision: A Computational Approach*, Freeman & Co. San Francisco.

McCoy, R. T., Yao, S., Friedman, D., Hardy, M. & Griffiths, T. L. (2023), 'Embers of Autoregression: Understanding Large Language Models Through the Problem They are Trained to Solve'. https://doi.org/10.48550/arXiv.2309.13638

Meng, K., Bau, D., Andonian, A. & Belinkov, Y. (2022), 'Locating and editing factual associations in gpt'. *Advances in neural information processing systems*, 35, 17359–17372.

Meyes, R., Lu, M., de Puiseau, C. W. & Meisen, T. (2019), 'Ablation Studies in Artificial Neural Networks'. https://doi.org/10.48550/arXiv.1901.08644

Millière, R. (2024), 'Philosophy of Cognitive Science in the Age of Deep Learning', *WIREs Cognitive Science* n/a(n/a), e1684.

Millière, R. & Buckner, C. (2024), 'A Philosophical Introduction to Language Models—Part I: Continuity with Classic Debates'. https://doi.org/10.48550/arXiv.2401.03910

Mirchandani, S., Xia, F., Florence, P., Ichter, B., Driess, D., Arenas, M. G., Rao, K., Sadigh, D. & Zeng, A. (2023), 'Large Language Models as General Pattern Machines'. https://doi.org/10.48550/arXiv.2307.04721

Mollo, D. C. & Millière, R. (2023), 'The Vector Grounding Problem'. https://doi.org/10.48550/arXiv.2304.01481

Nanda, N., Chan, L., Lieberum, T., Smith, J. & Steinhardt, J. (2022), 'Progress Measures for Grokking via Mechanistic Interpretability', in *The Eleventh International Conference on Learning Representations*. arXiv:2301.05217v3

Nanda, N., Lee, A. & Wattenberg, M. (2023), 'Emergent Linear Representations in World Models of Self-Supervised Sequence Models'. https://doi.org/10.48550/arXiv.2309.00941

Olsson, C., Elhage, N., Nanda, N., Joseph, N., DasSarma, N., Henighan, T., Mann, B., Askell, A., Bai, Y., Chen, A., Conerly, T., Drain, D., Ganguli, D., Hatfield-Dodds, Z., Hernandez, D., Johnston, S., Jones, A., Kernion, J., Lovitt, L., Ndousse, K., Amodei, D., Brown, T., Clark, J., Kaplan, J., McCandlish, S. & Olah, C. (2022), 'In-context Learning and Induction Heads', *Transformer Circuits Thread*. https://transformer-circuits.pub/2022/in-context-learning-and-induction-heads/index.html

OpenAI (2023), 'GPT-4 Technical Report'. https://doi.org/10.48550/arXiv.2303.08774

Piccinini, G. (2020), *Neurocognitive Mechanisms: Explaining Biological Cognition*, Oxford University Press, New York.

Plaut, D. C. & McClelland, J. L. (2010), 'Locating Object Knowledge in the Brain: Comment on Bowers's (2009) Attempt to Revive the Grandmother Cell Hypothesis', *Psychological Review* 117(1), 284–288.

Ravfogel, S., Elazar, Y., Gonen, H., Twiton, M. & Goldberg, Y. (2020), 'Null It Out: Guarding Protected Attributes by Iterative Nullspace Projection'. https://doi.org/10.48550/arXiv.2004.07667

Ravfogel, S., Prasad, G., Linzen, T. & Goldberg, Y. (2021), 'Counterfactual Interventions Reveal the Causal Effect of Relative Clause Representations on Agreement Prediction', in *Proceedings of the 25th Conference on Computational Natural Language Learning*, Association for Computational Linguistics, Online, pp. 194–209.

Räz, T. & Beisbart, C. (2024), 'The importance of understanding deep learning'. *Erkenntnis* 89(5), 1823–1840.

Rumelhart, D. E., Mcclelland, J. L. & Group, P. R. (1987), *Parallel Distributed Processing, Volume 1: Explorations in the Microstructure of Cognition: Foundations*, MIT Press, Cambridge, MA.

Sejnowski, T. J. & Rosenberg, C. R. (1987), 'Parallel Networks that Learn to Pronounce English Text', *Complex System* 1, 145–168.

Smolensky, P. (1986), 'Neural and Conceptual Interpretation of PDP Models', in *Parallel Distributed Processing: Explorations in the Microstructure, Vol. 2: Psychological and Biological Models*, MIT Press, Cambridge, MA, USA, pp. 390–431.

Smolensky, P. (1988), 'On the Proper Treatment of Connectionism', *Behavioral and Brain Sciences* 11(1), 1–23.

Syed, A., Rager, C. & Conmy, A. (2023), 'Attribution Patching Outperforms Automated Circuit Discovery'. https://doi.org/10.48550/arXiv.2310.10348

Templeton, A., Conerly, T., Marcus, J., Lindsey, J., Bricken, T., Chen, B., Pearce, A., Citro, C., Ameisen, E., Jones, A., Cunningham, H., Turner, N. L., McDougall, C., MacDiarmid, M., Freeman, C. D., Sumers, T. R., Rees, E., Batson, J., Jermyn, A., Carter, S., Olah, C. & Henighan, T. (2024), 'Scaling Monosemanticity: Extracting Interpretable Features from Claude 3 Sonnet', *Transformer Circuits Thread*. https://transformer-circuits.pub/2024/scaling-monosemanticity/index.html

Touvron, H., Martin, L., Stone, K., Albert, P., Almahairi, A., Babaei, Y., Bashlykov, N., Batra, S., Bhargava, P., Bhosale, S., Bikel, D., Blecher, L., Ferrer, C. C., Chen, M., Cucurull, G., Esiobu, D., Fernandes, J., Fu, J., Fu, W., Fuller, B., Gao, C., Goswami, V., Goyal, N., Hartshorn, A., Hosseini, S., Hou, R., Inan, H., Kardas, M., Kerkez, V., Khabsa, M., Kloumann, I., Korenev, A., Koura, P. S., Lachaux, M.-A., Lavril, T., Lee, J., Liskovich, D., Lu, Y., Mao, Y., Martinet, X., Mihaylov, T., Mishra, P., Molybog, I., Nie, Y., Poulton, A., Reizenstein, J., Rungta, R., Saladi, K., Schelten, A., Silva, R., Smith, E. M., Subramanian, R., Tan, X. E., Tang, B., Taylor, R., Williams, A., Kuan, J. X., Xu, P., Yan, Z., Zarov, I., Zhang, Y., Fan, A., Kambadur, M., Narang, S., Rodriguez, A., Stojnic, R., Edunov, S. & Scialom, T. (2023), 'Llama 2: Open Foundation and Fine-Tuned Chat Models'.

Voita, E., Talbot, D., Moiseev, F., Sennrich, R. & Titov, I. (2019), 'Analyzing Multi-Head Self-Attention: Specialized Heads Do the Heavy Lifting, the Rest Can Be Pruned'. https://doi.org/10.48550/arXiv.1905.09418

Weiss, G., Goldberg, Y. & Yahav, E. (2021), 'Thinking Like Transformers', in *Proceedings of the 38th International Conference on Machine Learning*, PMLR, pp. 11080–11090.

Woodward, J. (2005), *Making Things Happen: A Theory of Causal Explanation*, Oxford University Press, USA, New York.

Wu, Z., Geiger, A., Potts, C. & Goodman, N. D. (2023), 'Interpretability at Scale: Identifying Causal Mechanisms in Alpaca'. *Advances in neural information processing systems* 36, 78205–78226.

Yildirim, I. & Paul, L. A. (2023), 'From Task Structures to World Models: What Do LLMs Know?', *Trends in Cognitive Sciences* 28(5), 404–415.

Yousefi, S., Betthauser, L., Hasanbeig, H., Millière, R. & Momennejad, I. (2024), 'Decoding In-Context Learning: Neuroscience-inspired Analysis of Representations in Large Language Models'. https://arxiv.org/abs/2310.00313

Zhang, C., Bengio, S., Hardt, M., Recht, B. & Vinyals, O. (2021), 'Understanding Deep Learning (Still) Requires Rethinking Generalization', *Communications of the ACM* 64(3), 107–115.

Zhang, F. & Nanda, N. (2023), 'Towards Best Practices of Activation Patching in Language Models: Metrics and Methods'. https://doi.org/10.48550/arXiv.2309.16042

11 From cognitive semantics to neurosemantics

The neuroscience turn in the empirical study of word meaning

Fabrizio Calzavarini

A long tradition in the philosophy of language of the last century regards word meaning as a phenomenon that can be accounted for without taking into consideration the processes and representations implemented in the minds of language users (for discussion, see Gasparri & Marconi 2024). In truth-conditional semantics, for instance, both in its extensional (Tarski 1956; Davidson 1967) and intensional (Carnap 1952; Montague 1976) versions, the semantic values of words are typically exhausted by their referents in the world, without the intervention of mediating cognitive contents (Kripke 1980; Putnam 1975). Traditionally, this approach is non-psychological, focusing on the logical and formal properties of language rather than on how language is processed or represented in the mind.[1] A different anti-cognitivist approach, following Wittgenstein (1953), emphasizes word meaning as arising from inferential patterns (Dummett 1976) or norms of use (Brandom 1994) in social linguistic practices. Within this view, the essence of linguistic meaning can be fully understood without recourse to the psychological or neurological processes underlying language use but instead giving priority to publicly observable linguistic behavior and conventions.

An alternative line of thought, which has established itself in the last 50+ years, identifies word meanings with the semantic content of mental representations (Fodor 1975; Jackendoff 1983; Johnson-Laird 1983) or at least considers semantics as intimately connected with the psychological study of language understanding (Marconi 1997).[2] This view shares affinities with Chomsky's I-language framework, which emphasizes the internal, computational properties of language over externalized (E-language) notions like conventions, public usage, or referential relations to world entities (Chomsky 1986). The term *cognitive semantics*, in its broad sense, covers "research in philosophy, psychology, neural science, artificial intelligence, and other subject fields in cognitive science that takes the relationship between meaning and mind as the main object of study" (Li 2021). Cognitive semantic approaches are superficially diverse, being based on

DOI: 10.4324/9781003458531-11

mental representations of meaning postulates, semantic networks, lists of semantic features, or, more recently, distributional models and embodied representations (for discussion, see Section 11.3). Nevertheless, they reflect shared preoccupations, such as the assumption that semantic analysis can and should be informed by empirical data from cognitive science.

The cognitive semanticists of 1970–1980 took neuroscience to have very little relevance. The empirical basis of these theories, where present, was mostly made up of quantitative data about linguistic behavior, such as the reaction times of the experimental subjects, the accuracy of their verbal responses, and the errors they made. From around the mid-1980s, however, this situation changed drastically. The advent of new non-invasive (e.g., fMRI) or only moderately invasive (e.g., PET) neuroimaging techniques and the refinement of more traditional brain investigation tools (e.g., EEG) made the attempt to answer questions about the neural organization of semantics at least plausible. For example, what happens in our brain when we understand a fragment of natural language? Which brain areas and which neural operations participate in understanding, and to what extent? How long does it take our brain to build a semantic representation of a word or sentence? What do neurobiological data tell us about the nature of language comprehension? Thanks to these new methods, there was a cognitive neuroscience turn in cognitive semantics, which marks a transition from cognitive semantics to *neurosemantics* (Pulvermüller 2012; Panizza 2012; for discussion, see Calzavarini 2020).

The fundamental theoretical assumption of contemporary neurosemantics is that understanding the meaning of a word or a sentence ultimately consists in the activation and processing of neural representations, that is, semantically relevant brain patterns. Consequently, from this perspective, the central task of semantics consists in empirically studying the set of brain mechanisms that underlie our linguistic, and more specifically semantic, abilities (see, for instance, Pulvermüller 2012; Panizza 2012). This is consistent with the emergent general cognitive neuroscience account of the mind that explains cognition directly in terms of neural computations and representations, leading to a radical rethinking of levels, computation, representation, psychological explanation, and the relation between psychology and neuroscience (see Bechtel 2008; Milkowski 2013; Piccinini 2020; Buckner, 2024).

This chapter aims to analyze the neuroscience turn in the empirical study of word meaning.[3] The main claim of the chapter is that the advent of neuroimaging techniques has revolutionized research in this field, providing essential information that had not been obtained with previous experimental procedures. This claim openly contradicts Max Coltheart's (2006, 2013) famous critique that neuroimaging has not contributed to our understanding of cognition beyond the knowledge garnered from other methodologies. I will proceed as follows. After briefly describing the transition from traditional cognitive semantics to neurosemantics, I will consider some empirical generalizations about the neural substrates of word meaning that derive almost entirely from research using neuroimaging techniques. As we will see, one of these empirical generalizations is the *sensorimotor hypothesis*, namely the hypothesis that the cognitive representation of word meaning is partly distributed in perceptual and motor systems. Finally, I will provide some general considerations about the impact of these findings on the computational modeling of semantic memory. I will claim that the neuroscience turn has pushed the field toward the designing of partially embodied and embedded semantic AI systems (see Piccinini 2022), namely AI systems that integrate linguistic (e.g., patterns of words co-occurrences) and perceptual (e.g., image databases) information and representations.

11.1 From cognitive semantics to neurosemantics

For many decades, at least until the 1980s, the few empirical data relating to the neural organization of language understanding came almost exclusively from lesion studies, that is, studies of the behavior of subjects suffering from brain pathologies, such as vascular damage (ischemic or hemorrhagic strokes), tumors, or dementia, and on the inferential technique known as *double dissociation*, which infers the existence of independent cognitive systems in the presence of complementary cognitive deficits (for discussion, see Shallice 1988). The lesion studies of those years, however, being often conducted on single patients and without the aid of a control group or sophisticated statistical analyses (see Plaut 2003), had rather limited heuristic power and poor external validity. In other words, the conclusions obtained from those studies could hardly be generalized to the general population (see Morgan & Morgan 2001). Furthermore, the analysis of brain lesions—often very extensive and undifferentiated—could most often only take place *post-mortem*, drastically reducing the possibility of forming neuroanatomical hypotheses about the localization of semantics in the brain.

In the late 1980s, with the emergence of neuroimaging techniques, the situation in cognitive semantics changed dramatically. In 1988, Petersen and colleagues used PET to investigate the neural correlates of single-word comprehension. From 1991 to 1993, research groups led by McCarthy, Hinke, Binder, and Small used fMRI for the same purpose. Both neuroimaging techniques, by analyzing the variation of blood flow in the neuronal tissue between different experimental conditions, allowed researchers to visualize the functional activity of the brain *in vivo* during the execution of cognitive tasks, not only in brain-damaged subjects but also in neurotypical

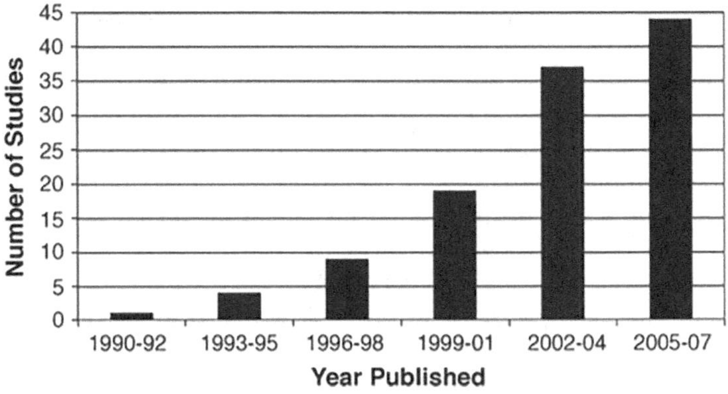

Figure 11.1 The temporal distribution of the studies analyzed by Binder and colleagues.

Source: Binder and colleagues (2009)

ones. In the space of a few years, research in neurolinguistics based on neuroimaging techniques intensified enormously. In a 2009 meta-analysis, for example, Binder and colleagues analyzed the results of 120 fMRI and PET studies (167 contrasts) dedicated specifically to the neural substrates of semantic processes. The temporal distribution highlighted in the study shows a significant increase in publications from the 1990s to the early 2000s, with few studies published in the two-year period 1990–1992 (less than five studies) and more than 40 studies published in the two-year period 2005–2007 (see Figure 11.1).

The data that emerges from this first phase of neuroimaging research is that language understanding is a highly distributed faculty, which involves a dense and highly interconnected network of brain areas, localized not only in the high-level association cortices (in the frontal, temporal, and parietal lobes) but also in the perceptual and motor systems. Furthermore, using the traditional subtractive method (Donders 1969), PET and fMRI studies have made it possible to identify with relative precision the neural networks selectively involved in phonological, morphosyntactic, and semantic processing (see, e.g., Vigneau et al. 2006), as well as in the processing of different classes of words, such as concrete and abstract words (e.g., Wang et al. 2010), or words that refer to entities with strong visual properties (for a review, see Kemmerer 2010) or acoustic properties (Kiefer et al. 2008).

Traditional neuroimaging studies aimed not only at *localizing* sub-processes of language understanding—that is, searching for increasingly refined correlations between the cognitive functions that are supposed to be involved in a given linguistic task, such as phonological or semantic processes, and brain structures (this is what is usually called *forward inference*; cf. Poldrack & Yarkoni 2016). Rather, the central assumption that guided much of the experimental research during those years—and which still guides much contemporary research—is that the activation of brain areas in a given study can provide clues to the correct functional decomposition of the cognitive task under examination (this is what is usually called *reverse inference*; see Calzavarini & Cevolani 2022). From this perspective, for example, the unexpected activation of visual areas during a linguistic task may suggest that, in the understanding of words with concrete referents, an imagery component is involved (Marconi et al. 2013).

As regards fMRI, until the early 2000s, the vast majority of studies used the *univariate* analysis technique, which limits itself to analyzing the average amount of neural activation recorded in a given brain region. Over the last 15 years, there has been a progressive interest in a class of more sophisticated neuroimaging techniques, called *multivariate pattern analysis*, which are able to extract information from the pattern of responses observed in the voxels distributed in a certain area, regardless of their average level of activation (for reviews, see Norman et al. 2006; Kriegeskorte & Kievit 2013; Naselaris et al. 2011). In particular, the focus has shifted to three classes of neuroimaging techniques:

1) Methods that exploit machine learning algorithms to make predictions on experimental stimuli starting from brain activity (*decoding*, or *multivoxel pattern classification*, MVPA).
2) Techniques that analyze the similarity matrices between neuronal patterns and between neuronal patterns and other types of data, such as behavioral or neurophysiological measures, to analyze the information encoded in brain areas (*representational similarity analysis*, RSA).
3) Methods that compare the predictions on neuronal activation obtained from a computational model (e.g., feature-based) of the stimuli, elaborated a priori, and the actual neuronal activation recorded in the experimental conditions (*voxel-wise encoding* or, more simply, *encoding*).

These neuroimaging methods have transformed cognitive neuroscience, correcting some limitations of traditional univariate techniques and significantly expanding the range of hypotheses that can be generated from experimental data. This change in perspective, which marks the transition from an *activation logic* to an *information logic* (Kriegeskorte & Bandettini 2007), also occurred within neurosemantics. Consequently, the number of multivariate studies devoted to the processing of linguistic meaning has increased considerably in recent years. For example, using MVPA, it has been shown that semantic brain areas encode the difference between lexical categories even when linguistic stimuli are presented to subjects subliminally, supporting the traditional view of comprehension as an automatic and partly unconscious process (Sheikh et al. 2019). Or, using the RSA technique, it has been shown that the patterns of semantic similarity between word classes are reflected in the patterns of neuronal activity recorded in the visual cortices of the ventral temporal lobe, suggesting a cortical organization "by semantic category" of these areas (Borghesani et al. 2016). Or again, in one of the first applications of the encoding technique, Mitchell et al. (2008) showed how a computational model that exploits the statistical information (co-occurrences between words) present in large textual databases is able to predict the neural activity generated by single words.

It has often been argued that neuroimaging techniques, however sophisticated, have a correlational nature. In other words, according to this view these techniques can at most highlight a covariation between some characteristic of the experimental task and the activation of a specific brain region, but they are unable to determine the causal role of this region (see Poldrack & Yarkoni 2016). This assumption has been questioned by some further recent developments in neuroimaging, such as techniques based on *effective connectivity*, which integrate information relating to the co-activation patterns of brain areas with formal causal models (see Stephan & Friston 2010). In neurosemantics, this technique has recently been used to formulate a model of the hypothetical functional-causal relationships between the neural components that support linguistic comprehension processes (Kuhnke et al. 2020, 2021). Neuroimaging techniques (structural fMRI) are also standardly used in modern neuropsychological studies to make more robust causal claims based on precise lesion-symptom mappings (Wilson 2017). For instance, Mesulam and colleagues (2015) have recently analyzed the lesion patterns of 72 aphasic patients using MRI, concluding that the classical Wernicke's area is not involved in language understanding.

Another rather widespread assumption in cognitive neuroscience is that fMRI is unable to detect the temporal dynamics of cognitive and neural processes. This assumption has also been partly called into question by recent developments, such as the analysis techniques and experimental paradigms of *fast fMRI* (see Yang & Lewis 2021), which have recently been used to study, with very high temporal resolution, the neural components involved in word reading (Malins et al. 2016). That said, more traditional neuroimaging methodologies, such as electroencephalography (EEG) or magnetoencephalography (MEG)—which analyze, respectively, the electrical and magnetic activity of the brain—remain the most suitable tools for this purpose (see Lopes da Silva 2013). Clearly, temporal information is also crucial in neurosemantics (Hauk 2016). The most robust data that emerges from a long tradition of EEG and MEG studies is that semantic processes are activated very rapidly, between 200 and 600 milliseconds after listening to a word, with a peak around 400 msec (see Kutas & Federmeier 2011). Furthermore, some EEG studies have reported even faster semantic effects, within 200 msec (see Pulvermüller et al. 2009).

To summarize, as can be seen from this brief review (which for obvious reasons cannot be exhaustive), contemporary neurosemantics has a wide range of experimental methods and techniques available to investigate the brain bases of language understanding, each of which has specific advantages and disadvantages. We will now try to understand what these tools can tell us about the nature of linguistic understanding.

11.2 Two empirical generalizations

As we have seen, an array of disparate neuroimaging techniques is currently employed in cognitively oriented semantics to investigate the neural underpinnings of language understanding. Thanks to the amount of neurobiological information provided by these techniques, today it is almost unimaginable to provide a cognitive theory of word meaning without considering neuroimaging data. *De facto*, cognitive semantics has irreversibly turned into neurosemantics.

What do neuroimaging data tell us about the nature of semantic processes? Today there are at least two empirical generalizations that are essential for the study of word meaning and have been established almost entirely based on neuroimaging techniques. The first generalization is the *sensorimotor hypothesis*: the cognitive representation of word meaning is partly distributed in perceptual and motor systems. This is the core tenet of the so-called *simulation theory*, according to which language understanding is a form of perceptual simulation (see, for instance, Barsalou 1999, 2016; Prinz 2002; Kemmerer 2010, 2022; Pulvermüller 2013a). This hypothesis was "in the air" before the advent of PET and fMRI (see Rogers & McClelland 2004). Wernicke himself advanced this hypothesis on a purely speculative basis (see Gage & Hickok 2005). Similarly, the neuropsychologists Warrington and Shallice (1984) proposed the existence of a "perceptual semantic system" to explain the pathological behavior of four patients with herpes encephalitis, who exhibited a semantic deficit selective for animal words (also McCarthy & Warrington 1984). Brain lesions in these patients, however, were widespread and included mainly non-sensory temporal regions. More generally, no clear case of patients with focal lesions in the sensorimotor cortices and a corresponding language impairment for concrete, high imageable words was reported before the 1990s (for discussion, see Calzavarini 2020, 2021b).

Starting from the late 1990s, the sensorimotor hypothesis has been increasingly supported by PET and univariate fMRI results. Univariate neuroimaging studies have shown that understanding concrete words correlates with activation of high-level sensorimotor areas, such as the secondary visual (see Kemmerer 2010) or auditory cortex (Kiefer et al. 2008), as well as with primary perceptual areas, such as the olfactory cortex (Barrós-Loscertales et al. 2012) and the gustatory cortex (González et al. 2006). To cite a representative example, in the fMRI study conducted by Hauk and colleagues (2004), subjects were asked to read words that refer to actions performed with the tongue (*lick*), fingers (*pick*), and legs (*kick*). Results of the experiment showed that simply reading these words activates the sections of the motor cortex that are involved, respectively, in the movement of the tongue, fingers, and legs. In an associated EEG experiment (Hauk et al. 2004), it was also shown that activation of the motor cortex occurs in the very first moments after the presentation of a word, between 100 and 200 msec. In light of this temporal dynamic, it is plausible that motor activation during linguistic tasks does not reflect post-semantic cognitive processes, such as the involvement of conscious motor imagination, but has an active functional role in language understanding.

The idea that language understanding is a form of perceptual simulation is also supported by multivariate neuroimaging studies. For example, using MVPC, Borghesani and colleagues (2016) showed that a machine learning algorithm that takes as input neural activity in six regions of interest along the ventral visual pathway, from the occipital lobe to the anterior temporal lobe (ATL), can predict the semantic category (animal vs. tool) of the word that the experimental subject is processing. In the same study, using the RSA technique, Borghesani and colleagues also showed how patterns of semantic similarity between classes of concrete words (computed through behavioral measures) are reflected in the patterns of neural activity recorded in the visual cortices, suggesting a cortical organization "by semantic category" of these areas. Further confirmation comes from studies that have used the encoding technique. For example, Fernandino and colleagues (2016) demonstrated that a semantic model based on five sensorimotor attributes (shape, color, movement, sound, and action) is able to predict the neural activity generated by concrete words in a brain network associated with semantic processes (61% average accuracy), suggesting that "the brain represents concepts as multimodal combinations of perceptual and motor representations" (2016, p. 1123). In a subsequent study, the same authors observed how a similar sensorimotor model formulates much more accurate predictions than semantic models that did not rely on sensorimotor information (Fernandino et al. 2022). Specifically, the sensorimotor model, which incorporates experiential dimensions such as sensory, motor, and affective features, outperformed both taxonomic models, which organize concepts based on hierarchical category structures, and distributional models, which rely on statistical patterns of word co-occurrence in language (see the next section for details).

There is no doubt that the neuroimaging data mentioned above support the sensorimotor hypothesis. Nevertheless, there are also empirical data that support an additional hypothesis. The second empirical generalization that has been established almost entirely based on neuroimaging techniques goes in the opposite, *amodal* direction: the cognitive representation of word meaning is partly distributed in a network of high-order associative, amodal neural systems, or amodal "hubs." Again, this hypothesis was "in the air" before the advent of PET and fMRI. Apparently, it is consistent with, for instance, Fodor's (1975) claim that the conceptual system is detached from the sensorimotor systems and is composed of amodal representations.[4] It is also consistent with early neuropsychological observations (Mesulam 1969) of patients with focal lesions in high-order, associative regions, distant and independent from primary sensorimotor cortices, and corresponding impairments in semantic abilities with words—this is the so-called *semantic variant of primary progressive aphasia*, also known as *semantic dementia*. Nevertheless, as is known, Fodor's amodal hypothesis was based almost exclusively on theoretical argumentation (see, for instance, Fodor 1975), while patient data on semantic dementia pointed only to a single amodal structure, the ATL.

From the early 1990s, PET and univariate fMRI studies have uncovered several additional amodal hubs involved in language understanding. For example, I mentioned the meta-analysis by Binder and colleagues (2009), who analyzed the results of 120 fMRI and PET studies dedicated to the neural substrates of semantic processing (Figure 11.2). Results of this study clearly show that language understanding, in its semantic aspects, involves an extensive network of brain areas whose main nodes are distributed in the associative, non-sensorimotor cortices of the brain, in the frontal, temporal, and parietal lobes. As Binder observes:

> [T]he nodes that make up this network, without exception, are high-level multimodal/supramodal areas distant from the primary sensory cortices and motor cortices. . . . The fact that they are activated by a wide range of relevant stimuli regardless of their sensorimotor and emotional content suggests that the information processed in these regions is not strictly tied to a particular perceptual or motor modality.
>
> (2016, p. 1097)

Independent evidence contributes to strengthening the supramodal nature of many of the nodes involved in this semantic network. For example, lesion studies have shown that ATL atrophy, one of the areas reported in Binder and colleagues' (2009) meta-analysis, is typically associated with

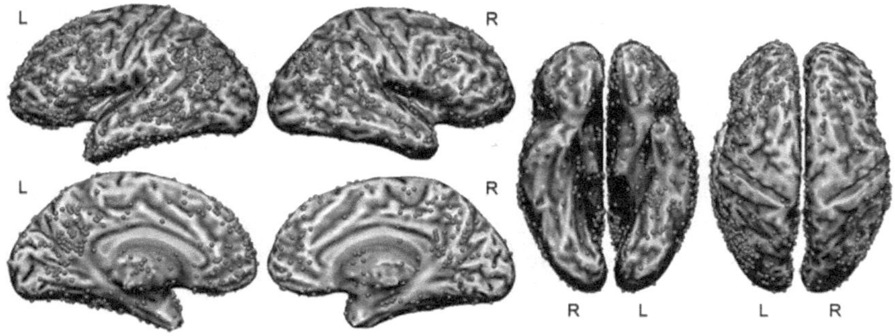

Figure 11.2 Results of the meta-analysis of Binder and colleagues.

Source: Binder and colleagues (2009)

semantic dementia syndrome, a disorder characterized by "multimodal" or "amodal" semantic deficits (see Lambon Ralph et al. 2017), affecting not only written and oral comprehension but also the recognition of perceptual stimuli in all modalities. Multivariate analyses also support the non-sensorimotor hypothesis, showing that associative regions encode semantic information for both concrete and abstract words (Anderson et al. 2021).

Based on this situation—that is, that today both the sensorimotor and the amodal hypotheses are well supported by neuroimaging data—the majority of scholars in neurosemantics have moved toward *dual* or *hybrid* accounts, according to which the cognitive representation of word meaning involves both amodal and sensorimotor components (Garagnani & Pulvermüller 2016; Lambon Ralph et al. 2017; Kuhnke et al. 2021). Examples are the LASS theory (*Language and Situated Simulation Theory*) of Barsalou and colleagues (2008), the WAT (*Words As Social Tools*) theory of Borghi and colleagues (Borghi & Binkofski, 2014; Borghi & Cimatti, 2009), the more recent variants of Pulvermüller's embodied theory (e.g., Pulvermüller 2013a, 2013b), the "interdependence of symbols" hypothesis by Louwerse (2008; see also Louwerse & Jeuniaux 2010), or the representational pluralism proposed by Guy Dove (2009, 2011). Hybrid theories of semantics are consistent with neurocognitive models of the *hub and spokes* type (Lambon Ralph et al. 2017). According to these models, the brain bases of semantics are constituted both by a set of modality-specific representations (the "spokes"), and by one or more amodal brain systems (i.e., not sensitive to the perceptual modality), such as the ATL, which are located in the association cortex (the "hubs").

Hybrid models of semantics are inspired (either implicitly or explicitly) by Paivio's *dual code* theory (Paivio 1986), according to which lexical semantic representations in the brain have a dual format, one verbal, the other perceptual. The perceptual component is multimodal, consistent with the key tenet of the simulation theory. The verbal component is amodal and consists of a cognitive representation of the phonetic form of words; each phonetic form is connected in a rich associative network with other phonetic forms (see Pulvermüller 2013b for a possible implementation of this verbal network). Note that this semantic architecture is significantly different from the Language of Thought hypothesis as classically conjectured by Fodor (1975). In Paivio's model—like in most hybrid models developed in contemporary neurosemantics—it is not necessary to postulate an additional system of propositional representations distinct from and functionally independent of the phonetic and orthographic forms, as Fodor's Mentalese. Direct neural associations between verbal representations, biologically established through Hebbian-like or similar learning mechanisms (for discussion, see Pulvermüller 2013a; Pulvermüller et al. 2021, 2023) are enough to sustain the amodal components of lexical semantic competence.[5]

What is critical here is that hybrid accounts of language understanding have been formulated before the advent of neuroimaging techniques, mainly based on theoretical argumentation or behavioral data. The dual coding theory by Paivio, for instance, was primarily supported by behavioral data, including experiments on memory and imagery. These experiments often involved tasks such as paired-associate learning or free recall, showing that participants performed better (i.e., better accuracy and shorter reaction times) when they could leverage both verbal and sensorimotor representations. Neuroimaging techniques have not only provided empirical support for these ideas but have also allowed the development of precise neurocognitive models of how amodal and sensorimotor semantic components interact in the semantic network. For example, Kuhnke et al. (2020) used the effective connectivity technique to analyze the presumed functional-causal connections between the components of this network, concluding that "the functional coupling between multimodal and modality-specific [conceptual] areas strongly depends on the type of task performed, and involves a flow of information in both a *top-down* and *bottom-up manner*" (Kuhnke et al. 2020, p. 3475). No such analysis would have been possible without the advent of neuroimaging techniques.

11.3 Implications for computational models of semantics

To summarize, I argued that the advent of neuroimaging techniques has revolutionized research in cognitive semantics, providing essential information that had not been obtained with previous experimental procedures. This claim openly contradicts Coltheart's famous critique that fMRI has not "contributed to our understanding of cognition beyond the knowledge garnered from other methodologies" (2013, p. 99). Critically, the information provided by neuroimaging techniques has also significantly influenced work on the computational modeling of semantic knowledge, suggesting new lines of research in this field (for reviews and discussions, see Rogers & McClelland 2004; Kumar 2021; Davis & Yee 2021; Kemmerer, *manuscript*).

Early attempts to model semantic knowledge in Artificial Intelligence and Knowledge Representation were largely based on logical-propositional representations such as meaning postulates (see Marconi 1997; Calzavarini 2020). Meaning postulates are used in the class of formal systems that have their roots in the structured inheritance networks, such as KL-ONE. More recently, similar systems have been proposed as logical formalism, under the umbrella term *Description Logics* (Baader et al. 2003). Alternatively, semantic knowledge has been represented as a semantic network (e.g., Collins & Quillian 1969; Steyvers & Tenenbaum 2005), as a list of the properties defining a particular concept, as in decompositional (e.g., Kamp & Reyle 1993) or feature-based models (e.g., Vinson & Vigliocco 2008), or else as particular data structures called frames (e.g., Minsky 1967). It is often remarked that, in the traditional propositional view, semantic representations are amodal and arbitrary, "because the way in which [such representations] are formed and deployed, along with their characteristics, bears no relationship to the physical and functional features of the referents" (Wilson & Foglia 2017). For instance, the concept CAT is linked to (or includes) the amodal, semantic component IT MEOWS, rather than the modal acoustic representation of a cat meowing.

A different approach to modeling semantic knowledge that has become popular in recent years is so-called *distributional semantics* (for reviews, see Lenci & Sahlgren 2023).[6] The approach is based on the assumption that semantic representations of words can be built by analyzing their statistical distribution in large text corpora. The resulting models are known as vectorial, co-occurrence, or corpus-based models. Two of the earliest models were latent semantic analysis (LSA) (Landauer & Dumais 1997) and Hyperspace Analogue of Language (Lund & Burgess 1996), which were followed by a plethora of models using different mathematical techniques. In recent times, distributional semantic models (DSMs) have started to exploit machine learning algorithms, in particular neural networks, to improve the extraction of lexical relations from large text corpora. For instance, Word2Vec (Mikolov et al. 2013) and GloVe (Pennington et al. 2014) use sophisticated vector-based approaches to capture similarities based on word co-occurrences in a vast array of contexts, improving performance in tasks such as text summarization. Moreover, transformer models like GPT (https://openai.com) have revolutionized the field by integrating deep learning to provide context-sensitive semantic representations, making significant advancement in natural language processing capabilities.

It is quite clear that both the propositional and the distributional AI approaches to semantic knowledge are consistent with one of the two empirical generalizations described in the previous paragraph, that is, the amodal hypothesis, according to which the cognitive processes underlying language understanding strongly relies on representations with an amodal, language-like format. Critically, recent fMRI studies have directly supported the empirical validity of DSMs by demonstrating that patterns of brain activation correlate with the semantic similarities predicted by these models (for reviews, see Kemmerer, *forthcoming*). For example, Carota and colleagues (2017) have demonstrated that words sharing similar linguistic contexts evoke comparable neural patterns of activation in regions like the inferior frontal gyrus and the posterior middle temporal gyrus. In particular, in a couple of studies using the Word2Vec (Carota et al. 2021) and GloVe (Carota et al. 2023) models, the same scholars found significant correlations between the semantic distances derived from these distributional models and the similarity of fMRI activation patterns observed during language tasks. This evidence supports the idea that distributional semantics can capture aspects of how the brain represents word meanings based on linguistic co-occurrences.

Not surprisingly, however, propositional and distributional approaches have been consistently criticized by the advocates of the simulation framework, who are against the thesis that semantic representations are abstract and amodal symbols like words in a natural language (Barsalou 2018; Borghi et al. 2023). As we have seen, neuroimaging studies have conclusively established that perceptual and motor processes possess a critical functional role in language understanding, and thus semantic knowledge cannot be completely detached from sensorimotor information. In particular, the distributional AI approach has attracted criticisms for deriving semantic representations exclusively from linguistic information and thus being affected by the symbol grounding problem (Harnad, 1990; Searle, 1980), that is, the fact that such representations are not grounded in the environment through perception and action.

In accordance with the sensorimotor hypothesis, the field is progressively moving toward designing partially "embodied" semantic AI systems, namely AI systems that integrate amodal (e.g., patterns of words co-occurrences) and perceptual semantic information (for discussion, see Kumar 2021; Kemmerer, forthcoming). The necessity of grounding distributional models of semantic memory, for instance, has led to renewed interest in feature-based models such as those proposed by McRae et al. (1997) and Smith et al. (1974), which represent words through collections of binary features (e.g., birds have wings). According to some authors, these models enable the direct encoding of sensory experiences, such as the color yellow for a banana, providing a more interpretable and grounded representation of meaning (see Baroni & Lenci, 2008). For instance, Andrews et al. (2009) used a Bayesian probabilistic topic model to combine experiential feature-based data with linguistic co-occurrence data, enhancing the richness of semantic representations. Studies by Jones et al. (2012) and Hoffman et al. (2018) further demonstrated that integrating perceptual information with linguistic data can account for a wide variety of semantic behaviors, bridging the gap between abstract linguistic models and embodied cognition.

Another method for grounding distributional AI models is to rely on computer vision algorithms to extract visual features from large image databases, combining them with linguistic data to create richer semantic representations, an approach known as *multimodal* distributional semantics (for reviews, see Bruni et al. 2014; Davis 2021). For instance, Bruni et al. (2014) used the "bag-of-visual-words" technique to discretize visual images into units comparable to words, which were then integrated with textual data through singular value decomposition. Convolutional neural networks have played a pivotal role in this multimodal approach by automatically extracting semantic features from images, videos, and text sequences. Kiela and Bottou (2014) demonstrated that concatenating image vectors derived from convolutional neural networks with word2vec vectors significantly improves semantic representations, outperforming models based purely on text. Similarly, Lazaridou et al. (2015) proposed a multimodal word2vec model that exploits both visual and amodal semantic representations, showing that abstract concepts can be grounded in concrete visual scenes. This multimodal approach is further exemplified by research integrating word-association norms, co-occurrence information, and phonological similarity into semantic networks (Stella et al., 2017).

The grounding of distributional models in neurosemantics parallels the situatedness of neural representations as characterized by Piccinini (2022), emphasizing the necessity for AI models to be embedded within real-world contexts to acquire meaningful semantic content. As observed by Piccinini, neural representations in biological organisms arise through dynamic interactions between the neurocognitive system, the body, and the environment, reflecting the organism's needs and experiences. Consequently, grounding AI models involves, *inter alia*, integrating such models with perceptual and sensorimotor data to enrich their internal representations, ensuring they are not just abstract symbols but tied to real-world experiences. In this sense, integrating distributional models with sensory data can be seen as a way to increment the *embodiment* and the *embeddedness* (Piccinini 2022) of these models. Furthermore, the need of organisms for situated learning, where neural networks adapt through interactions with their environment, somehow mirrors the way in which advanced multimodal distributional models use data from diverse sources to refine their semantic accuracy. Thus, in both biological and artificial systems, semantic representations of words emerge from their embedded, embodied, and interactive nature, highlighting the importance of more human-like and flexible AI models.

As a concluding remark, there is a growing tendency in neurosemantics to search for more neurobiologically realistic AI models of semantic architectures (see Piccinini 2022; Pulvermüller et al. 2021; Pulvermüller 2023). This has led researchers to progressively move away from localistic AI models where discrete symbols stand in for single concepts, akin to nodes in classical logic systems, semantic networks, or frames. As we have seen, neuroimaging techniques have firmly established that word meaning is represented across widely dispersed, interconnected neural systems rather than being confined to discrete, localized regions. In general, distributed AI models, such as deep learning neural networks, appear to offer a closer approximation to the brain's distributed processing of meaning by capturing complex patterns of activation across interconnected units. Current AI models, however, even the distributed ones, rarely respect biological constraints such as neuronal dynamics, plasticity, local circuit inter actions, and large-scale brain connectivity (for discussion, see Pulvermüller et al. 2021, 2023). This highlights the need for future AI models to ensure they reflect the dynamic and interconnected nature of neural systems, having the potential to bridge the gap between artificial and natural cognition.

11.4 Conclusions

To conclude, in this chapter, I analyzed the turn from traditional cognitive semantics to contemporary neurosemantics, which has marked a significant evolution in understanding word meaning. In traditional cognitive semantics, word meanings were explored without considering their neural implementation. Due to advances in neuroimaging techniques, research has shifted to examining the neural correlates of semantic processes. This turn has allowed researchers to discover that semantic processing is widely distributed in the human brain, with both associative and sensorimotor cortices involved. Thanks to techniques like fMRI, EEG, and machine learning it has been possible to develop multilevel neurocognitive architectures that integrate amodal, linguistic components (i.e., the hubs), with grounded, embodied neural layers (i.e., the spokes). As we have seen, these developments have influenced the design of AI systems in this field, which was traditionally characterized by propositional or language-like formalisms like semantic networks and distributional models. The integration of linguistic co-occurrences and perceptual information have improved the embodiment and the embeddedness of distributional models, enhancing their relevance and applicability in cognitive science and artificial intelligence. Needless to say, the neuroscience turn in the empirical study of word meaning has just begun, and much more theoretical and experimental work is needed in the following years.

Notes

1 This anti-cognitivist stance is the natural heritage of Frege's antipsychologism about logic and was defended by Montague, among others. For Montague, the main goal of (formal) semantics is to provide an effective recursive procedure to associate natural language expressions with proper truth conditions, and in this sense his approach was mathematical rather than psychological (Montague 1976). Already at the beginning of the development of truth-conditional semantics for natural language, however, scholars like Partee or Bach started to maintain a different, cognitively oriented approach, which was strongly influenced by the Chomskian view of linguistics as a branch of psychology (for discussion, see Partee 2010; Calzavarini 2020, 2021a). Within this view, the aim of (formal) semantics is to provide a psychological account of the semantic component of the speaker's faculty of language, that is, semantic competence. Nowadays, "truth-conditional cognitivism" (Calzavarini 2020, 2021a) is explicitly endorsed in most formal semantics introductory textbooks, both in Montagovian (e.g., Chierchia & McConnell-Ginet 2000) and Davidosnian (e.g., Larson & Segal 1995) traditions, as well as in several contemporary discussions in philosophy of language and linguistics (e.g., Borg 2004; Carston 2002).

2 According to Fodor, for instance, the semantics of natural language is fundamentally dependent on, and ultimately collapses into, the semantics of mental representations:

> It may be that Mentalese is the only language that needs a semantics; for example, that words and sentences of English have the contents they do because they are used to express the content of the corresponding words and sentences of Mentalese.
>
> (2007, p. 8)

Similarly, according to Jackendoff, "to study semantics of natural language *is* to study cognitive psychology" (1983, p. 3). In its strong sense, this view is intended to provide both a descriptive and a foundational semantic account in Stalnaker's sense (1997; see Kaplan 1989, on semantics vs. metasemantics). Descriptively, it specifies the semantic values of individual words in terms of the mental representations or concepts they evoke, rather than in terms of their referents (e.g., the semantic content of "cat" is the concept CAT). As observed by Gärdenfors (1999, p. 20), "the core idea of this approach is that meanings of expressions are mental. A semantics is seen as a mapping from the linguistic expressions to cognitive structures." Foundationally, it takes an internalist position (see Portner 2005; Riemer 2015), positing that semantic properties are grounded in internal cognitive and neural systems, rather than in external social or environmental relations. A more nuanced cognitive approach, held for instance by Marconi (1997), does not directly identify lexical meanings with cognitive entities but instead focuses on individual semantic competence with words, that is, lexical competence. This view is compatible with word meaning having also a normative aspect (for discussion, see Calzavarini 2020).

3 Note that the chapter is intended to remain neutral on foundational debates in semantics, such as semantic internalism versus externalism or cognitivism versus

anti-cognitivism. Defenders of the internalist/cognitivist perspective will likely find the chapter directly pertinent to semantics, given its focus on mental processes and representations. For externalists/anti-cognitivists, the discussion remains relevant insofar as it addresses the mechanisms underlying our ability to grasp word meaning. To ensure accessibility across these perspectives, I have adopted terminology like "knowledge of meaning," "language understanding," "semantic competence," and "mental/neural representations of semantic properties," avoiding direct identification of word meanings with mental/neural representations.

4 Critically, this does not mean that the Fodorian Language of Thought hypothesis is supported by neuroimaging data. The fact that the representations are amodal does not entail that they are language-like, let alone that they constitute the kind of language-like code posited by Fodor. As observed by Piccinini (unpublished manuscript), the Language of Thought hypothesis in its classical, Fodorian terms is a highly demanding thesis. This is not just the claim that cognitive states represent their targets in a language-like way (*representational LOT*) and are computations (*computational LOT*). *Classical LOT* "adds that the machinery that processes mentalese structures is like a digital computer. This, in turn, implies that, like the digital representations processed by digital computers . . ., mentalese structures are concatenations of finitely many types of digits" (Piccinini, unpublished manuscript). This stronger claim does not appear to have strong empirical support.

5 According to some scholars, such as Borghi and colleagues (2017), emphasizing the necessity of verbal representations does not mean embracing an amodal perspective on semantic competence. Linguistic experience, in fact, is ultimately a sensorimotor or embodied experience, since words are physical phenomena. They are produced by a phonatory system and processed by an acoustic system or a visual system. On the other hand, other scholars, such as Dove (2014) or Binder (2016) argue that phonetic/verbal information *per se* is not enough for sustaining semantic competence: additional processors are needed, outside phonology, that implement semantic processes such as compositionality, semantic inference, and the like. These mechanisms are located in areas of the brain that are not functionally dedicated to sensorimotor or phonetic processes, so it is hard to call these components "modal" in any clear sense. Anyway, even among these latter scholars, no one appears to be particularly fond of the Language of Thought hypothesis in its classical form. *Inter alia,* this is demonstrated by the fact that, as we will see in the following section, the reference AI architectures in this field are word embeddings, distributional semantics, large language models, and the like. Not classical digital computers.

6 Distributional AI models work very differently from classical, propositional ones, such as those based on Description Logics. For distributional models are based on vectorial representations and statistical algorithms, rather than discrete symbols and explicit logical rules (for discussion, see Lenci & Sahlgren 2023). In addition, distributional models are now frequently implemented by neural networks rather than classical digital computers.

References

Anderson, A. J., Kiela, D., Binder, J. R., Fernandino, L., Humphries, C. J., Conant, L. L., Raizada, R. D. S., Grimm, S., & Lalor, E. C. 2021. "Deep Artificial Neural Networks Reveal a Distributed Cortical Network Encoding Propositional Sentence-Level Meaning." *The Journal of neuroscience: the official journal of the Society for Neuroscience* 41(18): 4100–4119. https://doi.org/10.1523/JNEUROSCI.1152-20.2021

Andrews, M., G. Vigliocco, and D. Vinson. 2009. "Integrating Experiential and Distributional Data to Learn Semantic Representations." *Psychological Review* 116 (3): 463–98.

Baader, F., D. Calvanese, D. L. McGuinness, D. Nardi, and P. F. Patel-Schneider. 2003. *The Description Logic Handbook: Theory, Implementation, and Applications*. Cambridge: Cambridge University Press.

Baroni, M., and A. Lenci. 2008. "Concepts and Properties in Word Spaces." *Italian Journal of Linguistics* 20 (1): 55–88.

Barrós-Loscertales, A., J. González, F. Pulvermüller, et al. 2012. "Reading Salt Activates Gustatory Brain Regions: fMRI Evidence for Semantic Grounding in a Novel Sensory Modality." *Cerebral Cortex* 22 (11): 2554–63.

Barsalou, L. W. 1999. "Perceptual Symbol Systems." *The Behavioral and Brain Sciences* 22 (4): 577–609. discussion: 610–60.

Barsalou, L. W. 2016. "On Staying Grounded and Avoiding Quixotic Dead Ends." *Psychonomic Bulletin and Review* 23: 1122–42.

Barsalou, L. W. 2018. "Challenges and Opportunities for Grounding Cognition." *Journal of Cognitive Modeling* 7 (1): 23–33.

Barsalou, L. W., A. Santos, K. W. Simmons, et al. 2008. "Language and Simulations in Conceptual Processing." In *Symbols, Embodiment and Meaning*, edited by M. De Vega, A. M. Glenberg, and A. C. Graesser, 245–84. Oxford: Oxford University Press.

Bechtel, W. 2008. *Mental Mechanisms: Philosophical Perspectives on Cognitive Neuroscience*. New York: Routledge.

Binder, J. R. 2016. "In Defense of Abstract Conceptual Representations." *Psychonomic Bulletin & Review* 23: 1096–108.

Binder, J. R., R. H. Desai, W. W. Graves, and L. L. Conant. 2009. "Where is the Semantic System? A Critical Review and Meta-Analysis of 120 Functional Neuroimaging Studies." *Cerebral Cortex* 19 (12): 2767–96.

Borg, E. 2004. *Minimal Semantics*. Oxford: Oxford University Press.

Borghesani, V., F. Pedregosa, M. Buiatti, A. Amadon, E. Eger, and M. Piazza. 2016. "Word Meaning in the Ventral Visual Path: A Perceptual to Conceptual Gradient of Semantic Coding." *NeuroImage* 143: 128–40.

Borghi, A. M., and F. Binkofski. 2014. *Words as Social Tools: An Embodied View on Abstract Concepts*. New York: Springer.

Borghi, A. M., F. Binkofski, C. Castelfranchi, et al. 2017. "The Challenge of Abstract Concepts." *Psychological Bulletin* 143 (3): 263–92.

Borghi, A. M., and F. Cimatti. 2009. "Words as Tools and the Problem of Abstract Word Meanings." In *Proceedings of the 31st Annual Conference of the Cognitive Science Society*, edited by N. Taatgen and H. van Rijn, 2304–09. Amsterdam: Cognitive Science Society.

Borghi, A. M., C. De Livio, and F. Mannella. 2023. "Exploring the Prospects and Challenges of Large Language Models for Language Learning and Production." *Sistemi Intelligent* 35 (2): 361–77.

Brandom, R. B. 1994. *Making It Explicit: Reasoning, Representing, and Discursive Commitment*. Cambridge: Harvard University Press.

Bruni, E., N. K. Tran, and M. Baroni. 2014. "Multimodal Distributional Semantics." *Journal of Artificial Intelligence Research* 49: 1–47.

Buckner, D. 2024. *From Deep Learning to Rational Machines. What the History of Philosophy Can Teach Us about the Future of Artificial Intelligence*. Oxford University Press.

Calzavarini, F. 2020. *Brain and the Lexicon*. New York: Springer.
Calzavarini, F. 2021a. "Truth-Conditional Cognitivism and the Lexical Problem." *Topoi* 40: 43–54.
Calzavarini, F. 2021b. "The Empirical Status of the Pictorial View of Meaning." *Journal of Consciousness Studies* 11–12: 33–59.
Calzavarini, F., and G. Cevolani. 2022. "Abductive Reasoning in Cognitive Neuroscience: Weak and Strong Reverse Inference." *Synthese* 200: 70.
Carnap, R. 1952. "Meaning Postulates." *Philosophical Studies* 3 (5): 65–73.
Carota, F., N. Kriegeskorte, H. Nili, and F. Pulvermüller. 2017. "Representational Similarity Mapping of Distributional Semantics in Left Inferior Frontal, Middle Temporal, and Motor Cortex." *Cerebral Cortex* 27 (1): 294–309.
Carota, F., H. Nili, N. Kriegeskorte, and F. Pulvermüller. 2023. "Experientially-Grounded and Distributional Semantic Vectors Uncover Dissociable Representations of Conceptual Categories." *Language, Cognition and Neuroscience*, 1–25.
Carota, F., H. Nili, F. Pulvermüller, and N. Kriegeskorte. 2021. "Distinct fronto-temporal Substrates of Distributional and Taxonomic Similarity Among Words: Evidence from RSA of BOLD Signals." *NeuroImage* 224: 117408.
Carston, R. 2002. *Thoughts and Utterances*. Oxford: Blackwell.
Chierchia, G., and S. McConnell-Ginet. 2000. *Meaning and Grammar: An Introduction to Semantics*. Cambridge, MA: MIT Press.
Chomsky, N. 1986. *Knowledge of Language*. Santa Barbara: Prager.
Collins, A. M., and M. R. Quillian. 1969. "Retrieval Time from Semantic Memory." *Journal of Verbal Learning and Verbal Behavior* 8 (2): 240–47.
Coltheart, M. 2006. "What Has Functional Neuroimaging Told Us About the Mind (So Far)?" *Cortex* 42 (3): 323–31.
Coltheart, M. 2013. "How Can Functional Neuroimaging Inform Cognitive Theories?" *Perspectives on Psychological Science* 8 (1): 98–103.
Davidson, D. 1967. "Truth and Meaning." *Synthese* 17 (1): 304–23.
Davis, M. H., and E. Yee. 2021. "Neuroimaging and the Computational Modeling of Semantics." *Trends in Cognitive Sciences* 25 (6): 453–64.
Dove, G. 2009. "Beyond Perceptual Symbols: A Call for Representational Pluralism." *Cognition* 110: 412–31.
Donders, F. C. 1969. "Over de snelheid van psychische processen [On the Speed of Psychological Processes]." In *Attention and Performance: II,* translated and edited by W. Koster. Amsterdam: North-Hollan. (Original work published 1868).
Dove, G. 2011. "On the Need for Embodied and Disembodied Cognition." *Frontiers in Psychology* 1: 242.
Dove, G. 2014. "Thinking in Words: Language as an Embodied Medium of Thought." *Topics in Cognitive Science* 6: 371–89.
Dummett, M. 1976. "What is a Theory of Meaning? (II)." In *Truth and Meaning*, edited by G. Evans & J. McDowell, 67–137. Oxford University Press.
Fernandino, L., J. R. Binder, R. H. Desai, et al. 2016. "Concept Representation Reflects Multimodal Abstraction: A Framework for Embodied Semantics." *Cerebral Cortex* 26 (5): 2018–34.
Fernandino, L., J. Q. Tong, L. L. Conant, C. J. Humphries, and J. R. Binder. 2022. "Decoding the Information Structure Underlying the Neural Representation of Concepts." *Proceedings of the National Academy of Sciences of the United States of America* 119 (6): e2108091119.
Fodor, J. A. 1975. *The Language of Thought*. Cambridge, MA: Harvard University Press.

Fodor, J. A. 2007. "Semantics: An Interview with Jerry Fodor." *ReVEL* 5 (8): 1–12.
Gage, N. Y. Y., and Hickok, G. 2005. "Multiregional Cell Assemblies, Temporal Binding and the Representation of Conceptual Knowledge in Cortex: A Modern Theory by a 'Classical' Neurologist, Carl Wernicke." *Cortex* 41 (6): 823–32.
Gasparri, L., and D. Marconi. 2024. "Word Meaning." In *The Stanford Encyclopedia of Philosophy* (Summer Edition), edited by Edward N. Zalta and Uri Nodelman. https://plato.stanford.edu/archives/sum2024/entries/word-meaning/.
Garagnani, M., and F. Pulvermüller. 2016. "Conceptual Grounding of Language in Action and Perception: A Neurocomputational Model of the Emergence of Category Specificity and Semantic Hubs." *European Journal of Neuroscience* 43: 721–37.
Gärdenfors, P. 1999. "Some Tenets of Cognitive Semantics." In *Cognitive Semantics: Meaning and Cognition*, edited by J. Allwood and P. Gärdenfors, 19–35. Amsterdam: John Benjamins Publishing Company.
González, J., A. Barros-Loscertales, F. Pulvermüller, et al. 2006. "Reading Cinnamon Activates Olfactory Brain Regions." *NeuroImage* 32 (2): 906–12.
Harnad, S. 1990. "The Symbol Grounding Problem." *Physica D: Nonlinear Phenomena* 42 (1–3): 335–46.
Hauk, O. 2016. "Only Time Will Tell: Why Temporal Information is Essential for Our Neuroscientific Understanding of Semantics." *Psychonomic Bulletin & Review* 23 (4): 1072–79.
Hauk, O., I. Johnsrude, and F. Pulvermüller. 2004. "Somatotopic Representation of Action Words in Human Motor and Premotor Cortex." *Neuron* 41 (2): 301–07.
Hoffman, P., M. A. Lambon Ralph, and T. T. Rogers. 2018. "Semantic Diversity: A Measure of Semantic Ambiguity Based on Variability in the Contextual Usage of Words." *Behavior Research Methods* 50 (3): 1183–96.
Jackendoff, R. 1983. *Semantics and Cognition*. Cambridge, MA: MIT Press.
Johnson-Laird, P. N. 1983. *Mental Models*. Cambridge: Cambridge University Press.
Jones, M. N., B. T. Johns, and G. Recchia. 2012. "The Role of Semantic Diversity in Lexical Organization." *Canadian Journal of Experimental Psychology* 66 (2): 115–24.
Kamp, H., and U. Reyle. 1993. *From Discourse to Logic: Introduction to Model-Theoretic Semantics of Natural Language, Formal Logic and Discourse Representation Theory*. New York: Springer.
Kaplan, D. 1989. "Afterthoughts." In *Themes from Kaplan*, edited by J. Almog, J. Perry, and H. Wettstein, 565–614. Oxford: Oxford University Press.
Kemmerer, D. 2010. "How Words Capture Visual Experience: The Perspective from Cognitive Neuroscience." In *Words and the Mind: How Words Capture Human Experience*, edited by B. Malt and P. Wolff, 287–327. Oxford: Oxford University Press.
Kemmerer, D. 2022. *Cognitive Neuroscience of Language*. New York: Psychology Press.
Kemmerer, D. Forthcoming. *The Neuroscience of Word Meaning*. Cambridge: Cambridge University Press.
Kiefer, M., E. J. Sim, B. Herrnberger, et al. 2008. "The Sound of Concepts: Four Markers for a Link Between Auditory and Conceptual Brain Systems." *The Journal of Neuroscience* 28 (47): 12224–30.

Kiela, D., and L. Bottou. 2014. "Learning Image Embeddings Using Convolutional Neural Networks for Improved Multi-Modal Semantics." In *Proceedings of the 2014 Conference on Empirical Methods in Natural Language Processing (EMNLP)*, 36–45.

Kriegeskorte, N., and P. Bandettini. 2007. "Combining the Tools: Activation- and Information-Based fMRI Analysis." *NeuroImage* 38 (4): 666–68.

Kriegeskorte, N., and R. A. Kievit. 2013. "Representational Geometry: Integrating Cognition, Computation, and the Brain." *Trends in Cognitive Sciences* 17 (8): 401–12.

Kripke, S. A. 1980. *Naming and Necessity*. Cambridge, MA: Harvard University Press.

Kuhnke, P., M. Kiefer, and G. Hartwigsen. 2020. "Task-Dependent Recruitment of Modality-Specific and Multimodal Regions During Conceptual Processing." *Cerebral Cortex* 30 (7): 3938–53.

Kuhnke, P., M. Kiefer, and G. Hartwigsen. 2021. "Task-Dependent Functional and Effective Connectivity During Conceptual Processing." *Cerebral Cortex* 31 (7): 3475–93.

Kumar, A. A. 2021. "Semantic Memory: A Review of Methods, Models, and Current Challenges." *Psychonomic Bulletin & Review* 28 (1): 40–80.

Kutas, M., and K. D. Federmeier. 2011. "Thirty Years and Counting: Finding Meaning in the N400 Component of the Event-Related Brain Potential (ERP)." *Annual Review of Psychology* 62: 621–47.

Lambon Ralph, M. A., E. Jefferies, K. Patterson, et al. 2017. "The Neural and Computational Bases of Semantic Cognition." *Nature Reviews Neuroscience* 18: 42–55.

Landauer, T. K., and S. T. Dumais. 1997. "A Solution to Plato's Problem: The Latent Semantic Analysis Theory of Acquisition, Induction, and Representation of Knowledge." *Psychological Review* 104 (2): 211–40.

Larson, R. K., and G. Segal. 1995. *Knowledge of Meaning: An Introduction to Semantic Theory*. Cambridge, MA: MIT Press.

Lazaridou, A., N. T. Pham, and M. Baroni. 2015. "Combining Language and Vision with a Multimodal Skip-gram Model." In *Proceedings of the 2015 Conference of the North American Chapter of the Association for Computational Linguistics: Human Language Technologies*, 153–63. Association for Computational Linguistics.

Lenci, A., and M. Sahlgren. 2023. *Distributional Semantics*. Cambridge: Cambridge University Press.

Li, F. 2021. "Cognitive Semantics." In *Oxford Bibliographies Online in Linguistics*. https://doi.org/10.1093/obo/9780199772810-0279.

Lopes da Silva, F. 2013. "EEG and MEG: Relevance to Neuroscience." *Neuron* 80 (5): 1112–28.

Louwerse, M. M. 2008. "Embodied Relations Are Encoded in Language." *Psychonomic Bulletin & Review* 15: 838–44.

Louwerse, M. M., and P. Jeuniaux. 2010. "The Linguistic and Embodied Nature of Conceptual Processing." *Cognition* 114: 96–104.

Lund, K., and C. Burgess. 1996. "Producing High-Dimensional Semantic Spaces from Lexical Co-occurrence." *Behavior Research Methods, Instruments, & Computers* 28 (2): 203–24.

Malins, J. G., N. Gumkowski, B. Buis, et al. 2016. "Dough, Tough, Cough, Rough: A 'Fast' fMRI Localizer of Component Processes in Reading." *Neuropsychologia* 91: 394–406.

Marconi, D. 1997. *Lexical Competence*. Cambridge, MA: MIT Press.

Marconi, D., R. Manenti, E. Catricalà, et al. 2013. "The Neural Substrates of Inferential and Referential Semantic Processing." *Cortex* 49: 2055–66.

McCarthy, R., and Warrington, E. K. 1984. "A Two-Route Model of Speech Production." *Evidence from aphasia. Brain* 107 (Pt 2): 463–85.
McRae, K., V. R. de Sa, and M. S. Seidenberg. 1997. "On the Nature and Scope of Featural Representations of Word Meaning." *Journal of Experimental Psychology: General* 126 (2): 99–130.
Mesulam, M. M. 1969. "From Sensation to Cognition." *Brain* 121 (6): 1013–52.
Mesulam, M. M., C. K. Thompson, S. E. Weintraub, and E. J. Rogalski. 2015. "The Wernicke Conundrum and the Anatomy of Language Comprehension in Primary Progressive Aphasia." *Brain* 138 (8): 2423–37.
Mikolov, T., K. Chen, G. Corrado, and J. Dean. 2013. "Efficient Estimation of Word Representations in Vector Space." *arXiv preprint arXiv:1301.3781.*
Milkowski, M. 2013. *Explaining the Computational Mind.* Cambridge, MA: MIT Press.
Minsky, M. 1967. *Computation: Finite and Infinite Machines.* Englewood Cliffs, NJ: Prentice-Hall.
Mitchell, T. M., S. V. Shinkareva, A. Carlson, et al. 2008. "Predicting Human Brain Activity Associated with the Meanings of Nouns." *Science* 320 (5880): 1191–95.
Montague, R. 1976. "The Proper Treatment of Quantification in Ordinary English." In *Formal Philosophy: Selected Papers of Richard Montague*, edited by R. Thomason, 247–70. New Haven: Yale University Press.
Morgan, D. L., and R. K. Morgan. 2001. "Single-Participant Research Design: Bringing Science to Managed Care." *The American Psychologist* 56 (2): 119–27.
Naselaris, T., K. N. Kay, S. Nishimoto, and J. L. Gallant. 2011. "Encoding and Decoding in fMRI." *NeuroImage* 56 (2): 400–10.
Norman, K. A., S. M. Polyn, G. J. Detre, and J. V. Haxby. 2006. "Beyond Mind Reading: Multi-Voxel Pattern Analysis of fMRI Data." *Trends in Cognitive Sciences* 10 (9): 424–30.
Paivio, A. 1986. *Mental Representations.* Oxford: Oxford University Press.
Panizza, D. 2012. "Formal Neurosemantics: Logic, Meaning and Composition in the Brain." *Journal of Neurolinguistics.* https://doi.org/10.1016/j.jneuroling.2010.11.001.
Partee, B. H. 2010. "Formal Semantics: Origins, Issues, Early Impact." *Baltic International Yearbook of Cognition, Logic and Communication* 6 (1): 13.
Pennington, J., R. Socher, and C. D. Manning. 2014. "GloVe: Global Vectors for Word Representation." In *Proceedings of the 2014 Conference on Empirical Methods in Natural Language Processing (EMNLP)*, 1532–43. Association for Computational Linguistics.
Petersen, S. E., P. T. Fox, M. I. Posner, M. Mintun, and M. E. Raichle. 1988. "Positron Emission Tomographic Studies of the Cortical Anatomy of Single-Word Processing." *Nature* 331 (6157): 585–89.
Piccinini, G. 2020. *Neurocognitive Mechanisms: Explaining Biological Cognition.* Oxford: Oxford University Press.
Piccinini, G. 2022. "Situated Neural Representations: Solving the Problems of Content." *Frontiers in Neurorobotics* 16: 846979.
Piccinini, G. "Neural Hardware for the Language of Thought." Unpublished manuscript.
Plaut, D. C. 2003. "Interpreting Double Dissociations in Connectionist Networks." *Cortex* 39 (1): 138–41.
Poldrack, R. A., and T. Yarkoni. 2016. "From Brain Maps to Cognitive Ontologies: Informatics and the Search for Mental Structure." *Annual Review of Psychology* 67: 587–612.
Portner, P. 2005. *What is Meaning? Fundamentals of Formal Semantics.* Oxford: Blackwell.

Prinz, J. 2002. *Furnishing the Mind: Concepts and Their Perceptual Basis*. Cambridge, MA: MIT Press.

Pulvermüller, F. 2012. "Meaning and the Brain: The Neurosomatics of Referential, Interactive, and Combinatorial Knowledge." *Journal of Neurolinguistics* 25: 423–59.

Pulvermüller, F. 2013a. "How Neurons Make Meaning: Brain Mechanisms for Embodied and Abstract-Symbolic Semantics." *Trends in Cognitive Sciences* 17 (9): 458–70.

Pulvermüller, F. 2013b. "Semantic Embodiment, Disembodiment or Misembodiment? In Search of Meaning in Modules and Neuron Circuits." *Brain and Language* 127 (1): 86–103.

Pulvermüller, F. 2023. "Neurobiological Mechanisms for Language, Symbols and Concepts: Clues from Brain-Constrained Deep Neural Networks." *Progress in Neurobiology* 230: 102511.

Pulvermüller, F., Y. Shtyrov, and O. Hauk. 2009. "Understanding in an Instant: Neurophysiological Evidence for Mechanistic Language Circuits in the Brain." *Brain and Language* 110 (2): 81–94.

Pulvermüller, F., R. Tomasello, M. R. Henningsen-Schomers, and T. Wennekers. 2021. "Biological Constraints on Neural Network Models of Cognitive Function." *Nature Reviews Neuroscience* 22 (8): 488–502.

Putnam, H. 1975. "The Meaning of 'Meaning'." In *Mind, Language and Reality Philosophical Papers*, Vol. 2, 215–71. Cambridge: Cambridge University Press.

Riemer, N. 2015. "Internalist Semantics: Meaning, Conceptualization and Expression." In *The Routledge Handbook of Semantics*, 30–47. New York: Routledge.

Rogers, T. T., and J. L. McClelland. 2004. *Semantic Cognition: A Parallel Distributed Processing Approach*. Cambridge, MA: MIT Press.

Searle, J. R. 1980. "Minds, Brains, and Programs." *Behavioral and Brain Sciences* 3 (3): 417–24.

Shallice, T. 1988. *From Neuropsychology to Mental Structure*. Cambridge: Cambridge University Press.

Sheikh, U. A., M. Carreiras, and D. Soto. 2019. "Decoding the Meaning of Unconsciously Processed Words Using fMRI-Based MVPA." *NeuroImage* 191: 430–40.

Smith, E. E., E. J. Shoben, and L. J. Rips. 1974. "Structure and Process in Semantic Memory: A Featural Model for Semantic Decisions." *Psychological Review* 81 (3): 214–41.

Stalnaker, R. 1997. "Reference and Necessity." In *A Companion to the Philosophy of Language*, edited by C. Wright and B. Hale, 534–54. Oxford: Blackwell.

Stella, M., N. M. Beckage, M. Brede, and M. De Domenico. 2017. "Multiplex Model of Mental Lexicon Reveals Explosive Learning in Humans." *Scientific Reports* 7 (1): 2256.

Stephan, K. E., and K. Friston. 2010. "Analyzing Effective Connectivity with Functional Magnetic Resonance Imaging." *Wiley Interdisciplinary Reviews: Cognitive Science* 1 (3): 446–59.

Steyvers, M., and J. B. Tenenbaum. 2005. "The Large-Scale Structure of Semantic Networks: Statistical Analyses and a Model of Semantic Growth." *Cognitive Science* 29 (1): 41–78.

Tarski, A. 1956. *Logic, Semantics, Metamathematics*. Oxford: Clarendon Press.

Vigneau, M., V. Beaucousin, P. Y. Hervé, et al. 2006. "Meta-Analyzing Left Hemisphere Language Areas: Phonology, Semantics, and Sentence Processing." *NeuroImage* 30 (4): 1414–32.

Vinson, D. P., and G. Vigliocco. 2008. "Semantic Feature Production Norms for a Large Set of Objects and Events." *Behavior Research Methods* 40 (1): 183–90.

Wang, J., J. A. Conder, D. N. Blitzer, and S. V. Shinkareva. 2010. "Neural Representation of Abstract and Concrete Concepts: A Meta-Analysis of Neuroimaging Studies." *Human Brain Mapping* 31 (10): 1459–68.

Warrington, E. K., and T. Shallice. 1984. "Category Specific Semantic Impairments." *Brain* 107: 829–53.

Wilson, R. A., and L. Foglia. 2017. "Embodied Cognition." In *The Stanford Encyclopedia of Philosophy* (Fall Edition), edited by E. N. Zalta. https://plato.stanford.edu/archives/fall2017/entries/embodied-cognition/.

Wilson, S. M. 2017. "Lesion-Symptom Mapping in the Study of Spoken Language Understanding." *Language, Cognition and Neuroscience* 32 (7): 891–99.

Wittgenstein, L. 1953. *Philosophical Investigations*. Translated by G. E. M. Anscombe. Oxford: Basil Blackwell.

Yang, Z., and L. D. Lewis. 2021. "Imaging the Temporal Dynamics of Brain States with Highly Sampled fMRI." *Current Opinion in Behavioral Science* 40: 87–95.

12 Talking to ourselves

Inner speech and natural language as a language of thought

Wade Munroe

We talk to ourselves. Sometimes we do so out loud. However, frequently, we do so without making a sound. We use inner speech. In this chapter, I argue that inner speech modulates ongoing mnemonic, attentional, and sensorimotor processing and, in certain cases, serves as the necessary vehicle for conceptual processing. Language isn't merely for communication; it's for cognition (writ large). Additionally, I argue that recent work on inner speech can supplement embodied/grounded approaches to conceptual processing (GACP), especially in their explanations of abstract concepts (Borghi, 2023; Dove, 2022).

Although GACP are a diverse lot, they constitute a programmatic study of conceptual processing in relation to sensorimotor systems, the body, and the (physical/social) environment.[1] According to GACP, conceptual processing is (in part) functionally reliant on sensorimotor, affective, and interoceptive systems. GACP stand in stark contrast to classical, amodal views of conceptual processing and, in particular, the language of thought hypothesis (LOTH) (Fodor, 1975; Quilty-Dunn et al., 2023). Traditionally, advocates of LOTH posit (what I will call) a Classical LOT: a *lingua mentis* that possesses a grammar distinct from natural language. Instead of functionally relying on sensorimotor systems, Classical LOT operates in a central processing unit, separable from memory units and "sandwiched" (Hurley, 2001) between systems that subserve perception and action.

While Classical LOT is wanting for empirical evidence, there is now considerable evidence for GACP (Barsalou, 2008). However, GACP have a harder time accounting for the explanatory bread and butter of Classical LOT: (i) abstract conceptual processing and the much discussed (ii) systematicity and (iii) productivity of thought (Fodor & Pylyshyn, 1988). As I argue in the first half of the chapter, insofar as GACP are wanting in their explanation of (i)–(iii), recent work on inner speech can pick up the explanatory slack (however, for the sake of space, I focus on (i) as opposed to the well-trodden (ii) and (iii)). Assuming there is an explanatory need for a LOT, natural language suffices.

DOI: 10.4324/9781003458531-12

In the second half, I critically examine recent neurocognitive models of inner speech. While Classical LOT lacks a plausible implementation story, with LOTH advocates traditionally arguing for the autonomy of implementation-level and computational-/algorithmic-level explanations of cognition, covert speech can be situated within neurocognitive models of overt speech. However, as I demonstrate, these models still largely assume that language plays the solely communicative function of mapping nonlinguistic, conceptual representations onto (motor commands for) acoustic blasts and vice versa. In addition, extant models fail to jointly account for two empirical phenomena I detail in Section 12.4. If we are to cast off Classical LOT in favor of a natural language LOT, a series of issues need to be resolved in current neurocognitive models.

The chapter is structured as follows. In Section 12.1, I discuss GACP, their relation to LOTH, and the *prima facie* struggle of GACP to explain abstract conceptual processing and the systematicity and productivity of thought. In Section 12.2, I discuss the nature of inner speech, and I detail how recent empirical work on inner speech challenges Classical LOT in a manner congenial to GACP. In Sections 12.3 and 12.4, I critically examine neurocognitive models of inner speech.

12.1 GACP and classical LOT

Classical LOT consists in (i) a set of atomic symbols and (ii) a combinatorial syntax governing the recursive composition of LOT symbols/expressions. Theories that posit Classical LOT take conceptual processing to be the application of a set of formal inference rules operating over the syntactic structure of LOT expressions, where, again, these expressions operate outside of sensorimotor systems and the language faculty. The relevant notion of "grounding" used by GACP is related to the symbol grounding problem for classical, amodal views of conceptual processing (Harnad, 1990). The symbol grounding problem is a challenge to explain how Classical LOT expressions map onto things in the extramental world, for example, in virtue of some transduction process at Classical LOT-sensorimotor interfaces.

In contrast, on GACP, the distinction drawn by sandwich models of cognition between the sensorimotor bread and conceptual meat of cognition evaporates. There are no separable central processing, memory, and perception and action units. Conceptual processing involves simulation or the "neural reuse" (Anderson, 2010) of modality-specific systems—systems that subserve perception and action. So, thinking about dogs, for instance, doesn't consist in tokening some amodal symbol with the semantic value, *Canis familiaris*, but the selective reactivation of various engrams/experiential traces of previous visual, auditory, and tactile experiences with dogs (as well as one's affective responses) realized in the same systems that undergird those experiences.

It's important to note that GACP—or at least "weak" grounding accounts (Meteyard et al., 2012)—don't hold that conceptual processing reduces to unimodal processing in primary sensory or motor cortices. Instead, conceptual processing relies on neural and functional hierarchies that run from modality-specific areas up through heteromodal areas in association cortices (Barsalou et al., 2003; Binder, 2016). Broadly speaking, sensory and motor cortices are arranged hierarchically at both neural and functional levels of description. As we travel up a sensory processing hierarchy, for instance, a neuron or neural ensemble will preferentially respond to a conjunction of inputs from lower levels (conjunctive coding; Binder, 2016). Functionally speaking, a sensory processing stage, P, will remain tolerant to variance in and, thus, abstract away from various sensory features encoded by the conjunction of lower-level inputs to P, thus allowing for increasingly more categorical representations.

The top of a processing hierarchy operates as a semantic "hub" that binds functionally relevant processing in modality-specific regions (the "spokes"; Patterson & Ralph, 2016). Neuroimaging studies and work on persons with semantic dementia provide significant evidence that the anterior temporal lobes (ATLs) house (or are at least one central house for) semantic hubs (Chen et al., 2020). Although semantic hubs can function like amodal symbols (Binder, 2016), hub-and-spoke models are a far cry from Classical LOT. In sum, according to GACP, conceptual processing is distributed throughout modality-specific spokes and heteromodal ATL hubs.

While hub-and-spoke models help to provide a grounded account of *abstraction*—e.g., how we acquire and utilize a concept, DOG, that is stable over contingent feature variation in token dogs—it is far less clear how GACP account for *abstract* concepts. The token objects/properties/relations that fall under abstract concepts—concepts like FREE WILL, JUSTICE, and PRIME NUMBER—are not straightforwardly perceivable or manipulatable and can be spatiotemporally unbounded, as in the case of mathematical objects (at least on certain realist metaphysics) (Borghi, 2023; Dove, 2022).[2] We can perceive and interact with token dogs, but we can't perceive or interact with the prime number, 523.[3]

In addition, we've said little about how concepts could compose to ground the productivity and systematicity of thought. Roughly, productivity is the competence to entertain an unbounded array of thoughts via conceptual composition (whereas actual performance will be constrained by, e.g., time and memory limits). Systematicity is the intrinsic interdependence of the capacities to entertain certain thoughts or perform certain inferences. For instance, the capacities to think (i) that John loves Mary and (ii) that Mary loves John are intrinsically interdependent; a cognitive system can't possess one without the other. In order to entertain (i) or (ii), a cognitive system needs to understand, for example, thematic roles and that John or Mary can play the subject or object roles.

A full discussion of how advocates of GACP attempt to account for abstract concepts or the systematicity and productivity of thought is beyond the scope of this chapter. Instead, I move to a discussion of how the cognitive functions of inner speech can supplement GACP while challenging Classical LOT.

12.2 The nature and functions of inner speech

12.2.1 The nature of inner speech

The expression, "inner speech," and co-referring expressions—e.g., the French, "parole intérieure" (Egger, 1904)—are terms of art introduced in experimental psychology's nascency to designate conscious, linguistic imagery (typically) in the auditory modality. Unfortunately, popular contemporary definitions of inner speech—like the following from Ben Alderson-Day and Charles Fernyhough (2015, p. 931)—inherit the historical focus on phenomenology,

Phenomenological Definition (PD): "the subjective experience of language in the absence of overt and audible articulation."

The problem with something like PD, qua definition of inner speech, is that it (i) precludes, by fiat, the possibility of non-conscious inner speech and (ii) requires that those who (veridically) don't report experiencing linguistic imagery, thereby, lack inner speech (Nedergaard & Lupyan, 2023).

It's beyond the scope of this chapter to enumerate the reasons why we ought not treat PD as a strict definition. Instead, it will suffice to note that popular experimental methods for investigating the cognitive functions of inner speech don't assume that inner speech must be accompanied by conscious imagery (nor do the methods control for whether study participants report experiencing any linguistic imagery). For instance, take popular verbal interference paradigms (VIP; Nedergaard et al., 2023). VIP work, roughly, as follows: participants are asked to perform some primary task, T, that may or may not be modulated by linguistic processing. Simultaneously, participants perform a "verbal interference" task, V, like repeating a word. If performance on T is disrupted by V more than by control, non-verbal interference tasks, the experimenters conclude that language modulates T. VIP assume nothing about the presence or absence of linguistic phenomenology. Instead, VIP assume that verbal interference disrupts inner speech by loading speech planning/comprehension processes, which further assumes that inner speech is subserved by similar speech planning/comprehension processes to those that subserve our abilities to parse and produce overt speech (an assumption we discuss in Section 12.3). Thus, work-a-day experimental methods don't equate inner speech with phenomenology but with the offline operations of the language faculty, that is, operations that (roughly) don't causally result in or result from overt speech.

For the sake of this chapter, I take inner speech to be the following,

Processing Definition: the offline activation of speech planning/comprehension processes.

More broadly, theorists like Bence Nanay (2023) rightly argue that an analogous definitional shift needs to occur for mental and motor imagery, writ large, and for similar reasons—e.g., work-a-day experimental methods for investigating the cognitive functions of mental and motor imagery equate the two with process types (that can operate non-consciously) as opposed to phenomenological types. While classical characterizations of mental and motor imagery—also rooted in experimental psychology's nascency—forefront phenomenology, Nanay opts to characterize the imagery types as, roughly, offline perceptual and motor processing, respectively.

Insofar as conceptual processing involves simulation or neural reuse of sensorimotor systems, conceptual processing will, in part, constitute mental and motor imagery, in Nanay's sense of the terms. However, we should keep in mind that mental and motor imagery *needn't be conscious* (nor require working memory resources or effortful, top-down activation). Thus, claiming that conceptual processing involves mental and motor imagery isn't to claim that conceptual processing is conscious. Similarly, in the next subsection, I argue that conceptual processing, in part, constitutes the offline activation of speech planning/comprehension processes (i.e., inner speech), which also isn't to say that conceptual processing need be conscious.

12.2.2 The cognitive functions of inner speech

Speech planning and comprehension processes are used offline (thus, constituting inner speech) to modulate ongoing (i) mnemonic, (ii) attentional, and (iii) sensorimotor processing. In the following, I detail (i)–(iii) in turn, focusing on how relevant empirical literature challenges Classical LOT and supplements GACP.

Inner speech figures centrally in accounts of working memory, and, in particular, the multicomponent model of Alan Baddeley and colleagues

(Baddeley, 2017). For instance, using VIP, Dymarska et al. (2022) and Banks and Connell (2023) demonstrate that when tasked with remembering, respectively, a set of pictured objects and experienced events (non-verbal stimuli) people spontaneously use language to encode the objects or events. In addition, inner speech plays a chunking function, extending the number of objects or events people can keep in working memory in comparison to the computationally more costly use of distributed sensorimotor simulations.

Inner speech's mnemonic effects extend to long-term memory as well. For example, numerous studies with bilingual immigrants demonstrate that autobiographical memory is better cued by and recalled in the language used at memory encoding (regardless of whether the language is a first or second language), for example, memories are recalled in greater detail and with increased emotional salience (Purpuri et al., 2024). If autobiographical memories are encoded in Classical LOT (Mahr & Schacter, 2023)—or in any non-natural-language medium for that matter—such that language merely serves as a means to communicate the memories, we wouldn't expect any linguistic effects on cuing or recall.

Inner speech is also spontaneously used in maintaining (Nedergaard et al., 2023) and shifting endogenous attention (Cragg & Nation, 2010). As a concrete example—and one that will naturally segue into a discussion of sensorimotor processing—imagine you are trying to find your misplaced keys. You can facilitate visual search by using inner speech as a cue—e.g., covertly telling yourself, "I need to find my keys"—as opposed to activating, say, a visual template of your keys in visuospatial imagery in a manner unmediated by language (Lupyan & Swingley, 2012). Generally speaking, linguistic cues activate more prototypical or categorical representations in sensorimotor and association cortices than other cue types (Edmiston & Lupyan, 2015); and this fits nicely with GACP that deny there is any one neurocognitive thing—e.g., expression in Classical LOT—that constitutes, say, a person's concept, DOG, and that is invariably activated in thinking about dogs (Connell & Lynott, 2014). Instead, different aspects of the information in semantic memory associated with a given category will be flexibly activated in a manner sensitive to task demands. Concepts (or, more accurately, conceptualizations) are constructed on the fly, and we use language to modulate the construction.

However, we don't merely use language as a covert cue to activate category information realized in brain regions outside of the language faculty—regions where *real* conceptual processing occurs. Instead, language is itself a proper part of conceptual processing. Your typical human experiences large quantities of overt language tokens embedded in broader linguistic contexts. Linguistic distributional statistics—roughly, linguistic-entity-to-linguistic-context co-occurrences, for example, lexical-item-to-lexical-item and lexical-item-to-discursive-context/topic, co-occurrences—can serve as rich sources of semantic information (e.g., information about word meanings) but also information about the extralinguistic world encoded in the utterances of language users. Distributional semantic models (DSM) rely on the idea that words with similar meanings tend to occur in similar contexts. DSM (e.g., GloVe and word2vec) abstract and generalize over linguistic distributional statistics to represent word meanings as vectors in a semantic space (Günther et al., 2019). Although DSM have been mischaracterized by some as ungrounded, amodal competitors to GACP, contemporary advocates of GACP frequently integrate DSM into their models of conceptual processing (Davis & Yee, 2021).

Insofar as our concepts are grounded in perception and action, we should expect them to be, in part, grounded in our experiences—both overt and covert—of parsing and producing language (Borghi, 2023; Dove, 2022). In turn, just as the conceptual processing of a category involves, in part, the offline use of modality-specific systems that subserve perceiving and interacting with category tokens, conceptual processing also involves, in part, the offline use of speech planning and comprehension systems that subserve parsing and producing utterances about category tokens and types. For instance, take the considerable overlap in color cognition of sighted and congenitally blind persons, for example, in their abilities to reason about the structure of color space, the color constancy of objects, or why certain objects possess particular colors (Bi, 2021). A variety of neuroimaging research shows that sighted persons engage both posterior visual areas and anterior language-related areas during color cognition tasks, whereas congenitally blind persons engage only the latter (Bi, 2021). Congenitally blind persons extract information about color from language and reason with the information using the same faculties involved in parsing and producing utterances about color.

Importantly, for congenitally blind persons, color concepts are abstract in the relevant sense: the properties that fall under the concepts are not perceivable, unlike, for example, shape properties that can be experienced haptically. Looking to more general work on abstract concepts, several meta-analyses find that language areas are the most likely to exhibit increased activation with abstract concepts (e.g., Binder et al., 2009; Wang et al., 2018).

In the limiting case, certain abstract concepts will rely on the language faculty out of psychological necessity because we lack the means to represent the category independent of language. For example, our ability to think about and operate over integer values and quantities necessarily relies on the language faculty (Munroe, 2023). Briefly, there is a growing consensus that humans and certain non-human animals share an innate number sense that includes (i) the ability to represent and perform operations with exact quantities up to about three or four and (ii) an analog representation system—the Approximate Number System (ANS)—for dealing with approximate quantity (Feigenson et al., 2004). ANS undergirds an animal's ability to *estimate* and represent *approximate* quantities. Non-human animals from pigeons and rats to rhesus monkeys and chimpanzees have the ability to represent the *approximate* cardinality of sets with ANS, but the ability of these animals to distinguish successfully between sets based on cardinality is proportional to set size (Brannon et al., 2010).

In order to represent the *exact* cardinality of a set, an animal needs the ability to place set members in one-to-one correspondence with positive integers—put simply, the animal needs the ability to count. ANS can't subserve counting, as it can't consistently map set cardinality onto a stable representation across contexts that would operate as an integer representation. Unless we humans acquire a numeric code to represent positive integers—e.g., a language with a generative number vocabulary that includes recursive rules for combining lexical primitives to generate expressions of successive integers—we will also lack the cognitive means to count and, thus, think about and operate over integer values or quantities (Frank et al., 2008).

Importantly, a numeric code, like English number words, doesn't merely function as a means of communicating mathematical knowledge. Instead, our thoughts about and operations over integers are vehicled by the offline use of the code. Evidence for this latter claim can be found in the extensive literature on the relation between aphasia (acquired language deficits) and acalculia (acquired mathematical deficits). Although aphasia can take a range of forms, and, in turn, the mathematical deficits associated with aphasia are diverse, it is consistently found that the severity of mathematical impairments associated with aphasia vary with the severity of the aphasia (Ardila & Rosselli, 2002). If Classical LOT vehicled mathematical cognition, we wouldn't expect any correlation between aphasia and acalculia.

Instead of appealing to Classical LOT, we can use a DSM to model our understanding of a given integer. For instance, a native English speaker's understanding of, say, 15 will be encoded in linguistic embeddings like the following:

1. The ordinal embedding of, "15," in a count list, which one learns to generate through acquiring the syntactic rules for recursively composing English number words, rules which embed the successor function (Chu et al., 2020).
2. The word's embedding in addition and multiplication tables, which one acquires through rote verbal memorization and encodes in a linguistic format (Lee & Kang, 2002).

The mention of syntactic rules for recursively composing English number words nicely segues into a general discussion of natural language syntax conceived as a set of recursive operations (operations that include self-calls, that is, operations that employ themselves as sub-operations) that generate hierarchical, recursive structures (structures of a type, e.g., noun phrase, which include simpler structures of that same type). Insofar as a cognitive system can,

3. map concepts (or proper parts of concepts) realized outside of the language faculty in grounded simulations onto items in the mental lexicon (in other words, the system can lexicalize its concepts),
4. compose the resultant lexical items (along with function or closed-class expressions) using the operations of natural language syntax into hierarchal syntactic structures, and
5. map the syntactic structures back onto grounded simulations realized outside of the language faculty, or
6. use the syntactic structures to generate further structures, which one might then be able to 5.

Then, syntax can provide a powerful, computational system for conceptual composition, a system that could undergird the systematicity and productivity of thought.

I am not claiming that all conceptual composition utilizes natural language syntax. Undoubtably, at least some composition can occur without recourse to syntax, and advocates of GACP have provided models of how this occurs (e.g., Lynott & Connell, 2010). More conservatively, the claim is that language provides the mind with a computational system that can facilitate and expand its representational and inferential capacities. For instance, a series of correlational and training studies suggest that language facilitates the development of *explicit* Theory of Mind.[4] It's been suggested that the acquisition of mental state terms and complement clause syntax, in particular, facilitates explicit perspective taking, metacognition of one's own ignorance, and the acquisition of false belief understanding (Kaltefleiter et al., 2021). However, for the sake of space, my comments about syntax, systematicity, and productivity, will have to remain schematic.

The idea that natural language can serve as a LOT figured prominently in the work of philosophers like Gilbert Harman (2015) and, arguably, throughout the history of generative linguistics (Hinzen & Sheehan, 2013). For instance, take Chomsky's claim that the "Basic Property" of language is the "generation of an unbounded array of hierarchically structured expressions mapping to the conceptual-intentional interface, providing a kind of 'language of thought'—and quite possibly the only such LOT" (Chomsky, 2015, p. 13). Arguably, the popularity of Classical LOT sidelined the idea of a natural language LOT in cognitive science for some time. However, recent work on the ability of inner speech to modulate mnemonic, attentional, and sensorimotor processing provides fresh empirical support for a natural language LOT that complements GACP. But we should be clear about the scope of the claim that natural language can serve as a LOT: in certain limiting cases of abstract concepts, for example, integers, inner speech is the sole medium of conceptual processing, but the claim is not that *all* conceptual processing reduces to inner speech (cf. Gauker, 2011). Instead, the claim is that—just as conceptual processing is, in part, functionally reliant on sensorimotor, affective, and interoceptive systems—conceptual processing is, in part, functionally reliant on the language faculty.

12.3 Neurocognitive models of covert speech

Neurocognitive models of covert speech are typically situated in models of overt speech, as there is considerable evidence that, as Oppenheim and Dell claim, "we produce inner speech [in much] the same way that we speak, except that articulation is not present" (2008, p. 529). Historically, there has been an intellectual distribution of labor in which different scientific traditions have tackled different aspects of the overt speech planning and production process (de Zubicaray, 2023); in particular,

1. A psycholinguistic tradition has generated hierarchical models of lexical planning. These models—and, in particular, the model offered by Levelt, Roelofs, and Meyer (LRM's model; 1999), which is widely assumed in the inner speech literature—provide an account of how a speaker moves from a conceptual content she intends to communicate to an articulatory score, roughly, a representation of a time series of vocal tract gestures necessary for articulation.
2. A predictive motor control tradition has generated models of articulatory planning and motor control in overt speech. These models—like the popular Directions into Velocities of Articulators model (DIVA; Tourville & Guenther, 2011)—posit a series of inverse and forward models and comparators (commonly invoked in the broader literature on motor control; Wolpert & Ghahramani, 2000) to explain our ability to rapidly modulate articulation in virtue of afferent feedback. Predictive motor control models typically account for the generation of the articulatory score and further aspects of motor programming, thus picking up where 1 leave off.

   A full neurocognitive account of overt speech will be an amalgam of 1 and 2, although marrying 1 and 2 proves tricky, given differences in the concepts employed in the two traditions.

   Covert speech is typically located in the processes of monitoring and controlling overt speech. We have good evidence for two monitoring channels that assist in overt speech control:

3. An external channel—typically addressed by 2—that compares our perception of our own speech stream with a sensory prediction of the reafferent stimuli. This sensory prediction is typically thought to provide the "sensory content" of inner speech (Scott, 2013), that is, the content that, if accessed in working memory, would result in the "voice in the head" phenomenology that accompanies some forms of conscious inner speech.
4. An internal channel—typically addressed by 1—that directly monitors the speech planning process without recourse to reafferent stimuli. This internal monitor is standardly thought to provide other aspects of inner speech's contents, for example, the semantic, lexical, and syntactic aspects.

In the following subsection, I discuss (i) LRM's model of lexical planning, (ii) predictive motor control in overt speech, and (iii) inner speech's position in (i) and (ii). Because of constraints on space, I can't provide an exhaustive discussion of these topics (this would be a book-length project, e.g., Kemmerer, 2022). Instead, my aims are to (iv) provide the reader with enough detail to understand how inner speech can carry contents about the extralinguistic world, for example, autobiographical memory, color space, and integers, and (v) pinpoint conceptual and empirical issues with extant neurocognitive models.

12.3.1 LRM's model, predictive motor control, and inner speech

It's generally agreed that speech planning is subdivided into levels in which the semantic content, lexical and syntactic structure, morphophonological structure, and articulatory score for a planned utterance (whether overt or covert) are generated. Although there are disputes regarding the amount of interaction between levels, there is good neuropsychological evidence (e.g., evidence of double dissociations in speech errors between planning levels; Buchwald & Miozzo, 2011) for discrete representations operating at each tier. Thus, speech planning does not operate with representations in which semantic, syntactic, and morphophonological information are inextricably intertwined. In the following, I offer a critical discussion of the planning levels of LRM's model before discussing the predictive motor control involved in overt articulation.

According to LRM, the first stage of planning a meaningful utterance involves selecting "lexical concepts" based on the speaker's communicative intention, where a communicative intention constitutes a nonlinguistic, conceptual representation of the content the speaker intends to convey. Immediately, conceptual issues abound: Willem Levelt and colleagues assume communicative intentions are vehicled by a "conceptual structure, cast in a propositional language of thought" (Levelt, 1992, p. 5). Of course, we could reject LRM's appeal to Classical LOT and argue that grounded simulations could vehicle communicative intentions, but this still leaves us with lexical concepts. The supposed need for lexical concepts is typically justified along the following lines:

> Outside of when we are preparing to speak—that is, "thinking for speaking," (Slobin, 1996)—conceptual processing operates largely independently of natural language. Given linguistic diversity—e.g., diversity in how lexicons carve up the world—our conceptual representations of the content we intend to convey in an utterance must be "tuned to" (Levelt, 1989, p. 71) the language we will use in speech. Lexical concepts play this tuning function (Kemmerer, 2022).

As David Kemmerer (2022) suggests, we could take lexical concepts to be semantic hubs that function as heteromodal integration points for the myriad semantic features that constitute the (represented) meaning of a particular word. But this approach still assumes that (i) conceptual processing, outside of "thinking for speaking," takes place largely independently of natural language and (ii) the semantic content of lexical items is decomposable into semantic features that are stored across disparate areas of cerebral cortex and function as meaning components. However, as I argued in the previous section, language is itself a proper part of conceptual processing, and our understanding of something like integers is better modeled with a DSM (as opposed to a semantic-feature-based analysis).

The broader issue is that LRM's model assumes that language plays the purely communicative function of mapping nonlinguistic, conceptual representations onto (motor commands for) acoustic blasts and vice versa. If we want an account of how inner speech can, for example, subserve thought in more abstract domains, we need to drop any sharp boundary between conceptual processing and the language faculty. Although several theorists working on inner speech have noted similar shortcomings with LRM's model (e.g., Borghi, 2023; Kompa, 2024), the model's assumption that language plays a purely communicative function is still adopted by contemporary neurocognitive models of inner speech. It's beyond the scope of this chapter to detail how we ought to update the model. For the sake of space, I proceed to discuss the further processing stages of LRM's model.

The next planning stage—the syntactic-lemma stage—involves the selection of word lemmas, which are placed into a hierarchical syntactic structure and then linearized for subsequent stages. According to LRM, lemmas intervene between semantics and phonology and constitute abstract lexical representations that encode morphosyntactic features, like grammatical category and gender. A range of neuroimaging, neuromodulation, and neuropsychological work suggest that lemma selection is subserved by the mid-to-posterior medial temporal gyrus (de Zubicaray, 2023).

At the morphophonological level, the linearized structure of abstract word representations chosen at the syntactic-lemma level is realized in morphemes and phonemes. Subsequently, phonemic segments are bundled into syllables. At the final stage—the phonetic stage—a context-dependent, phonetic representation is derived by looking up syllabic gestures in a "mental syllabary." These syllabic gestures encode, "the selected word[s] in its context as a motor program" (Levelt et al., 1999, p. 31).

Models of predictive motor control typically begin their explanation at the phonetic stage. According to the DIVA model, posterior portions of the left inferior frontal gyrus (LIFG) subserve a "speech sound map," which holds a repository of motor programs for frequently used syllables that corresponds to LRM's mental syllabary. This speech sound map projects to motor cortex where a motor command is generated. The motor command propagates through motor areas and eventually results in overt speech. In addition, a copy of the motor command, called an "efference copy," is also generated and fed to forward models of the speech apparatus. These forward models generate predictions, called "corollary discharge" (CD),[5] of the auditory and somatosensory feedback (reafferent stimuli) we would experience in executing the motor command. These predictions are realized in the corresponding sensory regions that subserve speech perception, for example, bilateral superior temporal gyrus (STG) and superior temporal sulcus (STS) for auditory imagery and supramarginal gyrus (SMG) for somatosensory imagery.

Classically, CD is theorized to attenuate or inhibit reafferent stimuli that matches the predicted stimuli embodied in CD. CD allows an organism to (i) distinguish reafferent from exafferent stimuli and, thus, determine whether stimuli result from the organism's own actions or from features of the external environment and (ii) monitor its actions for any deviations from the predictions embodied in CD. Focusing on overt speech, there is good evidence that, when we speak, neural responses in auditory cortices to our own voice are attenuated relative to, for example, hearing a recording of our voice (Houde et al., 2002).

When it comes to inner speech, the motor command that issues in overt speech is inhibited, thus, articulation doesn't occur. Nonetheless, CD is still generated. In support, a range of neuroimaging studies demonstrate that inner speech recruits similar language planning and comprehension to overt speech, for example, LIFG and bilateral auditory areas in STG (Perrone-Bertolotti et al., 2014). In addition, the CD affiliated with inner speech has similar attenuating affects to the CD affiliated with overt speech.

For instance, Jack et al. (2019) demonstrated that study participants' inner speech reduced the amplitude of the N100 event-related potential response (ERP) to a voice recording when the content and timing of the participants' inner speech matched that of the recorded speech. The auditory N100 ERP is taken to be an index of processing in auditory cortices (e.g., STG). Thus, Jack et al. provide evidence that inner speech is accompanied by a temporally precise and content-specific CD, much like overt speech.

However, Jack et al.'s findings raise a puzzle (Patel, 2022): CD in inner speech is also theorized to constitute auditory and somatosensory imagery that provides the sensory content of inner speech (Scott, 2013). Insofar as imagery and perception in a given sense modality exhibit similar neural dynamics, the CD affiliated with inner speech ought to *amplify*, as opposed to *attenuate,* neural activity in auditory cortices. In addition, there is good neuroimaging evidence that the activity in auditory cortices induced by inner speech can resemble the amplified activity induced by auditory perception (Tian & Poeppel, 2010). So, how do we square the opposing amplifying and attenuating effects of CD? It's important to note that, as Crapse and Sommer (2008) write, "[t]here is no single type of [CD]—rather, there are numerous subtypes that correspond both to anatomical levels of the source and the target and . . . to functional utilities." CD can be generated at numerous levels of motor planning and can modulate sensory processing in multiple ways, including facilitating or amplifying sensory processing. The results of Jack et al. and Tian and Poeppel illustrate how various CD affiliated with inner speech can differentially modulate sensory processing (again, inner speech modulates ongoing sensorimotor processing).[6]

In sum, inner speech is produced in much the same way as overt speech. Despite the inhibition of the motor commands that would issue in overt speech, CD at multiple levels of motor planning is still generated. Extant neurocognitive models agree that the sensory content of inner speech is given by CD. In the following section, we discuss how competing neurocognitive models account for the lexical, syntactic, and semantic aspects of inner speech's contents.

12.4 The multiple contents of inner speech

We need more than CD to explain how inner speech can possess lexical and syntactic information, let alone semantic information about the extralinguistic world. Neurocognitive models that attempt to explain the additional aspects of inner speech's content can be grouped into two camps, which I call speech comprehension accounts (SCAs) and forward model accounts (FMAs). I discuss each in turn.

12.4.1 Speech comprehension accounts

According to SCAs, speech comprehension processes operate over CD in assigning lemmas (and, thus, lexical content), syntactic structure, and semantic content to covert utterances just as if one were parsing another person's overt speech. Thus, SCAs' explanation of inner speech post CD will mirror models of overt speech comprehension (e.g., Hickok & Poeppel, 2007).

Peter Carruthers (e.g., 2018) and Keith Frankish (e.g., 2018) adopt SCAs. A similar SCA is assumed by Levelt and numerous advocates of the perceptual loop theory of speech monitoring. As previously stated, inner speech is typically situated in models of overt speech monitoring, and the perceptual loop theory is one such model. According to Levelt and colleagues (1999), the output of morphophonological planning is "fed into" or "attended to" by speech comprehension processes, which allow a speaker to monitor what she is planning to say without afferent feedback by syntactically and semantically parsing her inner speech.[7]

A central tenet of all SCAs is as follows:

Sensory: Inner speech must possess sensory content, which is processed in auditory/somatosensory cortices.

Although justifications for Sensory differ in the details, they all involve claims about the types of information accessible to working memory. Carruthers and Frankish argue that inner speech—qua representation active in working memory capable of facilitating executive and higher-cognitive functions—must possess sensory content because the operations of working memory are sensory based. Speech comprehension processes operate over CD to generate an "event-file" (Hommel, 2004) in which the disparate codes involved in representing the semantic, lexical, and syntactic aspects of (overt or covert) speech can be bound to the sensory content given by CD and, subsequently, made accessible to working memory. More abstract lexical, syntactic, and semantic content can only be accessed by working memory through this perceptual binding process.

In contrast, Levelt doesn't make general claims about the nature of working memory but, instead, the more specific claim that "[g]rammatical relations are not consciously accessible to linguistically untrained speakers, and probably not even to trained linguists when they are in the course of speaking" (1989, p. 285). Working memory is only able to access the outputs of planning stages after the syntactic-lemma level and, thus, inner speech (or at least conscious inner speech) must possess morphophonological content.

However, Levelt doesn't provide much explanation or context for his claim about grammatical relations and conscious access. Granted, it's widely accepted that many of the representations and operations of the language faculty are not introspectively accessible. For instance, the representations operative at the syntactic-lemma level involve hierarchical syntactic structures, where the terminal nodes of these structures will consist of lemmas or empty categories that lack phonological content and, thus, won't activate any morphophonological information at subsequent processing stages (e.g., PRO and WH-traces). Clearly, conscious inner speech doesn't make the number and type of empty categories in a covertly uttered sentence luminous for introspection; this information would have to be inferred on the basis of explicit (as opposed to tacit) linguistic knowledge. However, we know from work on tip-of-the-tongue states that certain grammatical features encoded in the syntactic-lemma level can be accessed without processing at the morphophonological level, for example, information about grammatical gender (Vigliocco et al., 1997). In addition, as Levelt admits, working memory can't access *all* information output at the morphophonological or phonetic levels of planning either. Thus, it's unclear why the inaccessibility of information regarding grammatical relations (e.g., subject, direct object) should make us claim that morphophonological planning is required for inner speech to enter working memory.

More generally, SCAs hinge on the plausibility of their claims about working memory access and can be challenged on the basis of evidence to the contrary.[8] However, in this chapter I push against SCAs on another front. According to SCAs, inner speech results from the coupling of speech production and comprehension. Yet, although rare, there are documented cases of production-comprehension dissociations. For instance, in pure word deafness (PWD), a person can exhibit impairment in spoken word comprehension—and phonological awareness in particular—despite exhibiting preserved hearing, spoken word production, speech monitoring, and, crucially, performance on various silent phonological judgment tasks that require inner speech. As a classic example, Marshall et al. (1985) describe a patient who couldn't phonologically parse speech, for example, she was unable auditorily to identify the number of syllables in a word or distinguish stress contrasts. However, she performed well on silent phonological judgments, like picking out visual stimuli whose names rhymed without overtly saying the names.

If (i) inner speech requires that CD be projected to STG, which constitutes a "phonological network" crucial for parsing a speech stream into segmental and syllabic structure, and (ii) PWD typically results from bilateral lesions of the superior temporal cortex (Poeppel, 2001), including STG, then PWD ought to be invariably associated with inner speech deficits and impaired performance on silent phonological judgment tasks. Put simply, if Sensory is true, it ought not be possible for inner speech and speech comprehension to dissociate in the ways exhibited in documented cases of PWD.

Outside of PWD, production-comprehension dissociations can exist at additional levels of the speech planning and comprehension hierarchies (Bormann & Weiller, 2012). Although there isn't much work directly testing how various production-comprehension dissociations affect inner speech, these dissociations can be induced via neuromodulation techniques (e.g., Boatman et al., 2000) and utilized as an important source of evidence in testing the plausibility of SCAs.

In sum, production-comprehension dissociations with preserved (aspects of) inner speech suggest that Sensory is false; speech comprehension processing isn't necessary for working memory access to inner speech. At least at times, we can access aspects of the outputs of speech planning without mediation by speech comprehension. In other words, the contents of inner speech can be determined in a more top-down manner by speech planning as opposed to a more bottom-up manner by speech comprehension operating over CD.

12.4.2 Forward model accounts

FMAs and SCAs agree that the sensory content of inner speech is given by CD, sensory predictions of the auditory and somatosensory feedback that would be experienced in producing a planned utterance. However, in contrast to SCAs, FMAs invoke additional pairs of forward and inverse models and comparators to explain the internal monitoring channel and, in turn, other aspects of inner speech's contents. As Vicente and Jorba write:

> [I]f a prediction of the sound of a speech act [that is, CD] is the inner voice, a prediction of the full semantics of a speech act will be the meaning/content of such a speech act. That is, a prediction of the full semantics of a speech act is nothing other than what we identify as a propositional content. . . . Carruthers' view is that we bind the content of what we have said in our inner voice after interpreting our inner utterance via the comprehension system. . . . Another possibility, however, is that our inner speech already "comes" with a meaning.
>
> (Vicente & Jorba 2019, pp. 745–746)

Not only are there forward models of our vocal apparatus that generate CD, but there are also more abstract forward models that generate predictions regarding the lexical, syntactic, and semantic content of planned utterances. In turn, these more abstract forward models provide the lexical, syntactic, and semantic content of inner speech without the need for speech comprehension operating over CD. Thus, FMAs extend the use of concepts central to models of predictive motor control to explain the more abstract levels of speech planning.

Although FAMA advocates (e.g., Gambi & Pickering, 2016; Grandchamp et al., 2019; Pickering & Garrod, 2013; Vicente & Jorba, 2019) typically see one another as offering commensurate accounts of inner speech (and the internal monitoring channel), there are unrecognized and substantive differences between extant FMAs. In the following, I critically examine the two most developed FMAs: Pickering's, Gambi's, and Garrod's model (PGG's model) and the Condensation, Dialogality, and Intentionality model (ConDialInt) developed by Grandchamp et al. (2019).

12.4.2.1 PGG's model

According to PGG (Gambi & Pickering, 2016; Pickering & Garrod, 2013), speech planning begins with a non-natural language, conceptual representation (or "production command" in their terminology) that serves as input to a "production implementer." This is a rebranding of the first stage of LRM's model in the verbiage of predictive motor control and inherits the problems of separating conceptual processing and the language faculty. However, PGG also theorize that the production command is fed to a forward production model—which is an internal model of the core speech planning process itself—that generates predictions of the semantic, syntactic, lexical, and morphophonological representations that the production implementer eventually generates. These predictions are subsequently fed to a forward comprehension model, which further predicts how a listener would parse the utterance that is predicted to be produced by the production implementer. The outputs of the forward comprehension model are generated and buffered before the outputs of the production implementer. Internal monitoring is performed by comparators that can detect discrepancies between, for example, the predicted syntactic structure generated by the pair of forward models and the actual planned syntactic structure generated by the production implementer (a boxological depiction of PGG's model is provided in Figure 12.1).

PGG note a serious objection to their model that is important for our concerns: the representations operative at each level of speech planning appear to be generated thrice: once by the production implementer, once by the forward production model, and once by the forward comprehension model. This threefold proliferation of operations and representations seems greatly unparsimonious. For instance, if the forward production model produces the same types of representations at each speech planning stage as the slower production implementer, then it's unclear why we need the production implementer at all. The forward production model could, itself, generate motor commands that issue in overt speech, thus, precluding the need for the production implementer. In response, PGG argue that the predictions issued by the forward models constitute easier-to-compute,

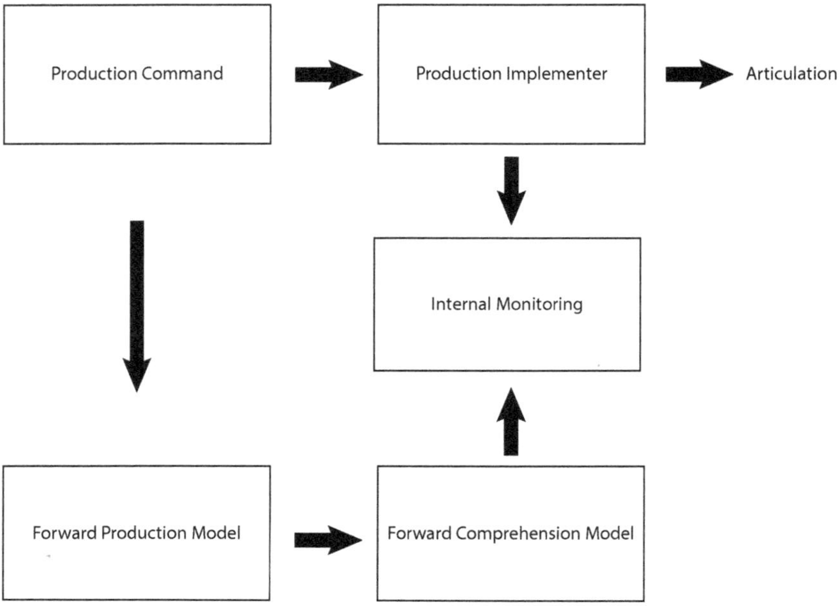

Figure 12.1 The internal monitoring channel of PGG's model. The external monitoring channel is not pictured.

"impoverished" representations that don't encode the details generated by the production implementer. This allows the outputs of the forward models to be computed quickly and buffered for comparison to the outputs of the production implementer, but it makes the forward models unsuitable for guiding articulation.

So, what gives inner speech its semantic, lexical, and syntactic contents on PGG's view? There seem to be three options: the production implementer, the forward production model, or the forward comprehension model. PGG initially claim that, ultimately, the forward comprehension model provides inner speech's content; however, they eventually introduce a fourth option:

> [W]hat we experience as inner speech is the outcome of a truncated language production process and, therefore, involves the activation of production representations within the implementer. . . . Such production representations are also accessible to the comprehension implementer [that is, core speech comprehension], at least at the phonological level.
> (Gambi & Pickering, 2016, p. 65)

In other words, PGG suggest that inner speech can work just like Levelt's internal perceptual loop. PGG don't indicate when inner speech's contents will result from forward modeling or core speech comprehension. I focus on forward modeling, as this is the only aspect of PGG's model that substantively differs from Levelt's SCA.

The problem for PGG is that if inner speech resulted from their forward models, inner speech couldn't play the cognitive roles discussed in Section 12.2.2. Again, PGG's forward production and comprehension models generate impoverished representations. Thus, if the contents of inner speech are given by forward models, the contents will be impoverished at every level. Pace Vicente and Jorba, the forward models posited by PGG don't generate a prediction of the *full semantics* of a speech act. At best, the forward models generate some partially filled semantic frame that lacks a determinate content. Therefore, it's unclear how inner speech could consistently vehicle something like exact calculation with integer values if its semantic content is necessarily impoverished.

12.4.2.2 The ConDialInt model

According to the ConDialInt model, speech planning begins with a desired language goal, that is, a representation of the content the speaker intends to convey vehicled by a non-natural language, conceptual medium. The desired language goal is converted into a desired message composed of lexical concepts by what Grandchamp et al. call the "conceptualization controller." The conceptualization controller is also a rebranding of the first stage of LRM's model in the verbiage of predictive motor control. Thus, the ConDialInt model also inherits the problems associated with assuming the independence of conceptual processing and the language faculty. The desired message is subsequently

> *sent back* as input to a conceptualization predictor [i.e., forward model], which predicts the language goal that would derive from it. Desired and predicted language goals can thus be compared, provided that the desired goal is buffered, so that desired and predicted signals are temporally aligned. . . . Any error at this early monitoring stage can be corrected for, by sending an error signal to the conceptualization controller and by delaying lower level processes.
>
> (Grandchamp et al., 2019, p. 7, emphasis mine)

As depicted in Figure 12.2, the model continues in an analogous fashion for each level of speech planning. At a given level, *n*, two representations will be compared as part of the internal monitoring channel—a "desired" and "predicted" representation. At the lexical concept, syntactic-lemma, morphophonological, and phonetic levels, the desired and predicted representations result, respectively, from two computational procedures, controllers and predictors, which operate over the *desired representations* (this

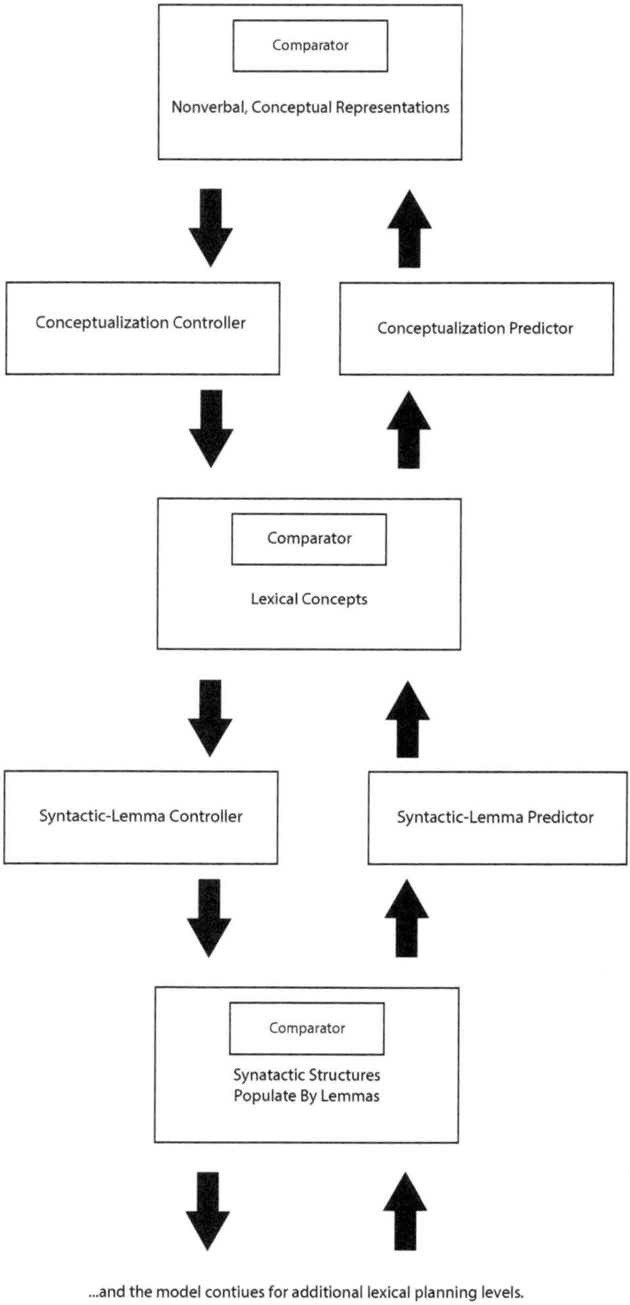

Figure 12.2 A portion of the ConDialInt model. Not all aspects of the model are depicted, as not all aspects are relevant to our discussion.

will be important for later discussion), respectively, at planning levels $n - 1$ and $n + 1$.

In contrast to PGG, on the ConDialInt model the predictors don't model the core speech planning process; thus, they don't predict *what will happen* at later stages in speech planning. Instead, they "predict" *what happened* at the previous planning stage. There is nothing like PGG's forward production model operating in the ConDialInt model. Insofar as we want to say Grandchamp et al.'s predictors are really predicting anything, they are predicting *what speech comprehension processes would* output if fed the desired representation operative at a given level.

So, what gives inner speech its semantic, lexical, and syntactic contents on the ConDialInt model? There seem to be two options: the controllers or predictors. Grandchamp et al. aren't particularly clear on the issue; however, given they claim that the sensory content of inner speech is provided by forward models and CD as previously described, I interpret the ConDialInt model as claiming that the semantic, lexical, and syntactic contents of inner speech are given by the associated predictors.

Because the ConDialInt model's predictors are not like PGG's forward production model, which needs to operate faster than the core speech planning process, Grandchamp et al. aren't forced to claim that the predictors merely generate impoverished representations. Thus, the ConDialInt model avoids the worries I raise for PGG's model. However, unlike PGG who explicitly claim that their forward models constitute special-purpose prediction mechanisms distinct from core speech production and comprehension (Pickering & Clark, 2014), Grandchamp et al. aren't explicit about whether their predictors are distinct computational procedures or, more parsimoniously, just core speech comprehension. For all we are told, the predictor operating over the desired syntactic structure, for instance, is just the semantic parsing stage of core speech comprehension "attending to" the syntactic structure generated by speech planning. If this is the case, the ConDialInt model would be a modification of Levelt's SCA that allows speech comprehension processes to access all stages of speech planning (as opposed to just the morphophonological stage).

Observed dissociations between comprehension and speech monitoring may push Grandchamp et al. to distinguish their predictors from core speech comprehension (Butterworth & Howard, 1987), but they are then left with a rather unparsimonious theory. Predictors and core speech comprehension would seem to be duplicate computational procedures, operating over and generating the same types of representations based on the same, tacit linguistic knowledge.

An additional worry arises from work on verbal transformations (VTs) in inner speech. VTs are illusions typically studied in overt speech that occur when hearing the rapid and continuous repetition of a speech form.

For instance, repeated, acoustically identical utterances of, "stress," may eventually be heard as, "dress." Hearing an utterance of, "stress," as, "dress," involves segregating /s/ from the speech stream and a shift in phonetic processing that mistakenly switches the alveolar plosives /t/ and /d/, but it also involves shifts at lexical and semantic levels. Hearing an utterance of, "stress," as, "dress," isn't a mere phonological/phonetic shift but also constitutes a shift in the meaningful English word identified in the speech stream.

VTs can be detected in conscious inner speech as well (Smith et al., 1995). However, it's implausible that the detection of, "dress," in one's inner speech stream when tasked with repeatedly and covertly uttering, "stress," is caused by a shift in core speech planning, for example, morphophonological planning spontaneously selecting /dress/ for utterance after several cycles of selecting /stress/. Instead, it's more plausible that the detection of, "dress," occurs after CD is projected to sensory areas at which point a shift in sensory-phonetic processing occurs, which subsequently causes speech comprehension processes to assign the "dress" lemma to the covert utterance. Covert VTs provide evidence for the claim that the content of inner speech can, at least at times, be determined in a more bottom-up fashion by speech comprehension processes operating over CD.

The ConDialInt model doesn't provide explicit room for a more bottom-up determination of inner speech's content in virtue of the fact that predictors always operate over *desired*, as opposed to *predicted*, representations. For instance, if phonological planning outputs /stress/, which is the desired representation that serves as input to the syntactic-lemma predictor that determines the lexical content of an inner speech utterance, there's no room for downstream sensory-phonetic processing of CD to propagate its way back up the speech hierarchy to fix the lexical content of the utterance as "dress."

Summarizing the claims in this section, extant neurocognitive models of inner speech inherit conceptual issues with LRM's model of lexical planning: the models assume that inputs to speech planning—communicative intentions—constitute non-natural language, conceptual representations that need to be mapped onto lexical concepts prior to the operations of the language faculty proper. This further assumes that conceptual processing and our representations of the semantic content of lexical items operate independently of natural language. But these assumptions are incompatible with the evidence reviewed in Section 12.2.2.

In addition, extant neurocognitive models fail to jointly account for two empirical phenomena:

1. observed production-comprehension dissociations in overt speech with preserved (aspects of) inner speech;
2. verbal transformations in inner speech.

The issue is that 1 and 2 pull in opposite directions: while 1 suggests that aspects of inner speech's contents can be determined in a top-down manner by core speech planning or forward modeling, 2 suggests that aspects of inner speech's contents can be determined in a more bottom-up fashion by speech comprehension. However, this shouldn't be too surprising, as most linguistic processing is a weighted amalgamation of top-down and bottom-up processing. For instance, depending on the context and acoustic properties of an overt utterance, our ability to parse the utterance will rely to a greater or lesser extent on bottom-up speech comprehension and top-down predictions of various linguistic features of the utterance issued by speech planning (Ryskin & Nieuwland, 2023). The story for inner speech is likely the same, but this still leaves us with plenty of work to be done in generating an adequate neurocognitive model of inner speech.

12.5 Conclusion

I've argued that recent work on inner speech can supplement GACP in providing grounded accounts of abstract conceptual processing and the systematicity and productivity of thought. Insofar as there is an explanatory need for a LOT, natural language suffices. In addition, while Classical LOT lacks a plausible implementation story, inner speech can be situated within neurocognitive models of overt speech. However, I detailed a series of outstanding issues with extant models of inner speech. If we want to trade in Classical LOT for a natural-language LOT that serves as a proper part of conceptual processing, our neurocognitive models of inner speech need to drop any sharp boundary between conceptual processing and the operations of the language faculty and allow for an amalgamation of top-down and bottom-up processing in the determination of inner speech's contents.

Notes

1 A host of theories fall under the broad GACP banner. See Meteyard et al. (2012) for further discussion.
2 See Borghi et al. (2017) for the abstraction/abstract distinction.
3 The problem is analogous to Benacerraf's problem for causal accounts of knowledge (1973).
4 Explicit Theory of Mind designates the Type-2 attitudes and inferences about minds and mental states to which we have access. In contrast, Implicit Theory of Mind is tracked by, for example, look-times or anticipatory look patterns.
5 The expressions "efference copy" and "corollary discharge" are used differently by different theorists. I follow (Grandchamp et al., 2019) in my usage.
6 See Ma and Tian (2019) for further discussion.
7 As I read Levelt, he uses, "inner speech," to refer to the representations and operations of *speech planning*. Thus, the speech comprehension representations activated as part of internal monitoring aren't inner speech. Instead, these speech comprehension representations constitute us "overhearing" our inner speech.

See Gauker (2018) for a similar use of the expression, "inner speech." In addition, in contrast to Carruthers and Frankish, Levelt doesn't talk of efference copies, forward models, or CD; however, Levelt is operating in a psycholinguistic tradition that doesn't utilize the vocabulary of predictive motor control. Levelt doesn't specify how speech comprehension "attends to" the output of speech planning. It could very well be that this "attending to" happens through efference copies, forward models, and CD.

8 For instance, evidence of unsymbolized thought (Hurlburt & Akhter, 2008).

References

Alderson-Day, B., & C. Fernyhough (2015). Inner speech: Development, cognitive functions, phenomenology, and neurobiology. *Psychological Bulletin, 141*(5), 931–965.

Anderson, M. L. (2010). Neural reuse: A fundamental organizational principle of the brain. *Behavioral and Brain Sciences*, *33*(4), 245–266.

Ardila, A., & Rosselli, M. (2002). Acalculia and dyscalculia. *Neuropsychology Review*, *12*(4), 179–231.

Baddeley, A. (2017). *Exploring Working Memory: Selected Works of Alan Baddeley*. London: Routledge.

Banks, B., & Connell, L. (2023). Access to inner language enhances memory for events. *Journal of Experimental Psychology: Learning, Memory, and Cognition, 50*(10), 1592–1615.

Barsalou, L. W. (2008). Grounded cognition. *Annual Review of Psychology*, *59*(1), 617–645.

Barsalou, L. W., Simmons, W. K., Barbey, A. K., & Wilson, C. D. (2003). Grounding conceptual knowledge in modality-specific systems. *Trends in Cognitive Sciences*, *7*(2), 84–91.

Benacerraf, P. (1973). Mathematical truth. *The Journal of philosophy*, *70*(19), 661–679.

Bi, Y. (2021). Dual coding of knowledge in the human brain. *Trends in Cognitive Sciences*, *25*(10), 883–895.

Binder, J. R. (2016). In defense of abstract conceptual representations. *Psychonomic Bulletin & Review*, *23*, 1096–1108.

Binder, J. R., Desai, R. H., Graves, W. W., & Conant, L. L. (2009). Where is the semantic system? A critical review and meta-analysis of 120 functional neuroimaging studies. *Cerebral Cortex*, *19*(12), 2767–2796.

Boatman, D., Gordon, B., Hart, J., Selnes, O., Miglioretti, D., & Lenz, F. (2000). Transcortical sensory aphasia: Revisited and revised. *Brain*, *123*(8), 1634–1642.

Borghi, A. M. (2023). *The freedom of words: Abstractness and the power of language*. Cambridge University Press.

Borghi, A. M., Binkofski, F., Castelfranchi, C., Cimatti, F., Scorolli, C., & Tummolini, L. (2017). The challenge of abstract concepts. *Psychological Bulletin*, *143*(3), 263.

Bormann, T., & Weiller, C. (2012). "Are there lexicons?" A study of lexical and semantic processing in word-meaning deafness suggests "yes". *Cortex*, *48*(3), 294–307. https://doi.org/10.1016/j.cortex.2011.06.003

Brannon, E. M., Jordan, K. E., & Jones, S. M. (2010). Behavioral signatures of numerical cognition. In M. Platt & A. Ghazanfar (Eds.), *Primate neuroethology* (pp. 144–159). Oxford University Press.

Buchwald, A., & Miozzo, M. (2011). Finding levels of abstraction in speech production: Evidence from sound-production impairment. *Psychological Science*, *22*(9), 1113–1119.
Butterworth, B., & Howard, D. (1987). Paragrammatisms. *Cognition*, *26*(1), 1–37.
Carruthers, P. (2018). The causes and contents of inner speech. In P. Langland-Hassan & A. Vicente (Eds.), *Inner speech: New voices* (pp. 31–52). Oxford University Press.
Chen, Y., Huang, L., Chen, K., Ding, J., Zhang, Y., Yang, Q., . . . Guo, Q. (2020). White matter basis for the hub-and-spoke semantic representation: Evidence from semantic dementia. *Brain*, *143*(4), 1206–1219.
Chomsky, N. (2015). *What kind of creatures are we?* Columbia University Press.
Chu, J., Cheung, P., Schneider, R. M., Sullivan, J., & Barner, D. (2020). Counting to Infinity: Does learning the syntax of the count list predict knowledge that numbers are infinite? *Cognitive Science*, *44*(8), e12875.
Connell, L., & Lynott, D. (2014). Principles of representation: Why you can't represent the same concept twice. *Topics in Cognitive Science*, *6*(3), 390–406.
Cragg, L., & Nation, K. (2010). Language and the development of cognitive control. *Topics in Cognitive Science*, *2*(4), 631–642.
Crapse, T. B., & Sommer, M. A. (2008). Corollary discharge across the animal kingdom. *Nature Reviews Neuroscience*, *9*(8), 587–600.
Davis, C. P., & Yee, E. (2021). Building semantic memory from embodied and distributional language experience. *Wiley Interdisciplinary Reviews: Cognitive Science*, *12*(5), e1555.
de Zubicaray, G. I. (2023). The neural organization of language production: Evidence from neuroimaging and neuromodulation. *Language Production*, 97–142.
Dove, G. (2022). *Abstract concepts and the embodied mind: Rethinking grounded cognition*. Oxford University Press.
Dymarska, A., Connell, L., & Banks, B. (2022). Linguistic bootstrapping allows more real-world object concepts to be held in mind. *Collabra: Psychology*, *8*(1), 40171.
Edmiston, P., & Lupyan, G. (2015). What makes words special? Words as unmotivated cues. *Cognition*, *143*, 93–100.
Egger, V. (1904). *La parole intérieure: essai de psychologie descriptive*. Alcan.
Feigenson, L., Dehaene, S., & Spelke, E. (2004). Core systems of number. *Trends in Cognitive Sciences*, *8*(7), 307–314.
Fodor, J. A. (1975). *The language of thought*, Vol. 87. Harvard University Press.
Fodor, J. A., & Pylyshyn, Z. W. (1988). Connectionism and cognitive architecture: A critical analysis. *Cognition*, *28*(1–2), 3–71.
Frank, M. C., Everett, D. L., Fedorenko, E., & Gibson, E. (2008). Number as a cognitive technology: Evidence from Pirahã language and cognition. *Cognition*, *108*(3), 819–824.
Frankish, K. (2018). Inner speech and outer thought. In P. Langland-Hassan & A. Vicente (Eds.), *Inner speech: New voices* (pp. 221–243). Oxford University Press.
Gambi, C., & Pickering, M. J. (2016). Predicting and imagining language. *Language, Cognition and Neuroscience*, *31*(1), 60–72.
Gauker, C. (2011). *Words and images: An essay on the origin of ideas*. Oxford: Oxford University Press.
Gauker, C. (2018). Inner speech as the internalization of outer speech. *Inner Speech: New Voices*, 53.
Grandchamp, R., Rapin, L., Perrone-Bertolotti, M., Pichat, C., Haldin, C., Cousin, E., . . . Lœvenbruck, H. (2019). The ConDialInt model: Condensation,

dialogality, and intentionality dimensions of inner speech within a hierarchical predictive control framework. *Frontiers in Psychology*, *10*, 1–30. https://doi.org/10.3389/fpsyg.2019.02019

Günther, F., Rinaldi, L., & Marelli, M. (2019). Vector-space models of semantic representation from a cognitive perspective: A discussion of common misconceptions. *Perspectives on Psychological Science*, *14*(6), 1006–1033.

Harman, G. H. (2015). *Thought*. Princeton University Press.

Harnad, S. (1990). The symbol grounding problem. *Physica D: Nonlinear Phenomena*, *42*(1), 335–346. https://doi.org/10.1016/0167–2789(90)90087–6

Hickok, G., & Poeppel, D. (2007). The cortical organization of speech processing. *Nature Reviews Neuroscience*, *8*(5), 393–402.

Hinzen, W., & Sheehan, M. (2013). *The philosophy of universal grammar*. Oxford: OUP.

Hommel, B. (2004). Event files: Feature binding in and across perception and action. *Trends in Cognitive Sciences*, *8*(11), 494–500.

Houde, J. F., Nagarajan, S. S., Sekihara, K., & Merzenich, M. M. (2002). Modulation of the auditory cortex during speech: An MEG study. *Journal of cognitive neuroscience*, *14*(8), 1125–1138.

Hurlburt, R. T., & Akhter, S. A. (2008). Unsymbolized thinking. *Consciousness and Cognition*, *17*(4), 1364–1374.

Hurley, S. (2001). Perception and action: Alternative views. *Synthese*, *129*, 3–40.

Jack, B. N., Le Pelley, M. E., Han, N., Harris, A. W. F., Spencer, K. M., & Whitford, T. J. (2019). Inner speech is accompanied by a temporally-precise and content-specific corollary discharge. *Neuroimage*, *198*, 170–180. https://doi.org/10.1016/j.neuroimage.2019.04.038

Kaltefleiter, L. J., Sodian, B., Kristen-Antonow, S., Wiesmann, C. G., & Schuwerk, T. (2021). Does syntax play a role in Theory of Mind development before the age of 3 years? *Infant Behavior and Development*, *64*, 101575.

Kemmerer, D. (2022). *Cognitive neuroscience of language*, 2nd ed. New York: Routledge.

Kompa, N. A. (2024). *Language, cognition, and the way we think: An interdisciplinary approach*. Bloomsbury Publishing.

Lee, K.-M., & Kang, S.-Y. (2002). Arithmetic operation and working memory: Differential suppression in dual tasks. *Cognition*, *83*(3), B63-B68.

Levelt, W. J. (1989). *Speaking: From intention to articulation*. Cambridge, MA: MIT Press.

Levelt, W. J. (1992). Accessing words in speech production: Stages, processes and representations. *Cognition*, *42*(1–3), 1–22.

Levelt, W. J., Roelofs, A., & Meyer, A. S. (1999). A theory of lexical access in speech production. *Behavioral and Brain Sciences*, *22*(1), 1–38.

Lupyan, G., & Swingley, D. (2012). Self-directed speech affects visual search performance. *Quarterly Journal of Experimental Psychology*, *65*(6), 1068–1085.

Lynott, D., & Connell, L. (2010). Embodied conceptual combination. *Frontiers in psychology*, *1*, 212.

Ma, O., & Tian, X. (2019). Distinct mechanisms of imagery differentially influence speech perception. *Eneuro*, *6*(5).

Mahr, J. B., & Schacter, D. L. (2023). A language of episodic thought? *Behavioral & Brain Sciences*, *46*.

Marshall, R. C., Rappaport, B., & Garcia-Bunuel, L. (1985). Self-monitoring behavior in a case of severe auditory agnosia with aphasia. *Brain and Language*, *24*(2), 297–313.

Meteyard, L., Cuadrado, S. R., Bahrami, B., & Vigliocco, G. (2012). Coming of age: A review of embodiment and the neuroscience of semantics. *Cortex*, *48*(7), 788–804.

Munroe, W. (2023). Semiotics in the head: Thinking about and thinking through symbols. *Philosophy and Phenomenological Research*, *107*(2), 413–438.

Nanay, B. (2023). *Mental imagery: Philosophy, psychology, neuroscience*. Oxford University Press.

Nedergaard, J., & Lupyan, G. (2023). *Not everyone has an inner voice: Behavioral consequences of anendophasia*. Paper presented at the Proceedings of the Annual Meeting of the Cognitive Science Society.

Nedergaard, J., Skewes, J. C., & Wallentin, M. (2023). "Stay focused!": The role of inner speech in maintaining attention during a boring task. *Journal of Experimental Psychology: Human Perception and Performance*, *49*(4), 451.

Nedergaard, J. S., Wallentin, M., & Lupyan, G. (2023). Verbal interference paradigms: A systematic review investigating the role of language in cognition. *Psychonomic Bulletin & Review*, *30*(2), 464–488.

Oppenheim, G. M., & Dell, G. S. (2008). Inner speech slips exhibit lexical bias, but not the phonemic similarity effect. *Cognition*, *106*(1), 528–537.

Patel, S. (2022). Towards a conative account of mental imagery. *Philosophical Psychology*, 1–25.

Patterson, K., & Ralph, M. A. L. (2016). The hub-and-spoke hypothesis of semantic memory. In *Neurobiology of language* (pp. 765-775). Elsevier.

Perrone-Bertolotti, M., Rapin, L., Lachaux, J.-P., Baciu, M., & Lœvenbruck, H. (2014). What is that little voice inside my head? Inner speech phenomenology, its role in cognitive performance, and its relation to self-monitoring. *Behavioural Brain Research*, *261*, 220–239.

Pickering, M. J., & Clark, A. (2014). Getting ahead: Forward models and their place in cognitive architecture. *Trends in Cognitive Sciences*, *18*(9), 451–456.

Pickering, M. J., & Garrod, S. (2013). An integrated theory of language production and comprehension. *Behavioral and Brain Sciences*, *36*(4), 329–347.

Poeppel, D. (2001). Pure word deafness and the bilateral processing of the speech code. *Cognitive Science*, *25*(5), 679–693.

Purpuri, S., Vasta, N., Filippi, R., Wei, L., & Mulatti, C. (2024). Does language shape the way we think? A review of the foreign language effect across domains. *International Journal of Bilingualism*, *0*(0), 13670069231225374. https://doi.org/10.1177/13670069231225374

Quilty-Dunn, J., Porot, N., & Mandelbaum, E. (2023). The best game in town: The reemergence of the language-of-thought hypothesis across the cognitive sciences. *Behavioral and Brain Sciences*, *46*, e261.

Ryskin, R., & Nieuwland, M. S. (2023). Prediction during language comprehension: What is next? *Trends in Cognitive Sciences, 27*(11), 1032–1052.

Scott, M. (2013). Corollary discharge provides the sensory content of inner speech. *Psychological science*, *24*(9), 1824–1830.

Slobin, D. I. (1996). *From "thought and language" to "thinking for speaking"*. New York, NY, US: Cambridge University Press.

Smith, J. D., Wilson, M., & Reisberg, D. (1995). The role of subvocalization in auditory imagery. *Neuropsychologia*, *33*(11), 1433–1454.

Tian, X., & Poeppel, D. (2010). Mental imagery of speech and movement implicates the dynamics of internal forward models. *Frontiers in psychology*, *1*, 7029.

Tourville, J. A., & Guenther, F. H. (2011). The DIVA model: A neural theory of speech acquisition and production. *Language and Cognitive Processes*, *26*(7), 952–981.

Vicente, A., & Jorba, M. (2019). The linguistic determination of conscious thought contents. *Noûs*, *53*(3), 737–759.

Vigliocco, G., Antonini, T., & Garrett, M. F. (1997). Grammatical gender is on the tip of Italian tongues. *Psychological Science*, *8*(4), 314–317.

Wang, X., Wu, W., Ling, Z., Xu, Y., Fang, Y., Wang, X., . . . Bi, Y. (2018). Organizational principles of abstract words in the human brain. *Cerebral Cortex*, *28*(12), 4305–4318.

Wolpert, D. M., & Ghahramani, Z. (2000). Computational principles of movement neuroscience. *Nature Neuroscience*, *3*(11), 1212–1217.

13 Working memory and the neural basis of intention

Wayne Wu

13.1 Introduction

Working memory has been a major target of empirical work in cognitive neuroscience, with well-established experimental paradigms along with correlated neural data collected during the performance of these paradigmatic working memory tasks (see D'Esposito and Postle 2015; Miller, Lundqvist, and Bastos 2018; van Ede and Nobre 2023). At the core is the idea of working memory as a cognitive workspace that is dynamic, required to support a variety of cognitive activities (Diamond 2013).

Interestingly, working memory has been often ignored as a target of investigation in contemporary philosophical psychology (but see Carruthers [2015] for one exception). For example, there are no philosophical theories of working memory despite many philosophical accounts of perception, thought, and various propositional attitudes, including other forms of memory like episodic memory. Given this, it is worth noting an interesting bridge from working memory to philosophical concern suggested by this striking passage in *Plans and the Structure of Behavior* by Miller, Galanter, and Pribram:

> When we have decided to execute some particular Plan, it is probably put into some special state of place where it can be remembered while it is being executed. . . . [W]e should like to speak of the memory we use for the execution of our Plans as a kind of quick access, "working memory." . . . When a Plan has been transferred into the working memory we recognize the special status of its incompleted parts by calling them "intentions".
>
> (G. A. Miller, Galanter, and Pribram 1960, 65)

This is one of the first invocations of working memory in English language psychology. What is notable is that the stated function of working memory is to store the agent's plan so that "it can be remembered while it is

DOI: 10.4324/9781003458531-13

being executed." The plan that must still be completed is an intention. We have a link from memory to a psychological state of substantial philosophical concern: an agent's intending to act (this is developed in detail in Wu [2023, chapters 3 and 4]).

In this chapter, I will argue that neural data gathered during the performance of working memory tasks cast new light on intention and its properties. The activity at issue is *delay period activity,* correlated activity while the subject is maintaining memoranda for (potential) future tasks. The properties of this activity have interesting philosophical and psychological implications regarding intentions to act. This is surprising in that a link to intention is not the typical framing of working memory in contemporary psychology and neuroscience. If the considerations I adduce are compelling, they highlight that a primary function of working memory, as originally conceived by those who introduced the construct, has fallen by the wayside, forgotten as it were, perhaps due to other concerns such as pervasive interest in measuring the limits of working memory capacity and its correlation with measures of intelligence (on capacity, see Cowan 2012).

I argue that the dynamic aspect of working memory reveals the unexplored dynamics of intention. I will then examine the significance of sensory recruitment models of working memory that have reoriented views about the neural substrates of working memory buffers, in many cases placing the buffers squarely in early sensory areas. I will discuss this in terms of the sensory aspects of intention. Accordingly, I suggest two ways that current models of working memory might lead to revamping how we see intention: intention is dynamic, not static, and it is, in some cases, partly sensory. This provides a framework for how neuroscience can lead to reconceptualizing well-established psychological categories.

13.2 A very brief primer on working memory

My claims regarding working memory will not depend on the truth of a specific overarching model (for eight distinct characterizations of working memory, see Cowan 2017). For example, I will not tie myself down to a specific working memory architecture—of which there are a variety, most famously that of Alan Baddeley and Graham Hitch (Baddeley and Hitch 1974). That said, most models of working memory involve a basic division between (1) a central executive and (2) memory stores. I will work with this common structure as a basis for understanding certain working memory *tasks*. The memory stores are by wide agreement capacity limited while the central executive functions to regulate and modulate the stores. We can treat this memory architecture as a functional model for understanding short-term remembering "for work." To speak of memory *for work* is to speak of the subject's holding some item in mind in order to use it to guide subsequent action, say when subjects retain spans of symbols such as the sequence of numbers in a phone number to dial or write down.

Cognitive science has probed working memory in a variety of tasks where working memory is activated by asking subjects to retain memoranda for some short duration before putting these items, potentially, to work in executing a task. There are a variety of working memory tasks, but a notable paradigm that I shall come back to is *retro-cueing* (Landman, Spekreijse, and Lamme 2003; Griffin and Nobre 2003). In this paradigm, subjects are asked to maintain several memoranda that are putatively task relevant. After the stimuli are removed and during the *delay period* where (representations of) the targets are maintained in working memory before executing a task, subjects receive a cue, a "retro-cue," that identifies one of the remembered targets as the likely target for subsequent test (e.g., the cue might be 80% valid in indicating the actual target). As in visual spatial cueing, retro-cueing improves performance such as reaction time (for a systematic review, see Souza and Oberauer 2016). Retro-cueing also demonstrates that working memory often involves manipulation of memoranda, in this case the prioritization of the cued target over any of the other remembered items.

Scientists have examined neural activity during the delay period and have frequently noted delay period activity that correlates with the retention of memoranda in working memory. That is, when nothing is before the eye, the brain maintains activity in relevant areas, initially noted and measured in prefrontal cortex (Goldman-Rakic et al. 1996; Funahashi 2017). A natural thought is that the delay period activity corresponds to the activation of relevant representations via the activation of their neural realizers. Notably, the content of working memory can often be decoded from this activity with appropriately trained decoders across different imaging modalities. While much of the focus on delay period activity during working memory focused on prefrontal cortex, recent decoding work has suggested that decodable delay period activity can be found throughout the cortex (Christophel et al. 2017). For example, in some visual working memory tasks, that is, tasks where the memoranda are previously presented visual stimuli, working memory contents can be decoded as early as primary visual cortex (V1)(Harrison and Tong 2009; Serences et al. 2009). While sensory recruitment remains controversial (for trenchant criticisms, see Xu 2017 [2020]), a growing number of working memory theorists, especially those investigating visual working memory, have endorsed it (for defense, see, among others, Gayet, Paffen, and Van der Stigchel 2018; Scimeca, Kiyonaga, and D'Esposito 2018; Iamshchinina et al. 2021). Readers can understand the main claim of this chapter to be conditional on the truth of sensory recruitment.

In what follows, we will focus on a subset of cases where, during a working memory task like retro-cueing, subjects must maintain and manipulate memoranda. Correlated with retention and modulation during a task-defined delay period is neural activity that can be instantiated in different parts of cortex depending on the nature of the task. As noted, in some of these cases, delay period activity occurs in early sensory areas, including the primary sensory cortex.

13.3 Sensory recruitment and reactivation

If delay period activity is observed during the maintenance period of a working memory task such that an appropriate decoder can decode the identity of the memoranda from this activity, one might hypothesize that this activity is the neural substrate of active working memory maintenance, namely the subject's retaining (representations of) specific targets for later work. John Jonides and coworkers give a nice expression of a *sensory recruitment* account of working memory maintenance: "the same brain regions that are involved in encoding information are those that are maintained in a temporarily active state while that information is retained for short intervals. That is, these same areas mediate working memory." Furthermore, sensory cortices provide the "first site of working memory storage" (Jonides, Lacey, and Nee 2005, 2–3)

One thing that continues to be unclear in the literature is the scope of quantification in a sensory recruitment account that holds that working memory maintenance is instantiated in early sensory areas. The strongest version of the view is that *every* working memory task draws on sensory substrates as the neural basis of working memory maintenance, say sensory substrates in early perceptual areas. This might seem too strong since sometimes, the contents of working memory come from long-term memory and it is not obvious that the latter must be sensory, and if not, then it is also not clear why working memory would need to use sensory substrates to maintain non-sensory long-term memories that are activated in working memory (for working memory as activated long-term memory, see Cowan 1995). A weaker, and perhaps more plausible, position is that only *some* working memory tasks draw on sensory substrates during maintenance, and in particular perceptual working memory tasks. Even here, one might wonder whether for sensory working memory tasks, the claim is that in performing these tasks, working memory maintenance *always* involve sensory substrates. Whether they do, of course, is an empirical issue.

For my purposes, it is enough to note this need for clarification but pivot to a weak version of the central hypothesis, one that is likely to attract the most adherents: at least for some working memory tasks, maintenance involves early sensory areas (D'Esposito and Postle 2015; Postle 2016). This is enough to cast new light on intention.[1]

13.4 Intention and memory for work

Earlier, I noted Miller, Galanter and Pribram's emphasis that working memory is for the storage of plans to regulate behavior. This suggests that in appropriate studies of working memory, we can glimpse the dynamics of intention understood as the agent's remembering what she intends to do, namely in maintaining a plan to execute.

Let us see matters from the other direction, from intention to memory. Michael Bratman (1987) once noted that intention has inertia, that it carries forward the upshot of one's practical reasoning and decision. This seems to me a memory function. So, having decided to act in a certain way, to then act in the future as one intends, one must remember what one decided to do. Forgetting is one way to fail to do what one intended. Let's say you suffer amnesia and forget what you planned to do. Yet the loss of such a memory is not cushioned by the additional presence of a distinct intention, as if you could lose one but not the other. If there were a distinct intention to fill in the memory blank, then the intention would just be performing a memory function. There would be a redundancy in remembering plans, a role that intentions fulfill. That said, I suggest that if one forgets one's plan, if one loses track of what one decided to do, there isn't an intention to fall back on: "Well, I don't remember what to do, but at least intention, not being memory, is not affected by my amnesia!" That seems implausible. The inertial properties of intention that Bratman emphasized just show that it is the relevant memory, the agent's retaining the action she previously decided to perform. Conversely, losing track of intention, an agent would no longer be able to access what she is presently doing. She would have forgotten. This supports the claim that intention is a type of memory, namely memory for work or as I say elsewhere, a *practical* memory (Wu 2023).

Contemporary research on working memory is consistent with this link between working memory and intention. Recent work has focused on the storage capacity of the working memory buffers and its capacity limitations as well as theorizing about the nature of such storage (e.g., see interesting work by Ma, Husain, and Bays [2014]). Drawing on the Baddeley and Hitch model, we can see this as emphasizing the properties of working memory buffers as primary targets of investigation.

Yet consider why those targets are held in working memory in the first place. They are not stored for no reason. Rather, they are targets of action. As such, they are targets of intention, for in holding them in mind, I remember them so I can act on them as I intend. The buffers are, after all, *for work*. Subjects maintain relevant targets in mind for subsequent task performance. Here, the central executive reasserts itself, for the items held in working memory are not there randomly (though the idea of spontaneous encoding into working memory is underexplored, see Dube and Golomb

2021). In experimental contexts, as well as in life, we hold things in mind because they are relevant to our present tasks. We hold them according to a plan, something that arguably must either direct the central executive's regulatory power, in which case a homunculus worry arises (who is using this plan to control the central executive?), or better, because the plan is held *in* the central executive, this module can directly regulate specific targets in the memory buffers (for more discussion on the regress worry and an argument that the central executive must store the agent's plan, see Wu [2023, Chapter 3]).

Retro-cueing shows that these remembered targets can dynamically change over time. In the ordinary course of the day, we might retain various targets in mind to guide near-future behavior, say a list of groceries as we enter the market. A text on our phone, a cue, from home might lead us to prioritize one item on the list ("Just get some diapers! Hurry!"). When this happens, our intentions become more fine-grained, in effect narrowing to more specific targets, diapers but not milk. Similarly, in retro-cueing, we prioritize one of the remembered targets for later performance. If we narrowly focus on the buffer component of working memory, we might see only the manipulation of stored items, but this would be to forget the central executive and the "for work" function of memory. After all, the remembered items are the targets of our plans.

This idea of working memory as memory for work (action) is a postulate highlighting the action function of working memory and can be motivated independently of the earlier quote from Miller et al. The postulate makes sense of what subjects are doing in the lab when they are asked to hold a set of objects during a delay. Reflection on ordinary and laboratory actions across time suggest that working memory serves memory for work (intention). That said, we also have a foundation from the earlier quote by Miller et al., who introduced the construct precisely to postulate a memory store to hold intentions that regulate our actions.

What this suggests, again at least in some cases, is that the dynamics of working memory as probed in a number of experimental paradigms such as retro-cueing reveal the dynamics of the corresponding intention. These dynamics concerns the updating of the targets of intention as the environment changes. In the laboratory, the dynamics of working memory in the retro-cueing task alter what the intention targets, moving from an intention that targets several potentially task-relevant stimuli to sharpening the intention to target a single (likely) stimulus, namely the item cued. Thus, one might shift from intending to act on A or B or C to intending to act specifically on (cued) A. Alternatively, one might shift from intending to act on A or B or C to prioritizing A over B and C though all three remain in memory. Accordingly, changes in what we intend over time, how we orient toward potential targets, is realized in the dynamics of working memory.

Seeing this possibility clearly only requires a reminder that regulating one's plans was why the working memory construct was established in the first place.[2]

13.5 Revamping intention in philosophical psychology

In this section, I consider two ways in which the cognitive neuroscience of working memory can change the way we think about intentions as psychological states. First, we have reason to treat intention not as static but dynamic. Second, we have reason to think of it as partly sensory. These are interesting claims that will meet resistance from philosophers and psychologists. That said, these claims identify plausible hypothesis that *reemerge* once we recall the originating function of the working memory construct and bring it up to date with recent research on sensory recruitment.

Intention in philosophical discussions is a canonical cause of intentional action. It is conceived of as a mental *state*, so in a sense static. Donald Davidson was troubled by invoking intention as a cause since on his view, events are causes but intentions, beliefs, and desires, the standard folk psychological causes of action, being static states, are not proper causes of action. A cause must be *occurrent,* not static. For this reason, he spoke of a specific event, the onslaught of a state, as how intentions and their ilk cause action (Davidson 1980). But are intentions static? The link between intention and working memory suggests that they are not.

Let's say that some intentions show a change in targeting over time, such as in narrowing the identity of the intended targets. This can be the result of practical reasoning. For example, one intends to have a snack but without there being a concrete thing to eat, the intention will always be just an abstract plan. So, one looks for a snack, and on seeing one, an apple, one intends to eat *that* thing. Here, the initial abstract representation of a target gets rendered more precise through practical reasoning: (a) I intend to eat a snack, (b) that (apple) is a snack, (c) I intend to eat that. Similarly in our grocery case, I intend to buy A, B, and C; Text: Buy just A!; I intend to buy just A. These are cases where the dynamics are relatively slow, but this can be sped up as the time between delay, the holding of the intention before action, and the time for action is compressed. We do not explicitly reason in this way during the flow of actions even as intentions have to update to keep track of changing targets (consider how the athlete's plan changes dynamically during the fast flow of a game).

Felipe de Brigard notes, however, that many philosophers will treat the transitions in the inference I noted to identify different intentions because each intention differs in content. An intention to eat a snack is different from an intention to eat an apple. In general, two mental states are distinct if they differ in content. This is a good point and suggests a potentially radical upshot of the view. If the content of intention tracks the content of the realizing working memory store and if we hold that the dynamics of the latter store support changes in the content of the *same* intention, then it looks like we must reject a venerable position in the philosophy of mind, namely, that we individuate states by their contents.

Yet I think the proposal is not actually radical in that way but in another. The principle at issue most naturally applies to states, but the radical claim to explore is that intention is not a state. A better model for intention as a dynamic mental phenomenon is visual experience over time, say the visual experience of motion. The content of this experience changes, say, in where it locates the object at each time during which the subject is experiencing motion. Here, we speak of the same visual experience as changing its content, and arguably, to understand seeing motion requires that we think of a single "state" with dynamic content rather than a number of states with distinct contents (paradoxes are in the offing). What the proposal of this chapter suggests is a reorientation in the philosophical psychology of intention: we should think of intention more on the experience model rather than the static state model.

So, there are dynamic changes in memory for work (intention). The proposal is that this dynamic has a basis in working memory, specifically in the modification of working memory stores over time, as illustrated in retro-cueing. The intention-in-action that is operative over the duration of a working memory task is itself dynamic. Accordingly, if *some* intentions have a dynamic aspect, then not all intentions are properly construed as *states*. They are better conceived, perhaps, as activities, a malleable aspect of mind that reflects how a plan, as remembered, changes online as action proceeds, updating as the world changes and new information is made available. The contents of the working memory buffer reflect the targets of the agent's intentions, and the dynamics of that buffer is the substrate for the dynamics of the content of intention. If we think of working memory as instantiating a dynamic state of mind, that being one of the distinct signatures of working memory as memory, then the psychological phenomena it constitutes is no less dynamic. This dynamic is, of course, reflected in delay period activity, the brain's actively maintaining representations for future work. So, a number of intentions, intentions-in-action, are not static. They are dynamic in that they are constantly updating as one acts.[3]

Now, if the targets of the agent's intentions are maintained in the working memory buffer, then the neural data that motivates a weak version of sensory recruitment points to the working memory buffer as realized in sensory neural substrates. So, in some cases where my intention is to act on *A* where *A* is the target remembered, then this target might be remembered in a sensory format as evidenced by sensory recruitment in delay period activity as the neural basis of memory maintenance. After all, in intending to act on *A*, an intention that is retained over a delay, *A* must be kept in mind for work. It is an empirical question where the substrates of such memory are, but if in the cases where memory maintenance is realized in early sensory areas, a reasonable hypothesis is that in intending to *A*, my intention involves a type of sensory-based targeting. That is, the representation of *A* is in that sense sensory (cf. Carruthers 2015).

Consider the idea of visual working memory, a memory state that from the get-go would be understood to be partly sensory since its content, what is remembered, has a sensory substrate. This is not, of course, an obligatory conclusion, but if neuroscience can motivate plausible hypotheses about the nature of mind, then the fact that certain memories have their basis in early sensory areas, even as early as primary sensory areas, should establish a hypothesis of literal sensory memory, that is, memory constituted by sensory representations. The invocation of sensory recruitment along with the original functional role of working memory when it was introduced now expands the hypothesis. It is not just visual working memory, with narrow focus on the visual memory buffer, that is sensory based, but a type of thought itself, namely memory for work (intention) that has sensory targets held in mind.

Philosophers might work up the point as follows: an agent intends to act on *that A*, say, an instance of *A* that is presented visually. Here, we begin with intention as a demonstrative thought where the demonstrative, *that A*, is a visual demonstrative that picks out a concrete, visible *A* (e.g., a specific apple). Thus, we have an intention to act on *that A* over a period of time while *A* is in view. Now, *A* disappears (e.g., in a retro-cueing experiment). An intention that began with the presence of a visual stimulus that fixed its content must be maintained despite the offset of that stimulus. The demonstrative can no longer be visual in the sense of relying on a perceptually mediated information link to a visible *A* since an *A* is no longer present before the eye. Rather, the intention relies on a *memory* demonstrative.

From the armchair, one might imagine that the neural realization of these demonstrative representations is based first in sensory areas and then in memory areas. That is, the neural basis of the representations shifts from a sensory area to a non-sensory (cognitive/memory) area. The neural data that has motivated many to endorse sensory recruitment speaks against this division, for the sensory and memory area in question is the same, collapsed in the brain. That is the core idea of sensory recruitment accounts of working memory. Sensory recruitment holds that in some cases, the neural basis of this transition from a perceptual to a memory demonstrative is seamless precisely because, as Jonides et al. hypothesized, the same neural basis of sensory experience of *A*, what was activated to allow the subject to see the target in order to demonstratively refer to it and to encode it, is the same neural basis for the memory retention of *A*. The striking consequence is that a cognitive state like intention might have a sensory core in the same way that visual working memory has a sensory core, driven by the nature of the neural basis of memory maintenance.

It would take a lot of work to substantiate the two hypotheses I've started to unpack, and this is not the place to do so (again, see Wu [2023, chapter 3] for initial discussion). We can also make progress by recovering hypotheses that are lost because a certain connection has been forgotten, namely the link between working memory and the retention of plans that Miller et al. emphasized decades ago. Once we restore that perspective and draw on the immense amount of work on the neural basis of working memory and performance in working memory tasks, understanding the brain through work in cognitive neuroscience casts a new light on the psychological state that motivated talk of working memory, namely, the agent's intention. From the perspective of this volume, the neural data, merged with concrete behavioral tasks and a functional characterization, percolates all the way up to the philosophical psychology of intention.

13.6 Concluding thoughts

Our understanding of the mind is enriched by multiple levels of analysis and investigation. A major challenge is to find ways to connect these levels. The cognitive neuroscience of working memory has begun to do this by tying neural activity to working memory through concrete experimental tasks and concurrent neuroimaging using various modalities. A further bridge can be made to psychological theory, namely, by remembering that the point of postulating working memory was to theorize about the role of remembered plans in regulating and structuring behavior. This allows us to bridge cognitive neuroscience to the theory of intention. The goal of this chapter has not been to definitively establish its main claims, but to uncover forgotten hypotheses and to elaborate them, providing some initial reasons to endorse them. In doing so, we can broaden philosophical and psychological perspectives on intention by drawing on cognitive neuroscience.[4]

Notes

1 Felipe de Brigard suggests some caution regarding a too easy invocation of decoding in support of inferences regarding representations. The caution is well taken. For example, he and coworkers have raised some questions regarding how much weight to put on multivariate pattern analysis (MVPA) in inferences about neural representations (Gessell, Geib, and De Brigard 2021). MVPA is the basis of some of the evidence in favor of sensory recruitment.

2 Another function of working memory is in setting attention. I discuss this function at length in Wu (2023, chapter 3) but see also the biased neural competition

theory of attention (Desimone and Duncan 1995). For additional empirical discussion, see Kiyonaga and Egner (2013).

3 No doubt, a philosopher who wants to have an ontology of only states and events can make sense of each updating as an event. This option is duly noted, but my goal is not to convince said philosopher that their approach is wrong. Rather, it is to suggest that in opening up philosophical concern to other explanatory schemes such as cognitive neuroscience, we might see something in those perspectives that will alter ours.

4 I am grateful to Gualtiero Piccinini for his patience and his comments on the manuscript and to Felipe de Brigard for helpful advice.

References

Baddeley, Alan D., and Graham J. Hitch. 1974. "Working Memory." In *The Psychology of Learning and Motivation*, edited by Gordon H. Bower, 8: 47–89. New York: Academic Press.

Bratman, M. 1987. *Intention, Plans and Practical Reason*. Cambridge, MA: Harvard University Press.

Carruthers, Peter. 2015. *The Centered Mind: What the Science of Working Memory Shows Us about the Nature of Human Thought*. Oxford: Oxford University Press.

Christophel, Thomas B., P. Christiaan Klink, Bernhard Spitzer, Pieter R. Roelfsema, and John-Dylan Haynes. 2017. "The Distributed Nature of Working Memory." *Trends in Cognitive Sciences* 21 (2): 111–24. https://doi.org/10.1016/j.tics.2016.12.007.

Cowan, Nelson. 1995. *Attention and Memory: An Integrated Framework*. Oxford: Oxford University Press.

Cowan, Nelson. 2012. *Working Memory Capacity*. Psychology Press.

Cowan, Nelson. 2017. "The Many Faces of Working Memory and Short-Term Storage." *Psychonomic Bulletin & Review; New York* 24 (4): 1158–70.

Davidson, Donald. 1980. "Actions, Reasons and Causes." In *Essays on Actions and Events*, 3–19. Oxford: Oxford University Press.

Desimone, Robert, and John Duncan. 1995. "Neural Mechanisms of Selective Visual Attention." *Annual Review of Neuroscience* 18 (1): 193–222. https://doi.org/10.1146/annurev.ne.18.030195.001205.

D'Esposito, Mark, and Bradley R. Postle. 2015. "The Cognitive Neuroscience of Working Memory." *Annual Review of Psychology* 66 (1): 115–42. https://doi.org/10.1146/annurev-psych-010814–015031.

Diamond, Adele. 2013. "Executive Functions." *Annual Review of Psychology* 64 (1): 135–68. https://doi.org/10.1146/annurev-psych-113011–143750.

Dube, Blaire, and Julie D. Golomb. 2021. "Perceptual Distraction Causes Visual Memory Encoding Intrusions." *Psychonomic Bulletin & Review* 28 (5): 1592–600. https://doi.org/10.3758/s13423-021-01937-6.

Ede, Freek van, and Anna C. Nobre. 2023. "Turning Attention Inside Out: How Working Memory Serves Behavior." *Annual Review of Psychology* 74 (1): 137–65. https://doi.org/10.1146/annurev-psych-021422–041757.

Funahashi, Shintaro. 2017. "Working Memory in the Prefrontal Cortex." *Brain Sciences* 7 (5): 49. https://doi.org/10.3390/brainsci7050049.

Gayet, Surya, Chris L. E. Paffen, and Stefan Van der Stigchel. 2018. "Visual Working Memory Storage Recruits Sensory Processing Areas." *Trends in Cognitive Sciences* 22 (3): 189–90. https://doi.org/10.1016/j.tics.2017.09.011.

Gessell, Bryce, Benjamin Geib, and Felipe De Brigard. 2021. "Multivariate Pattern Analysis and the Search for Neural Representations." *Synthese* 199 (5): 12869–89. https://doi.org/10.1007/s11229-021-03358-3.

Goldman-Rakic, P. S., A. R. Cools, K. Srivastava, A. C. Roberts, Trevor William Robbins, and Lawrence Weiskrantz. 1996. "The Prefrontal Landscape: Implications of Functional Architecture for Understanding Human Mentation and the Central Executive." *Philosophical Transactions of the Royal Society of London. Series B: Biological Sciences* 351 (1346): 1445–53. https://doi.org/10.1098/rstb.1996.0129.

Griffin, Ivan C., and Anna C. Nobre. 2003. "Orienting Attention to Locations in Internal Representations." *Journal of Cognitive Neuroscience* 15 (8): 1176–94. https://doi.org/10.1162/089892903322598139.

Harrison, Stephenie A., and Frank Tong. 2009. "Decoding Reveals the Contents of Visual Working Memory in Early Visual Areas." *Nature* 458 (7238): 632–35. https://doi.org/10.1038/nature07832.

Iamshchinina, Polina, Thomas B. Christophel, Surya Gayet, and Rosanne L. Rademaker. 2021. "Essential Considerations for Exploring Visual Working Memory Storage in the Human Brain." *Visual Cognition* 0 (0): 1–12. https://doi.org/10.1080/13506285.2021.1915902.

Jonides, John J., S. C. Lacey, and Derek Evan Nee. 2005. "Processes of Working Memory in Mind and Brain." *Current Directions in Psychological Science* 14 (1): 2–5.

Kiyonaga, Anastasia, and Tobias Egner. 2013. "Working Memory as Internal Attention: Toward an Integrative Account of Internal and External Selection Processes." *Psychonomic Bulletin & Review* 20 (2): 228–42. https://doi.org/10.3758/s13423-012-0359-y.

Landman, Rogier, Henk Spekreijse, and Victor A. F. Lamme. 2003. "Large Capacity Storage of Integrated Objects before Change Blindness." *Vision Research* 43 (2): 149–64. https://doi.org/10.1016/S0042–6989(02)00402–9.

Ma, Wei Ji, Masud Husain, and Paul M. Bays. 2014. "Changing Concepts of Working Memory." *Nature Neuroscience* 17 (3): 347. https://doi.org/10.1038/nn.3655.

Miller, Earl K., Mikael Lundqvist, and André M. Bastos. 2018. "Working Memory 2.0." *Neuron* 100 (2): 463–75. https://doi.org/10.1016/j.neuron.2018.09.023.

Miller, George A., Eugene Galanter, and Karl H. Pribram. 1960. *Plans and the Structure of Behavior*. New York: Holt, Rinehart and Winston.

Postle, Bradley R. 2016. "How Does the Brain Keep Information 'in Mind'?" *Current Directions in Psychological Science* 25 (3): 151–56. https://doi.org/10.1177/0963721416643063.

Scimeca, Jason M., Anastasia Kiyonaga, and Mark D'Esposito. 2018. "Reaffirming the Sensory Recruitment Account of Working Memory." *Trends in Cognitive Sciences* 22 (3): 190–92. https://doi.org/10.1016/j.tics.2017.12.007.

Serences, John T., Edward F. Ester, Edward K. Vogel, and Edward Awh. 2009. "Stimulus-Specific Delay Activity in Human Primary Visual Cortex." *Psychological Science* 20 (2): 207–14. https://doi.org/10.1111/j.1467–9280.2009.02276.x.

Souza, Alessandra S., and Klaus Oberauer. 2016. "In Search of the Focus of Attention in Working Memory: 13 Years of the Retro-Cue Effect." *Attention, Perception, & Psychophysics* 78 (7): 1839–60. https://doi.org/10.3758/s13414-016-1108-5.

Wu, Wayne. 2023. *Movements of the Mind*. Oxford: Oxford University Press.

Xu, Yaoda. 2017. "Reevaluating the Sensory Account of Visual Working Memory Storage." *Trends in Cognitive Sciences* 21 (10): 794–815. https://doi.org/10.1016/j.tics.2017.06.013.
Xu, Yaoda. 2020. "Revisit Once More the Sensory Storage Account of Visual Working Memory." *Visual Cognition* 28 (5–8): 433–46. https://doi.org/10.1080/13506285.2020.1818659.

14 Basic emotion theory meets the brain

Radicals and reformists in the arena of neuroscience

Marco Viola and Fausto Caruana

14.1 Introduction

Once upon a time, cognitive psychology used to conceive the mind as software, often disregarding the so-called implementation level. But the rise of new neuroscientific developments, and most prominently neuroimaging techniques, changed the face of cognitive psychology. The mainstream view is now that purely cognitive explanations—once the gold standard in psychology—are at best elliptical mechanistic explanations to be integrated with details that depend on the neural implementation (Piccinini & Craver 2011).

Several scholars propose that our cognitive ontology, that is, the catalogue of *bona fide* psychological constructs that can be included in our scientific theories, must be revised based on neuroscientific considerations (e.g., Anderson 2015; Poldrack 2010; Klein 2010; Poldrack & Yarkoni 2016; Price & Friston 2005; Viola 2017). For instance, in a proof of concept study, Poldrack's team scrutinized several constructs related to the domain of cognitive control (Lenartowicz et al. 2010). Noting that the neural activations correlating with "task switching" were poorly discriminable from those that correlate with other constructs, they suggest eliminating task switching from our cognitive ontology.

The proposal that neuroscientific data can arbitrate between psychological theories—and between the constructs they posit—has encountered criticism. Toward the beginning of the new century, some neuroskeptic cognitive psychologists challenged the usefulness of neuroscience, and especially of neuroimaging data from early fMRI experiments, in adjudicating psychological theories (Uttal 2001; Coltheart 2006; Loosemore & Harley 2010). Their skepticism largely hinged on the fact that psychological theories are underdetermined by neural data, since the same pattern of neural activity is consistent with multiple psychological interpretations.

If psychological debates were such that even large amounts of neural data favored none of the rival theories, this might seem to undermine the

DOI: 10.4324/9781003458531-14

very usefulness of spending so much money on fancy neuroimaging scanners. In this chapter, we will examine one such debate, namely, the debate regarding the six basic emotions posited by Paul Ekman (1992). Despite a wealth of data, the scientific community seems unable to reach a final verdict about their existence. Some scholars vocally argue that extant data militate against basic emotions. Others claim that they actually vindicate them. *Prima facie*, this disagreement could be taken as evidence that neuroskeptics were right: neuroscience cannot solve disputes about cognitive ontology. On the contrary, we will show that, although the same data may be consistent with multiple theoretical perspectives, whatever theoretical perspective one adopts, neuroscience definitely shows that the ontology of basic emotions posited by Ekman is wrong and must be amended.

We will proceed as follows. In the next section, we will briefly introduce Ekman's Theory of Basic Emotions and discuss some neuroimaging evidence about them as well as the different interpretations of this evidence (Section 14.2). We will then seek to explain what causes these divergent, even opposite, interpretations (Section 14.3). We will argue that the disagreement is rooted in three different approaches to cognitive ontology: conservative, radical, or reformist. Then, we will reassess the evidence on the neural correlates of basic emotions following the rules of each approach to see where they lead (Sections 14.4–14.6). Lastly, we will make a plea to complement the oft-cited neuroimaging evidence with electrophysiological and clinical data (Section 14.7). The final upshot is that, on any approach, there are no grounds for restricting our ontology of emotions to the six emotions included in Ekman's list. We will conclude that, whatever approach one adopts, neuroscience does make some difference in cognitive ontology. In this case, it dictates that we move past the myth of the canonical six (Section 14.8).

14.2 Basic emotions and their neural correlates

The label "Basic Emotion Theory" indicates several theories with family resemblance among them (see Ortony & Turner 1990). The best known version is arguably that developed by Paul Ekman and systematized during the 1990s (Ekman 1992). He construes each basic emotion as a sort of mental module, that is, a phylogenetically inherited hardwired mechanism working fast and independently from conscious beliefs (cf. Griffiths 1990). His list of six basic emotions—anger, disgust, fear, joy, sadness, and (less often) surprise—has become a sort of canon for the ontology of emotions within affective science, as well as in some applied field, such as in facial emotion recognition technology (see Viola 2024).

How do we identify basic emotions? Ekman's favorite marker was the distinctive facial expression that correlates with each basic emotion (Ekman

1992). However, Ekman predicted that other markers could be found, including neural correlates: "[t]here must be unique physiological patterns for each emotion, and these [Central Nervous System] patterns should be specific to these emotions not found in other mental activity" (1992, p. 183). During the following years, several neuroimaging laboratories searched for these patterns. Commenting on the proliferation of neuroimaging studies on the correlates of basic emotions, Scarantino (2009, p. 946) joked that "[t]he Holy Grail of contemporary affective neuroscience is the search for separate neural circuits corresponding to different discrete emotions." As a matter of fact, between 1990 and 2015, the quest for this Holy Grail resulted in dozens of univariate studies.[1] While each individual study could analyze but a specific facet of the neural correlates of emotion activity, depending on the specifics of the task and the stimuli, from 2002 to 2017, seven meta-analyses about specific emotion categories aimed to provide a more comprehensive picture (Phan et al. 2002; Murphy et al. 2003; Fusar-Poli et al. 2009; Vytal & Hamann 2010; Lindquist et al. 2012a; Wager et al. 2015; Kirby & Robinson 2017. See Table 14.1 for an overview).

These meta-analyses hoped to establish the neural correlates of each emotion category. They include studies based on hemodynamic neuroimaging (PET or fMRI), analyzed with univariate analyses, using healthy subjects.[2] With the exception of Fusar-Poli et al. (2009), who restrict their scope to the correlates of *seeing facial expressions* of Ekman's list of emotions (except surprise), other meta-analyses pool together activations from

Table 14.1 Seven meta-analyses on the neural correlates of emotion experience and/or recognition. See the main text for further details.

	No. of studies	*No. of subjects*	*Timespan*	*Main analysis techniques*
Phan et al. (2002)	55*	761	1990–2000	Based on 20 broad regions
Murphy et al. (2003)	106	1167	1994–2001	Based on eight broad regions
Fusar-Poli et al. (2009)	105 fMRI	1600	1990–2008	ALE (Laird et al. 2005)
Vytal and Hamann (2010)	83*	n/a	1993–2008	ALE (Laird et al. 2005)
Lindquist et al. (2012a)	91*	n/a	1993–2007	MKDA (Wager et al. 2007)
Wager et al. (2015)	148*	2519	1990–2011	BSPP (Kang et al. 2011)
Kirby and Robinson (2017)	27 to 141$	550 to 1426$	up to 2014	ALE (Laird et al. 2005)

* = only studies on discrete emotions included in this counting.
$ = depending on specific emotions, based on BrainMap database (as of 2014).

several experimental paradigms, often lumping together the perception and experience of emotions under the working hypothesis that perceptions and experiences of the same emotion share a common neural "core." This hypothesis is confirmed inasmuch as at least some loci of neural activations stably correlate with specific instances of a given emotion category irrespective of the paradigm—be them emotion perception tasks such as spotting facial expressions or emotion experiences elicited via several techniques, such as video-clips or recalling emotional memories.[3]

Despite the differences in techniques and inclusion criteria, all meta-analyses yield the same two results:

(i) First, there is no one-to-one mapping between neural areas and emotions, that is, "every region that was activated for one emotion category was activated for at least one other category" (Lindquist et al. 2012a, p. 129). In short, the mapping between regions and emotions is many-to-many.
(ii) Second, *many-to-many* does not amount to *any-to-any*. The mapping is *complex*, that is, neural activations corresponding to each emotion are scattered across the brain and partially overlap. But the mapping is not *random*. On the contrary, there is "a reliable brain basis for diagnosing instances of each emotion category . . . with moderate to high accuracy, depending on the category, in spite of substantial heterogeneity in the paradigms, imaging methods, and subject populations used" (Wager et al. 2015, p. 10).

However, somehow puzzlingly, different scholars draw *opposite* conclusions from (i-ii). For some, the evidence is favorable to basic emotions:

> [T]hese findings support the predictions of basic emotion theories by demonstrating that basic emotion states are associated with consistent patterns of brain activation and that these patterns differ significantly between emotion.
>
> (Vytal & Hamann 2010, p. 2, 879; see also Phan et al. 2002; Murphy et al. 2003; Fusar-Poli et al. 2009; Kirby & Robinson 2017)

For others, quite the contrary:

> Our findings do not support basic emotion theories, which are inspired by our phenomenology of distinct experiences of anger, sadness, fear, disgust and happiness that should be mirrored in distinct modules that cause each emotion.
>
> (Wager et al. 2015, p. 12; cf. Lindquist et al. 2012a)

Hence, rather than a Holy Grail, neural evidence turns out to be a Rubin vase, the bi-stable figure where you see either a goblet or two faces.

Hamann (2012) acknowledges that Lindquist and colleagues' (2012a) reach different conclusions from him (Vyal & Hamann 2010) based on similar data and suggests that one of these two interpretations might be ruled out by complementing data from *univariate* neuroimaging with data from experiments employing *multivoxel pattern analysis* (Haxby et al. 2014). Several such studies were performed during the last decade. For instance, Kragel and LaBar (2015) employed musical stimuli and short clips to elicit the "canonical six" plus contentness (a calmer version of happiness) in their subjects. Saarimäki et al. (2016) elicited the canonical six except surprise by means of self-elicitation and short clips. In both studies, multidimensional patterns of neural activity corresponding to each discrete emotion were employed to train a classifier, which proved capable of predicting the emotion based on the neuroimaging data with an accuracy significantly greater than chance. Predictions were robust across methods and subjects, that is, using neural correlates of emotion category E elicited via method M_1 on subject S_1 as training data yielded above-chance predictions when used for training the classifier to predict E from the neural correlates obtained either via a different method M_2 or on a different subject S_2 (for a recent overview of more studies, see Kragel et al. 2024).

In reviewing these studies, Kragel and LaBar (2016) concluded that "emotion categories are uniquely represented in the activity of distributed neural systems that span cortical and subcortical regions" (Kragel & LaBar 2016, abstract). In other words, even if one-to-one mappings between basic emotions and their neural correlates are nowhere to be found, there are neural fingerprints for each discrete basic emotion. But, again, where Kragel and LaBar saw the vase, others saw the faces. Clark-Polner and colleagues (2017) contend that, since the patterns associated with different emotions largely overlap, this undermines Basic Emotion Theory, whose tenets seem to imply that each Emotion has its own dedicated neural machinery.

Some may take this disagreement vis-à-vis the same evidential basis to support skepticism about the usefulness of neural data to arbitrate disputes in cognitive ontology. After all, they might ask, if different scientists agree on the structure of evidence (i-ii) but draw opposite conclusions about psychological theories, what are brain data good for? We resist this skeptical conclusion. What this controversy actually shows, instead, is that *data do not speak for themselves*: they are always interpreted in the light of some auxiliary hypotheses. A plurality of auxiliary hypotheses may be legitimate in principle. But are they also equally fruitful? To answer this question, in the next section, we will outline different ways to interpret the data and explore where they lead.

14.3 Understanding the disagreement

14.3.1 Three stances on cognitive ontology

In commenting on the inconclusiveness of neural evidence in the dispute for or against basic emotions, Loaiza (2020, p. 9) reminds us that, to draw any such conclusion, "we must have a set of criteria to map emotions onto brain structures." Which criteria one adopts depend on their approach to cognitive ontology. We will describe three approaches, building on Anderson (2015). These approaches can be placed on a continuum, varying in the degree of weigh that they assign neural evidence in assessing cognitive categories.

On one end of the spectrum, we find *conservatives*. Unlike neuroskeptics (e.g., Uttal 2001; Coltheart 2006), conservatives grant that neural evidence has something to say when it comes to cognitive ontology. Specifically, neural evidence plays the role of validating constructs if they map onto stable neural patterns across subjects and experimental conditions, and rejecting them if they do not. If some cognitive constructs do not withstand the scrutiny of neural data, new ones must be devised by cognitive theorizing based on methods from traditional, "boxological" cognitive psychology.

At the opposite, brain-friendliest end of the spectrum we find *radicals*. These are people willing

> to rethink the very foundations of psychology in light of evidence from neuroscience and evolutionary biology. They [think] that the required revisions might involve the construction of categories that divide the mind in very different ways than the current ontology does.
>
> (Anderson 2015, p. 70)

The radicals are not afraid to depart from extant taxonomies. Instead of trying to save the old categories inherited from pre-neuroscientific psychology, they advocate a bottom-up approach that starts from patterns in neural data and tries to figure out their psychological significance. As two fierce radicals put it:

> Rather than starting with a commonsense category and looking for its neural correlates, experiments would manipulate specific psychological operations in an a priori manner, and then examine their consequences for the content and type of resulting mental states.
>
> (Lindquist & Barrett 2012, p. 538)

The middle ground is occupied by the *reformists*.[4] For them, neural data are useful not only for assessing independently obtained cognitive constructs.

Instead, neural data go hand in hand with old-fashioned cognitive theorizing when it comes to revising cognitive ontology. Within this approach, many sources, including cognitive desiderata, neuroscientific findings such as neural dissociations, abductive reverse inferences (Calzavarini & Cevolani 2022), and failure to find neural correlates, can motivate "revisions that can include splitting categories once thought unified, merging categories between which the brain makes no distinction, and even jettisoning constructs altogether" (Anderson 2015, p. 70).

As anticipated in the introduction, we will now appraise the evidence on the neural correlates of emotions through the lens of all three approaches to see where each one leads and derive some theoretical considerations from that. Luckily for us, this does not require too much guesswork, because some scholars explicitly defend some of these positions.

14.3.2 The desiderata of the three stances

Everybody, be them conservative, radical, or reformist, would agree on the following claim:

> (N) a specific neural correlate consistently and selectively activated by instances of a putative psychological construct C is evidence that C exists.

For radicals, (N) is both necessary and sufficient to develop our cognitive ontology. For conservatives, (N) is only necessary: other psychological considerations must be checked before granting a psychological kind a place in our cognitive ontology. As to reformists, (N) is sufficient for *considering inclusion* in our Cognitive Ontology but does not warrant automatic inclusion: before that happens, other conditions pertaining to cognitive theorizing must also be met.

The approaches also differ with respect to what *kind* of neural correlate they require for a psychological construct to be a *bona fide* psychological kind. To begin with, we might distinguish at least two versions of (N):

> (N_R) a specific *brain region* R_x consistently and selectively activated by instances of a putative psychological construct C is evidence that C exists.
>
> (N_S) a specific *set* S_x *of brain regions (R1, R2, . . . &* R_n*)* consistently and selectively activated by instances of a putative psychological construct C is evidence that C exists.

Everybody, including radicals, would grant ontological reality to basic emotions were they to meet the condition posed by (N_R). But, as explained above (Section 14.2), meta-analyses of *univariate* data and more recent multivariate experiments conclusively show that no emotion can be mapped one-to-one to a specific neural region (and indeed, the same holds for any mental construct). What obtains instead is that specific emotions consistently activate specific *sets* of brain regions, although these sets overlap (i.e., some regional activations are common to multiple emotions). For reformists and conservatives, the consistent but non-selective mapping granted by (N_S) suffices to establish the existence of basic emotions. For radicals, it does not.

Why should (N_R) be stronger than (N_S)? Arguably, because neural regions can be reasonably considered *bona fide* neuroanatomical kinds. Their neuroanatomical reality may even justify merging putatively distinct psychological kinds that map on that region.[5] In contrast, when it comes to a set of regions corresponding to some emotion, the only glue keeping together brain regions R_1 and R_2 rather than R_1 and R_3 is the emotion they correlate with. For a reformist or a conservative, a set of regions consistently mapping to fear or anger is simply a fear *network* or an anger *network*. But for a radical, a conjunction of neural structures based on their correlation with pre-given psychological constructs rather than on neuroanatomical grounds is a non-starter. Indeed, granting psychological reality to whatever constructs meets a lax criterion such as having a stable neural correlate in some ad hoc network runs the risk of overinflating our cognitive ontology with arbitrarily chosen kinds (Krickel 2024).

That does not mean that radicals only allow for such a strict principle as (N_R). They are perfectly aware "that the field of cognitive neuroscience has been moving steadily toward the conclusion that psychological functions are best described at the level of networks than at the level of individual brain regions" (Lindquist et al. 2012b, p. 180). However, (N_S) is too lax a principle for them, as it overlooks the distinction "between a hard-wired, anatomical network and a dynamic functional network that is generated on the fly during an emotional experience or perception" (Lindquist et al. 2012b). Within the radicals' brain-driven approach to cognitive ontology, only psychological processes that correlate with what they call "intrinsic networks" deserve to be included in our cognitive ontology (Lindquist & Barrett 2012; Barrett & Satpute 2013). In other words, radicals are not against using brain networks as a basis for including psychological constructs in our cognitive ontology. What they ask is that the network be grounded in neuroanatomy. In their words:

> [I]f emotion categories were supported by anatomically given, inherited networks, then there should be intrinsic networks that correspond to the brain regions active during the experience or perception of instances of anger, sadness, fear, and other emotion categories.
>
> (Lindquist et al. 2012a, p. 141)

For now, let us take stock. The radicals are open to psychological constructs that consistently map onto intrinsic networks because, in their agenda, intrinsic networks are the solid foundation on which a renewed, brain-friendly cognitive ontology may be built. On the other hand, conservatives and reformists can settle for (N_S) because, as long as they correlate with some psychological construct posited by respectable theories, "dynamic networks constructed on the fly" are legitimate neural correlates.

Up to this point, we have merely provided a rational reconstruction of the debate. In the next three sections, we ask the following questions: Is Basic Emotion Theory solid enough for warranting a conservative defense of all and only the canonical six emotions? Does the radicals' quest for intrinsic networks provide solid ground for a brain-based cognitive ontology? Is the reformist open-ended investigation of the neural and nonneural hallmarks of emotions leading to substantial revisions to the canonical list? To what extent do methodological considerations about neuroscientific techniques (neuroimaging vs. other techniques) affect the philosophical debate?

14.4 Conservatives: the canonical six as a historical accident

For many decades, Ekman's favorite six basic emotions—anger, disgust, fear, happiness, sadness, and surprise—have enjoyed a privileged status in affective science, becoming a sort of canon. They figure in several psychology and neuroscience textbooks. They are more represented than other putative affect categories in the abstracts of affective neuroscience papers. The most popular standardized databases of facial expressions (Dawel et al. 2022; Viola 2024) include facial expressions for these six categories (plus or minus one). They are also the default set for most facial emotion recognition systems (see Cohn & De la Torre 2015; Huang et al. 2019). Such hegemony suggests that, whatever reasons there were for picking these very categories rather than others, they must be solid.

Alas, this is not the case.

Let us leap back in time to the Sixties. The young Paul Ekman was about to leave for New Guinea to test whether the nine affects posited by his mentor Silvan Tomkins' theory were universally recognized even by the Fore, a preliterate culture that had not interacted with westernized countries. Phoebe Ellsworth was collaborating with Ekman to help him gather a balanced set of pictures of facial expressions of *all* Tomkins' nine affects. Yet, by the time they had to leave for New Guinea, they managed to find enough stimuli for only six of them. According to Ellsworth, "With a few more weeks or months, the list of canonical emotions might have been longer. So, the canonical six were an accident. We simply fell short of Tomkins' canonical nine, which still constituted just one canon among many"

(Ellsworth 2014, p. 23). Notice that Tomkins' list of nine affects was not strongly validated either: these nine affects were hypothesized, which is legitimate, and some of them were proved to be universally recognizable by Ekman and other scholars' studies. Yet the list *per se* was just a working hypothesis, which provides no strong rationale for excluding emotion categories which did not figure therein (Leys 2017, ch. 1). How did these six emotion categories become reified into a canon (despite Ekman's own attempt to enlarge the list by adding contempt during the Eighties; Ekman & Friesen 1986)? While providing a historical explanation for the success of the canonical six goes beyond the scope of this chapter, we speculate that the fame and relative simplicity of Ekman's pioneering experiments prompted so many replications (reviewed by Elfenbein & Ambady 2002) that the protocol, including the emotion categories tested, became paradigmatic. Such paradigmatic protocols sometimes become so deeply embedded in the background assumptions of a research community that they are rarely questioned.

Another possibility, suggested by the radicals, is that Ekman's basic six are alluring because they are close to folk psychological concepts of emotion (Lindquist & Barrett 2012). For our purposes, it is enough to note that the main reason why scholars conducted more experiments involving the canonical six than any other emotions is the lack of sufficient effort at comparing these six emotions with other candidates. In other words, inertia. Of course, inertia is not a valid reason to resist the revision of some categories in our cognitive ontology. Any conservative wishing to preserve a cognitive ontology of emotions that sticks with the basic six should say why the list must include all and only those six rather than (say) love, lust, contempt, or envy. Here, some conservatives might be tempted to appeal to Ekman's gold standard for identifying emotion, that is, the presence of distinctive and universal facial expressions and reject other candidates on the grounds that, unlike the basic six, they do not map neatly onto universal facial expressions. As we will see in Section 14.6, this does not seem to be the case.

14.5 The radicals: basic emotions do not carve at the brains' joints

The radicals have a dream: rebuilding cognitive ontology by carving at the brain's natural joints. Sitting at the opposite end of the cognitive ontology spectrum, they wish to eliminate extant psychological categories in favor of categories that share a consistent and neuroscientifically solid neural basis, such as an anatomically confined brain region (N_R) or a set of regions kept together by some sort of neural glue, that is, "intrinsic networks." Given such a premise, the chance that constructs devised independently from neural data survive this severe scrutiny seems poor. Yet the canonical six (or at least some of them) might survive the severe threshold proposed by the radicals if they map regularly onto neurally respectable structures, that is, brain regions or intrinsic networks. We have already seen that the evidence is definitely against one-to-one mapping between emotion categories and brain regions (cf. Section 14.2). What about mappings between emotions and intrinsic networks?

To answer this question, we first need to grasp what the radicals mean by "intrinsic networks." Since Lindquist et al. (2012a) speak of "anatomical, hard-wired networks," we would expect that they have in mind the structural connectivity network revealed by techniques such as Diffusion Tensor Imaging.[6] However, their operationalization of intrinsic network refers to more indirect measures based on physiological data from fMRI. More specifically, they grant the honorary label of "intrinsic network" to *resting state networks*, that is, sets of regions that co-activate due to spontaneous fluctuations in hemodynamic activity when subjects lie in a scanner without any explicit task (Gessell 2024).

In slightly more formal terms, some radicals endorse the following principle:

> (N_{RS}) a specific *set* S_x *of brain regions* (R_1, R_2, . . . *&* R_n) consistently and selectively activated by instances of a putative emotion E is evidence that E exists *if and only if* S_x is an intrinsic network, that is, R_1, R_2, . . . *&* R_n are also consistently and selectively activated by resting state.

Thus, an emotion (or whatever cognitive construct) deserves admission in the cognitive ontology if and only if its activation pattern matches that of some resting state network.

But is that the case? A study by Touroutoglou and colleagues (2015) straightforwardly addresses this question. Touroutoglou and colleagues acquired resting state data for 89 subjects. Then, they picked the three or four largest activation peaks for each emotion category described by Vytal and Hamann's (2010) meta-analysis as seeds for the discovery of co-activation maps within the resting state networks of their subjects, that is, they clustered together those voxels that co-activate with those seeds. Why? In their words, because

> [i]f anatomically constrained networks for each emotion category exist in the intrinsic architecture of the human brain, as the basic emotion view predicts [they reasoned], then the meta-analytically derived seed regions for a given emotion category . . . should produce "discovery" maps whose spatial overlap reveals a network for that category.
>
> (Touroutoglou et al. 2015, p. 1259)

This condition does not obtain: co-activations during instances of different emotions cross-cut resting state networks. Therefore, Touroutoglou and colleagues conclude that "individual instances of anger, disgust, fear, etc., could be understood as high dimensional brain states reflecting neural assemblies within broadly distributed [intrinsic] networks, as well as the dynamic interaction of those assemblies" (Touroutoglou et al. 2015, p. 1264). This is enough to settle the matter concerning Ekman's six basic emotions: from a radical perspective, they are not vindicated by neural data.

However, we think that the principle (N_{RS}) defended by Touroutoglou and colleagues is problematic. Even if we set aside the issues facing all bottom-up reforms of cognitive ontology (see McCaffrey & Machery 2016; Francken et al. 2022; De Brigard & Gessell 2024), the special status accorded to resting state networks hinges upon assumptions like the following:

> Spontaneous activity observed when the brain is at rest reveals the basic building blocks of brain functioning.

In Touroutoglou and colleagues' words, "spontaneous neuronal activity within these networks reflects the intrinsic organization of the brain (Buckner et al., 2013), which in turn forms the functional architecture of the mind" (Touroutoglou et al. 2015, p. 1258). However, the very paper they cite to support their claim (Buckner et al., 2013) seems to go in the opposite direction: therein we read that "rest should be considered as just another arbitrary task state" (Buckner et al. 2013, p. 833). Finn (2021, p. 4) goes even further, noting that "rest is just as biased as any other task state," and "it is biased in ways that may be less easily measured or mitigated," and thus suggests downsizing the hype for collecting resting state data in favor of task-based paradigms.

Moreover, noting how identifying resting state networks hinges upon sampling neural activity at relatively long timescales, McCaffrey and Danks (2022) contend that allegedly unified networks might rather be artificial mixtures of functionally isolated sub-networks. This may explain why the psychological interpretation of these networks varies greatly from study to study (Andrews-Hanna et al. 2014). The radicals are eager to reinterpret these multiple functional attributions as different manifestations of a single domain-general operation whose functionality must be construed at a deeper, more abstract level (Lindquist & Barrett 2012; Barrett & Satpute 2013). These reinterpretations are the bread and butter of cognitive ontology reforms (Price & Friston 2005), but they may bring about functional labels that are too general to provide any constraint to psychological interpretation (Klein 2012).

Even worse, the hemodynamic activity behind resting state networks is not (totally) explained by cognitive activity but hinges in part upon non-cognitive physiological processes subserving brain homeostasis (Haueis 2018). In fact, since extant evidence suggests that hemodynamic activity at rest might be less driven by neural activity than task-evoked activity (e.g., Winder et al. 2017), investing resting state networks with psychological significance might be a mistake.

Besides, while a radical like Barrett makes very clear that we should exclude basic emotions and other old-fashioned psychological constructs from our cognitive ontology, she is way more elusive about how to replace them. According to older version of her theory (Barrett 2006), what we commonly call "emotions" are indeed the products of conceptual acts upon a primitive dimension called *Core Affect*. Core Affect is the process that monitors the subject's well-being, making it available to phenomenal consciousness through two dimensions: arousal (inactivated—activated) and valence (feeling bad—feeling good). However, neural correlates of subjective arousal and valence are even *more scattered* and *less predictive* than those of specific emotion categories (Kragel & LaBar 2015; Wager et al. 2015), making the ontological reality of core affective dimensions even less consistent with the radicals' desiderata of a brain-driven ontology (see also Scarantino 2009). This might explain why throughout the years, in Barrett's writings, valence and arousal become more peripheral than they used to be (for instance, in Barrett 2006). Initially, they are demoted to a "descriptive features of core affect that bear no resemblance to or inform about how affect is caused" (Barrett & Bliss-Moreau 2009, p. 188), until finally they disappeared from her later writings (e.g., Barrett 2017).

But what, then, are the neurally solid, ontologically serious constructs at play in her theory of emotions? A promising response points toward a large-scale intrinsic brain network subserving interoception and allostasis (Kleckner et al. 2017). Again, allostasis (i.e., an organism's adaptation to environment for the sake of promoting homeostasis) is so generic a task label that it may lack the specificity we would expect from a good cognitive label (cf. Klein 2012). In sum, the radical agenda may have a hard time revising our cognitive ontology so as to satisfy both their appetite for bottom-up, neuroscientifically robust psychological categories and the need to make them psychologically meaningful.

14.6 The reformists: emotions are welcome, but why six?

There is no place for basic emotions within the strongly neuro-inspired cognitive ontology dreamt of by the radicals. What if we relax the criterion for what shall count as a neural correlate from the criterion proposed by the radicals—that is, (N_{RS}) or (N_R)—to a looser criterion, namely, (N_S)? Recall that all (N_S) requires is that instances of a putative cognitive construct map to a relatively stable set of neural regions, often called a "functional network." Of course, if this lax neural criterion were the only desideratum for admitting a construct in our cognitive ontology, we would risk inflating it with arbitrary and suspicious categories. After all, although seeing Pokémons evokes consistent neural responses in the ventro-temporal cortex of former players of Pokémon video games (Gomez et al. 2019), we might want to resist including a construct like "representing Pokémons" in our cognitive ontology (cf. Krickel 2024). Luckily, for both reformists and conservatives, consistency of neural basis is merely a *necessary* condition for inclusion in cognitive ontology. The difference between reformists and conservatives is that, since reformists grant neural evidence a heuristic role in revising cognitive ontology, they have the onus to explain why we ought to reject a construct such as "representing Pokémons," whereas conservatives can simply shrug their shoulders and answer "simply because no cognitive theory posits Pokémons."

Nevertheless, neuroimaging studies have found not only stable neural correlates of Pokémons but also of *non-Basic emotions*. Recall Saarimäki and colleagues' (2016) multivariate study discussed in Section 14.2, which managed to get above-chance predictions of basic emotions based on neural activations even across different subjects or modalities. In a later opinion paper, the senior and the first author of that study noted:

> [S]howing that basic emotions can be classified from brain imaging data alone . . . by no means confirms that these would be the only discrete emotion systems in the brain, or that this classification scheme would be the optimal way to represent human emotions.
>
> (Nummenmaa & Saarimäki 2019, p. 5. See also Hamann 2012)

Their remark is substantiated by their follow-up study (Saarimäki et al. 2018). Instead of confining themselves to the canonical six, this new study also included eight less explored emotions such as "love," "pride," and "guilt"—plus a neutral condition. They used short emotion-inducing narratives to elicit different emotions in 25 subjects. Then, as in their previous study, they used these emotion categories to train a classifier. Again, the classifier proved capable of accurate prediction (significantly above the chance level of 6.7%) for all emotions except "shame" and "longing." However, mean accuracy greatly differs for the various categories, ranging from <10% for shame and longing to >25% for surprise, up to >30% for the neutral condition. While the average prediction accuracy of the canonical six taken as a group was greater than the average for non-basic emotions, some individual non-basic emotion performed far better than the alleged basic ones. For example, instances of love were predicted twice as accurately as instances of anger.

Similar findings are also reported by other neuroimaging studies employing elicitors (clips) codified for a broad range of emotions. Horikawa et al. (2020) found, for 27 emotions out of the 34 they tested, reliable correlates across subjects. Koide-Majima et al. (2020) found stable correlates for 25 categories out of 80 on average (depending on subjects). We emphasize that *not all the emotions tested in these studies mapped consistently with a given neural pattern*. Thus, even a principle as permissive as (N_S) is enough to reject some psychological constructs.

From a reformist's perspective, these findings may motivate revising the list of emotions to include in our cognitive ontology, provided we also have some non-neuroscientific reasons to grant them psychological reality. Do we? A scrutiny of recent psychological literature suggests that we might well do.

Let us set aside for a moment the discussion of neural correlates and focus instead on other markers of emotions, such as Ekman's classical criterion, namely the presence of universal facial expressions. In a nuanced recognition task based on computer-generated faces, Jack et al. (2016) reduced facial expressions to four core motifs, shaped by cultural variation. Other researchers leveraged (big) data-driven machine learning technologies in real-life contexts and found robust evidence for 16 (Cowen et al. 2021) or 35 (Srinivasan & Martinez 2018) universal patterns of facial movements associated with emotion.

The very scholars who nowadays present themselves as heirs to Ekman's Basic Emotion Theory explicitly argue for overcoming his face-centrism in favor of multimodal behavioral markers (Keltner et al. 2019). They have collected evidence that some emotions are universal by considering different channels, such as affective self-reports (Cowen et al. 2019a), music (Cowen et al. 2020), and non-verbal vocalizations (Cordaro et al. 2016; Cowen et al. 2019b). In all these studies, they report universal stable cross-cultural stimuli-emotion associations for many more emotions than the canonical six, some of which could not be spotted in facial expressions. While the list is still provisional, the final one is unlikely to converge on all and only the canonical six.

To sum up, for a reformist the only sensible way to preserve (at least some of) the canonical six within our cognitive ontology is to open its doors to other emotions as well (cf. Scarantino 2015; Scarantino & Griffiths 2011).

Rather than throwing the baby out with the bathwater, this conclusion should prompt us to roll up our sleeves and find more refined categories. In the end, that's exactly what happened in the case of memory, where neuroscientific research discovered a plurality of theoretically homogeneous and scientifically reliable kinds of memory: episodic, iconic, procedural, autobiographical, sensory, and working memory. Proceeding in a similar way, neuroscientific research will provide us with a new list of theoretically homogeneous kinds of emotions, where the canonical six might be split into sub-types of fear, anger, happiness, etc. Some research in the field of animal electrophysiology is already doing this. As one example among many, studies in rodents have shown that different types of fear are linked to different sub-sectors of the amygdala, alternately recruited depending on the type of threat and the elicited reaction to them (Gross & Canteras 2012). Instead of eliminating basic emotions, neuroscience might multiply them.

14.7 Cognitive ontology beyond neuroimaging

Until now, much of the scientific discussion about basic emotions was based on results from PET and fMRI studies (cf. Section 14.2; see also Viola 2024). Yet, if we aim to rethink our cognitive ontology on the basis of neuroscientific results, we should include all available neuroscientific evidence. The lack of neuroanatomically robust emotional fingerprints highlighted by the radicals, indeed, might be due to the fact that neuroimaging is not the best technique to tackle this issue. First, according to many scholars, emotions are context-dependent action tendencies, and neither real-life emotional stimuli nor real-life behavioral responses can be studied within fMRI scanners. Second, neuroimaging data are correlative in nature: they can reveal which brain regions are *involved* in emotion processing, but they cannot tell which brain regions are *required* for emotion generation or recognition (Caruana 2017).[7] Third, neuroimaging is known to have rather low temporal and spatial resolution compared with other neurophysiological techniques, and this heuristic weakness can greatly impact the ability to dissect the different contribution of adjacent regions (which is particularly true for small subcortical nuclei).

Electrophysiological and clinical data such as those obtained by electrical stimulation, brain lesions, and pathological conditions, are more rarely considered by cognitive neuroscientists than neuroimaging data. But such data allow to study how the activation or the inhibition of specific brain regions causally contribute to cognition and behavior. Let's consider the case of human electrical stimulation, routinely performed in surgical patients. While some emotions traceable to those identified by Basic Emotion Theory (in particular, fear, disgust, and happiness) have been elicited by electrically stimulating a limited number of brain regions, other basic emotions (anger, surprise, contempt—cf. fn. 1) along with non-basic ones (shame, embarrassment) have never been evoked. Fear and some related negative responses, such as panic attacks or anxiety, can be elicited from a limited number of interconnected areas, including the amygdala (in particular, the central nuclei and the laterobasal group; Zhang et al. 2023) or the dorsalmost sector of the anterior cingulate cortex (Caruana et al. 2018). Similarly, disgust is systematically elicited by the anterior insula (Krolak-Salmon et al. 2003), and happiness has been associated with anterior cingulate or temporal pole stimulation (Caruana et al. 2020). Different responses, including positive emotions elicited by the stimulation of the amygdala (Lanteaume et al. 2007) and the insula (Villard et al. 2023), are extremely rare and can be conceived as the exception that proves the rule.

The considerable number of stimulation data published to date converge, reassuring on the solidity of this trend. This fact may suggest that, from a neuroscientific point of view, some basic emotions are more basic than others. Moreover, emotions other than the canonical six can also be elicited by electrically stimulating cortical and subcortical regions, leading neurophysiologists to articulate a new set of basic emotions. The most famous case in neuroscience is probably that of Panksepp's (1998) set of basic emotions (which he called *Primary Process Affects*), obtained by a variety of electrophysiological (electrical/chemical stimulation) studies on rodents: seeking, lust, care, play, fear, rage, and panic.

The cognitive ontology that one can derive from clinical data is even more complex. For instance, epileptic seizures can induce rage, fear, disgust and, in the case of ecstatic seizures, well-being and intense serenity (Bartolomei et al. 2005; Pelliccia et al. 2022; Gschwind & Picard 2016). In addition, seizures can induce mood disorders, in particular depression, in the long run. Of course, interpreting the underpinnings of seizures is very challenging, as seizures may recruit both adjacent and distant regions. It is nonetheless curious that certain emotions are frequently elicited while others are not, and that the list of those frequently elicited does not match the one derived by electrical stimulation studies.

In fact, it may seem that different neuroscientific techniques support multiple Cognitive Ontologies. For now, all of them seem to be individually coherent but reciprocally inconsistent. The discrepancy between such different findings suggests that understanding the emotional brain is akin to deciphering a book composed in a polysemic language, much like James Joyce's Finnegans Wake. Contrary to polysemic books, however, the outcome of the different investigations finally deserves a cohesive interpretation, and this task is among the future challenges for the philosophy of neuroscience.

14.8 Conclusion

Let us briefly recap. We opened our chapter in Section 14.1 by noting that, apart from some neuroskeptics, many scholars believe that neuroimaging data can be purposefully employed to test whether psychological constructs deserve inclusion in our cognitive ontology. Among other constructs, this scrutiny applied to the canonical six emotions made popular by Paul Ekman, that is, anger, disgust, fear, happiness, sadness, and surprise. Yet this scrutiny led to mixed, even conflicting results, possibly revamping skepticism about the usefulness of neuroscience. In Section 14.2, we then reviewed some data about the neural correlates of these emotions. Meta-analyses of univariate studies and qualitative reviews of multivariate studies reveal that (i) no one-to-one mapping could be established between specific regions and specific emotions, but (ii) each emotion mapped quite consistently with a given set of neural areas or activation patterns (some of which were shared by multiple emotions).

Some scholars interpret these findings as evidence in favor of basic emotions. Others see them as evidence to the contrary. Why? The reason, we claimed in Section 14.3, is different approaches to neural-based reforms of cognitive ontology. We distinguished three approaches: conservatives, who maintain that neuroscientific data can at best validate psychological constructs; radicals, aiming for a brain-driven ontology; and reformists, who allow neuroimaging data to play a heuristic role in revising our Cognitive ontology. After sketching the rules of the game for each camp (in Section 14.3.2), we tried to play the game, that is, we assessed neural evidence according to each ruleset.

We started by asking, do conservatives have reasons to resist reform? Given that the success of the canonical six seems due in part to a "historical accident," as summarized in Section 14.4, the answer to this question is negative. Unsurprisingly, as we show in Section 14.5, abiding by the radicals' recipe implies jettisoning basic emotions from cognitive ontology, because the sets of areas correlating with them do not correspond to any robust construct from their perspective, such as a resting state network—although we also flag some reasons why mapping with a resting state network could be a poor guide to the reform of cognitive ontology. The reformist approach (examined in Section 14.6) provides a more promising ground for reshaping the field. However, since neuroscientific and behavioral evidence suggest that there may be many more discrete emotions than the canonical six, a reformist should consider enriching their ontology. Finally, we have shown that different techniques suggest different ontological landscapes. Although the debate on emotions is based almost exclusively on neuroimaging data, we highlighted that different ontologies can be derived by other neuroscientific techniques, such as electrophysiological and clinical data, posing new challenges for the philosophy of neuroscience (Section 14.7).

Radicals and reformists agree on two things: (a) a list including all and only the canonical six must be rejected, and (b) more neuroscience research will tell us more about the emotional brain. The only way to save all and only the canonical six would be, from a conservative standpoint, having good theoretical reasons to exclude other emotions. But such reasons are nowhere to be found. We conclude that, whichever perspective we adopt on cognitive ontology, if we follow it consistently, we will dismiss the canonical six—or at least downplay their special status. Hence, neuroskeptics are right that neural data *per se* underdetermine psychological theory. Yet they would be wrong in inferring that neural data are useless. *Pace* neuroskeptics, as soon as a set of working hypotheses is adopted and its implications are drawn consistently, neuroscientific data *do* bear fruits.

Acknowledgments

Among the many who commented on previous versions of this chapter, a special thanks goes to the audiences at the online seminars on the Problem of cognitive ontology in 2020, the SILFS conference in 2021, and the online conference on NeuroCognitive Foundations of Mind in 2022, and to Beate Krickel, Gualtiero Piccinini, and an anonymous reviewer for their insightful comments.

Notes

1 In early fMRI studies, the traditional approach for statistical analysis was univariate. In univariate analysis, each voxel is marked either active or inactive at each time depending on whether their signal reaches a certain threshold. During the 2000s, however, multivariate analysis gained traction. Unlike a univariate analysis, whose result is a 3D "map" of brain regions that are either active or inactive, the output of multivariate analysis is a N-dimensional matrix, with each dimension representing the activity of a different voxel. For an introduction to multivariate analysis, see Haxby et al. (2014). For a discussion of the shortcomings of univariate and multivariate analyses, see Klein (2010) and Ritchie et al. (2019), respectively.

2 Importantly, while Ekman's pivotal studies (Ekman et al. 1969; Ekman & Friesen 1971) and many of those inspired by them (reviewed in Elfenbein & Ambady 2002) had the merit of testing their generalizability on subjects from preliterate cultures, most of the neuroimaging studies included in these meta-analyses were performed on WEIRD subjects (Henrich et al. 2010). While not easy to overcome, this limitation constrains the generality of neuroscientific findings discussed throughout this article.

3 Vytal and Hamann (2010, p. 2880) explicitly investigate whether this working hypothesis is sound. They found that the neural activation patterns that correlate with experiencing five basic emotions (all except surprise) largely overlap those that correlate with the perception of emotional facial expressions.

4 While Anderson (2015) speaks of "moderates," we prefer the label "reformists" to emphasize that this approach might actually foster significant modifications to Cognitive Ontology.

5 In fact, common neural activation for allegedly distinct cognitive kinds T_1 and T_2 sometimes spur scientists to posit a common cognitive process C underlying both (Price & Friston 2005, but see Klein 2012 for a cautionary caveat).

6 Diffusion Tensor Imaging is a neuroimaging technique that tracks the diffusion of water molecules in biological tissues. Applied to the brain, it provides a tentative sketch of the architecture of axonal connections in the brain.
7 Here is how the father of affective neuroscience, Jaak Panksepp (2007, p. 282), puts it:

> With regard to modern brain imaging, simply consider the fact that the pseudocolor statistical maps of neural activity changes (rarely more than a few percentage points different than baseline levels) hide vast oceans of neurophysiological and neurochemical activities, with multiple functional systems interpenetrating, especially in subcortical regions, and with overlapping, interacting circuits generating affective mentality. Each imaged pseudocolor region of the human brain hides enormous complexities and individual variabilities that need to be considered. What we can surely say of the typical final products of most averaged results from human brain imaging studies, consisting of isolated islands of arousal, is as follows:
>
> [T]he brain does not work that way.

References

Anderson, M. L. (2015). Mining the brain for a new taxonomy of the mind. *Philosophy Compass*, *10*(1), 68–77.
Andrews-Hanna, J. R., Smallwood, J., & Spreng, R. N. (2014). The default network and self-generated thought: component processes, dynamic control, and clinical relevance. *Annals of the New York Academy of Sciences*, *1316*(1), 29.
Barrett, L. F. (2006). Are emotions natural kinds? *Perspectives on Psychological Science*, *1*(1), 28–58.
Barrett, L. F. (2017). The theory of constructed emotion: an active inference account of interoception and categorization. *Social Cognitive and Affective Neuroscience*, *12*(1), 1–23.
Barrett, L. F., & Bliss-Moreau, E. (2009). Affect as a psychological primitive. *Advances in Experimental Social Psychology*, *41*, 167–218.
Barrett, L. F., & Satpute, A. B. (2013). Large-scale brain networks in affective and social neuroscience: towards an integrative functional architecture of the brain. *Current Opinion in Neurobiology*, 23(3), 361–372.
Bartolomei, F., Trébuchon, A., Gavaret, M., Régis, J., Wendling, F., & Chauvel, P. (2005). Acute alteration of emotional behaviour in epileptic seizures is related to transient desynchrony in emotion-regulation networks. *Clinical Neurophysiology*, *116*(10), 2473–2479.
Buckner, R. L., Krienen, F. M., & Yeo, B. T. (2013). Opportunities and limitations of intrinsic functional connectivity MRI. *Nature Neuroscience*, *16*(7), 832–837.
Calzavarini, F., & Cevolani, G. (2022). Abductive reasoning in cognitive neuroscience: weak and strong reverse inference. *Synthese*, *200*(2), 70.
Caruana, F. (2017). What is missing in the "Basic emotion vs. Constructionist" debate? Pragmatist insights into the radical translation from the emotional brain. *Pragmatism Today*, *8*(1), 87–108.
Caruana, F., Avanzini, P., Pelliccia, V., Mariani, V., Zauli, F., Sartori, I., . . . & Rizzolatti, G. (2020). Mirroring other's laughter. Cingulate, opercular and temporal contributions to laughter expression and observation. *Cortex*, *128*, 35–48.
Caruana, F., Gerbella, M., Avanzini, P., Gozzo, F., Pelliccia, V., Mai, R., . . . & Rizzolatti, G. (2018). Motor and emotional behaviours elicited by electrical stimulation of the human cingulate cortex. *Brain*, *141*(10), 3035–3051.

Clark-Polner, E., Johnson, T. D., & Barrett, L. F. (2017). Multivoxel pattern analysis does not provide evidence to support the existence of basic emotions. *Cerebral Cortex*, *27*(3), 1944–1948.
Cohn, J. F., & De la Torre, F. (2015). Automated face analysis for affective computing. In Calvo, R. A., D'Mello, S., Gratch, J. M., & Kappas, A. (Eds.). *The Oxford handbook of affective computing*. Oxford Library of Psychology, 131–150.
Coltheart, M. (2006). Perhaps functional neuroimaging has not told us anything about the mind (so far). *Cortex*, *42*(3), 422–427.
Cordaro, D. T., Keltner, D., Tshering, S., Wangchuk, D., & Flynn, L. M. (2016). The voice conveys emotion in ten globalized cultures and one remote village in Bhutan. *Emotion*, *16*(1), 117.
Cowen, A. S., Elfenbein, H. A., Laukka, P., & Keltner, D. (2019b). Mapping 24 emotions conveyed by brief human vocalization. *American Psychologist*, *74*(6), 698.
Cowen, A. S., Fang, X., Sauter, D., & Keltner, D. (2020). What music makes us feel: at least 13 dimensions organize subjective experiences associated with music across different cultures. *Proceedings of the National Academy of Sciences*, *117*(4), 1924–1934.
Cowen, A. S., Keltner, D., Schroff, F., Jou, B., Adam, H., & Prasad, G. (2021). Sixteen facial expressions occur in similar contexts worldwide. *Nature*, *589*(7841), 251–257.
Cowen, A., Sauter, D., Tracy, J. L., & Keltner, D. (2019a). Mapping the passions: toward a high-dimensional taxonomy of emotional experience and expression. *Psychological Science in the Public Interest*, *20*(1), 69–90.
Dawel, A., Miller, E. J., Horsburgh, A., & Ford, P. (2022). A systematic survey of face stimuli used in psychological research 2000–2020. *Behavior Research Methods*, 1–13.
De Brigard, F., & Gessell, B. (2024). The mirage of big-data phrenology. *The British Journal for the Philosophy of Science*. https://doi.org/10.1086/732152
Ekman, P. (1992). An argument for basic emotions. *Cognition & Emotion*, 6(3–4), 169–200.
Ekman, P., & Friesen, W. V. (1971). Constants across cultures in the face and emotion. *Journal of Personality and Social Psychology*, *17*(2), 124.
Ekman, P., & Friesen, W. V. (1986). A new pan-cultural facial expression of emotion. *Motivation and Emotion*, *10*(2), 159–168.
Ekman, P., Sorenson, E. R., & Friesen, W. V. (1969). Pan-cultural elements in facial displays of emotion. *Science*, *164*(3875), 86–88.
Elfenbein, H. A., & Ambady, N. (2002). On the universality and cultural specificity of emotion recognition: A meta-analysis. *Psychological Bulletin*, *128*(2), 203–235.
Ellsworth, P. C. (2014). Basic emotions and the rocks of New Hampshire. *Emotion Review*, 6(1), 21–26.
Finn, E. S. (2021). Is it time to put rest to rest?. *Trends in Cognitive Sciences*, *25*(12), 1021–1032.
Francken, J. C., Slors, M., & Craver, C. F. (2022). Cognitive ontology and the search for neural mechanisms: three foundational problems. *Synthese*, *200*(5), 378.
Fusar-Poli, P., Placentino, A., Carletti, F., Landi, P., Allen, P., Surguladze, S., . . . & Politi, P. (2009). Functional atlas of emotional faces processing: a voxel-based meta-analysis of 105 functional magnetic resonance imaging studies. *Journal of Psychiatry & Neuroscience, 34*(6), 418–432.
Gessell, B. (2024). Resting-state fMRI and cognitive neuroscience. In Heinzelmann, N. (Ed.). *Advances in neurophilosophy*. Bloomsbury Publishing, 103–132.

Gomez, J., Barnett, M., & Grill-Spector, K. (2019). Extensive childhood experience with Pokémon suggests eccentricity drives organization of visual cortex. *Nature Human Behaviour*, *3*(6), 611–624.

Griffiths, P. E. (1990). Modularity, and the psychoevolutionary theory of emotion. *Biology and Philosophy*, *5*(2), 175–196.

Gross, C. T., & Canteras, N. S. (2012). The many paths to fear. *Nature Reviews Neuroscience*, *13*(9), 651–658.

Gschwind, M., & Picard, F. (2016). Ecstatic epileptic seizures: a glimpse into the multiple roles of the insula. *Frontiers in Behavioral Neuroscience*, *10*, 21.

Hamann, S. (2012). Mapping discrete and dimensional emotions onto the brain: Controversies and consensus. *Trends in Cognitive Sciences*, *16*(9), 458–466.

Haueis, P. (2018). Beyond cognitive myopia: a patchwork approach to the concept of neural function. *Synthese*, *195*(12), 5373–5402.

Haxby, J. V., Connolly, A. C., & Guntupalli, J. S. (2014). Decoding neural representational spaces using multivariate pattern analysis. *Annual Review of Neuroscience*, 37, 435–456.

Henrich, J., Heine, S. J., & Norenzayan, A. (2010). The weirdest people in the world?. *Behavioral and Brain Sciences*, *33*(2–3), 61–83.

Horikawa, T., Cowen, A. S., Keltner, D., & Kamitani, Y. (2020). The neural representation of visually evoked emotion is high-dimensional, categorical, and distributed across transmodal brain regions. *iScience*, *23*(5), 101060.

Huang, Y., Chen, F., Lv, S., & Wang, X. (2019). Facial expression recognition: a survey. *Symmetry*, *11*(10), 1189.

Jack, R. E., Sun, W., Delis, I., Garrod, O. G., & Schyns, P. G. (2016). Four not six: revealing culturally common facial expressions of emotion. *Journal of Experimental Psychology: General*, *145*(6), 708–730.

Kang, J., Johnson, T. D., Nichols, T. E., & Wager, T. D. (2011). Meta analysis of functional neuroimaging data via Bayesian spatial point processes. *Journal of the American Statistical Association*, *106*(493), 124–134.

Keltner, D., Sauter, D., Tracy, J., & Cowen, A. (2019). Emotional expression: advances in basic emotion theory. *Journal of Nonverbal Behavior*, 1–28.

Kirby, L. A., & Robinson, J. L. (2017). Affective mapping: an activation likelihood estimation (ALE) meta-analysis. *Brain and Cognition*, *118*, 137–148.

Kleckner, I. R., Zhang, J., Touroutoglou, A., Chanes, L., Xia, C., Simmons, W. K., . . . & Barrett, L. F. (2017). Evidence for a large-scale brain system supporting allostasis and interoception in humans. *Nature Human Behaviour*, *1*(5), 1–14.

Klein, C. (2010). Philosophical issues in neuroimaging. *Philosophy Compass*, *5*(2), 186–198.

Klein, C. (2012). Cognitive ontology and region-versus network-oriented analyses. *Philosophy of Science*, *79*(5), 952–960.

Koide-Majima, N., Nakai, T., & Nishimoto, S. (2020). Distinct dimensions of emotion in the human brain and their representation on the cortical surface. *NeuroImage*, *222*, 117258.

Kragel, P. A., & LaBar, K. S. (2015). Multivariate neural biomarkers of emotional states are categorically distinct. *Social Cognitive and Affective Neuroscience*, *10*(11), 1437–1448.

Kragel, P. A., & LaBar, K. S. (2016). Decoding the nature of emotion in the brain. *Trends in Cognitive Sciences*, *20*(6), 444–455.

Kragel, P. A., Sander, D., & LaBar, K. S. (2024). Can brain data be used to arbitrate among emotion theories?. In Scarantino, A. (Ed.). *Emotion theory: the Routledge comprehensive guide, Vol. i*. Routledge, 511–542.

Krickel, B. (2024). The new mechanistic approach and cognitive Ontology—Or: What role do (Neural) mechanisms play in cognitive Ontology? *Minds and Machines*, *34*(3), 17.

Krolak-Salmon, P., Hénaff, M. A., Isnard, J., Tallon-Baudry, C., Guénot, M., Vighetto, A., . . . & Mauguiere, F. (2003). An attention modulated response to disgust in human ventral anterior insula. *Annals of Neurology: Official Journal of the American Neurological Association and the Child Neurology Society*, *53*(4), 446–453.

Laird, A. R., Fox, P. M., Price, C. J., Glahn, D. C., Uecker, A. M., Lancaster, J. L., Turkeltaub, P. E., Kochunov, P., & Fox, P. T. (2005). ALE meta-analysis: controlling the false discovery rate and performing statistical contrasts. *Human Brain Mapping*, *25*(1), 155–164

Lanteaume, L., Khalfa, S., Régis, J., Marquis, P., Chauvel, P., & Bartolomei, F. (2007). Emotion induction after direct intracerebral stimulations of human amygdala. *Cerebral Cortex*, *17*(6), 1307–1313.

Lenartowicz, A., Kalar, D. J., Congdon, E., & Poldrack, R. A. (2010). Towards an ontology of cognitive control. *Topics in Cognitive Science*, 2(4), 678–692.

Leys, R. (2017). *The ascent of affect: genealogy and critique*. University of Chicago Press.

Lindquist, K. A., & Barrett, L. F. (2012). A functional architecture of the human brain: emerging insights from the science of emotion. *Trends in Cognitive Sciences*, *16*(11), 533–540.

Lindquist, K. A., Wager, T. D., Bliss-Moreau, E., Kober, H., & Barrett, L. F. (2012b). What are emotions and how are they created in the brain?. *Behavioral and Brain Sciences*, *35*(03), 172–202.

Lindquist, K. A., Wager, T. D., Kober, H., Bliss-Moreau, E., & Barrett, L. F. (2012a). The brain basis of emotion: a meta-analytic review. *Behavioral and Brain Sciences*, *35*(03), 121–143.

Loaiza, J. R. (2020). Emotions and the problem of variability. *Review of Philosophy and Psychology*, 1–23.

Loosemore, R., & Harley, T. (2010). brains and minds: on the usefulness of localization data to cognitive psychology. In Hanson, S., & Bunzl, M. (Eds.). *Foundational issues in human brain mapping*. MIT Press, 217–240.

McCaffrey, J., & Danks, D. (2022). Mixtures and psychological inference with resting state fMRI. *The British Journal for the Philosophy of Science*, *73*(3), 583–611.

McCaffrey, J. B., & Machery, E. (2016). The reification objection to bottom-up cognitive ontology revision. *Behavioral and Brain Sciences*, *39*, e125.

Murphy, F. C., Nimmo-Smith, I. A. N., & Lawrence, A. D. (2003). Functional neuroanatomy of emotions: a meta-analysis. *Cognitive, Affective, & Behavioral Neuroscience*, *3*(3), 207–233.

Nummenmaa, L., & Saarimäki, H. (2019). Emotions as discrete patterns of systemic activity. *Neuroscience Letters*, *693*, 3–8.

Ortony, A., & Turner, T. J. (1990). What's basic about basic emotions?. *Psychological Review*, *97*(3), 315–331.

Panksepp, J. (1998). *Affective neuroscience: the foundations of human and animal emotions*. Oxford University Press.

Panksepp, J. (2007). Neurologizing the psychology of affects: how appraisal-based constructivism and basic emotion theory can coexist. *Perspectives on psychological Science*, 2(3), 281–296.

Pelliccia, V., Del Vecchio, M., Avanzini, P., Revay, M., Sartori, I., & Caruana, F. (2022). 70 Years of human cingulate cortex stimulation. Functions and dysfunctions through the lens of electrical stimulation. *Journal of Clinical Neurophysiology*, 10–1097.

Phan, K. L., Wager, T., Taylor, S. F., & Liberzon, I. (2002). Functional neuroanatomy of emotion: a meta-analysis of emotion activation studies in PET and fMRI. *Neuroimage*, *16*(2), 331–348.

Piccinini, G., & Craver, C. (2011). Integrating psychology and neuroscience: functional analyses as mechanism sketches. *Synthese*, *183*(3), 283–311.

Poldrack, R. A. (2010). Mapping mental function to brain structure: how can cognitive neuroimaging succeed?. *Perspectives on Psychological Science*, *5*(6), 753–761.

Poldrack, R. A., & Yarkoni, T. (2016). From brain maps to cognitive ontologies: informatics and the search for mental structure. *Annual Review of Psychology*, 67, 587–612.

Price, C. J., & Friston, K. J. (2005). Functional ontologies for cognition: the systematic definition of structure and function. *Cognitive Neuropsychology*, *22*(3–4), 262–275.

Ritchie, J. B., Kaplan, D. M., & Klein, C. (2019). Decoding the brain: neural representation and the limits of multivariate pattern analysis in cognitive neuroscience. *The British journal for the Philosophy of Science*, 70(2), 581–607.

Saarimäki, H., Gotsopoulos, A., Jääskeläinen, I. P., Lampinen, J., Vuilleumier, P., Hari, R., . . . & Nummenmaa, L. (2016). Discrete neural signatures of basic emotions. *Cerebral Cortex*, *26*(6), 2563–2573.

Saarimäki, H., Ejtehadian, L. F., Glerean, E., Jääskeläinen, I. P., Vuilleumier, P., Sams, M., & Nummenmaa, L. (2018). Distributed affective space represents multiple emotion categories across the human brain. *Social Cognitive and Affective Neuroscience*, *13*(5), 471–482.

Scarantino, A. (2009). Core affect and natural affective kinds. *Philosophy of Science*, *76*(5), 940–957.

Scarantino, A. (2015). Basic emotions, psychological construction and the problem of variability. In Russell, J., & Barrett, L. F. (Eds.). *The psychological construction of emotion*. New York: Guilford Press, 334–376.

Scarantino, A., & Griffiths, P. (2011). Don't give up on basic emotions. *Emotion Review*, *3*(4), 444–454.

Srinivasan, R., & Martinez, A. M. (2018). Cross-cultural and cultural-specific production and perception of facial expressions of emotion in the wild. *IEEE Transactions on Affective Computing*, *12*(3), 707–721.

Touroutoglou, A., Lindquist, K. A., Dickerson, B. C., & Barrett, L. F. (2015). Intrinsic connectivity in the human brain does not reveal networks for 'basic' emotions. *Social Cognitive and Affective Neuroscience*, *10*(9), 1257–1265.

Uttal, W. R. (2001). *The new phrenology: the limits of localizing cognitive processes in the brain*. The MIT Press.

Villard, C., Dary, Z., Léonard, J., Villalon, S. M., Carron, R., Makhalova, J., . . . & Bartolomei, F. (2023). The origin of pleasant sensations: insight from direct electrical brain stimulation. *Cortex*, *164*, 1–10.

Viola, M. (2017). Carving mind at brain's joints. The debate on cognitive ontology. *Phenomenology and Mind*, (12), 162–172.

Viola, M. (2024). (Mis) decoding affect in the face and in the brain. In Ienca, M., & Starke, G. (Eds.). *Brains and machines: towards a unified ethics of AI and neuroscience*. Elsevier, 125–145.

Vytal, K., & Hamann, S. (2010). Neuroimaging support for discrete neural correlates of basic emotions: a voxel-based meta-analysis. *Journal of Cognitive Neuroscience*, *22*(12), 2864–2885.

Wager, T. D., Kang, J., Johnson, T. D., Nichols, T. E., Satpute, A. B., & Barrett, L. F. (2015). A Bayesian model of category-specific emotional brain responses. *PLoS Computational Biology*, *11*(4), e1004066.

Wager, T. D., Lindquist, M., & Kaplan, L. (2007). Meta-analysis of functional neuroimaging data: current and future directions. *Social Cognitive and Affective Neuroscience*, *2*(1/4), 150.

Winder, A. T., Echagarruga, C., Zhang, Q., & Drew, P. J. (2017). Weak correlations between hemodynamic signals and ongoing neural activity during the resting state. *Nature Neuroscience*, *20*(12), 1761–1769.

Zhang, H., Wang, D., Wei, P., Fan, X., Yang, Y., An, Y., . . . & Zhao, G. (2023). Integrative roles of human amygdala subdivisions: insight from direct intracerebral stimulations via stereotactic EEG. *Human Brain Mapping*, *44*(9), 3610–3623.

Index

Note: *Italic* page numbers refer to figures, **bold** page numbers refer to tables and page numbers followed by 'n' refer to notes.

activation patching 201–204, *203*
Alderson-Day, Ben 249
amodal hypothesis 230, 231, 234
amygdala 71–72, 303, 304
Anderson, M. L. 235, 293, 306n4
anterior cingulate cortex (ACC) 73, 304
anterior temporal lobes (ATLs) 229, 248
Approximate Number System (ANS) 253
architectural dogma 145, 147, 151, 152
artificial neural networks (ANNs) 2, 12, 94, 108, 144, 190
attribution methods 194–195
autointerpretability 213
autonomism 1–4

bacteria 170–171
Baddeley, Alan 250–251, 276, 279
Banks, B. 251
Bargh, J. A. 168
Barrett, L. F. 300
basal ganglia 72–73
basic emotions 289–291, 296, 301
Bayesian inference 177
Bayes' theorem 44, 45
behavioral homology 68
behaviorism 19
Bergeron, Vincent 67–73
Binder, J. R. 224, 225, *225*, 230, *231*, 238n5
birdsong syntax 74, 75
black boxes 190, 192–195, 200, 215
Block, Ned 101, 116n11
Blood Oxygen Level Dependent (BOLD) 85, 95n1
Boghossian, P. 167, 168
Boone, W. 48
Borghesani, V. 229
Borghi, A. M. 238n5
Bottou, L. 235
brains 9; basic emotions 289–292; function 96n5; neural correlates 289–292
Bratman, Michael 279
Bricken, T. 212
Brigard, Felipe de 281, 284n1, 285n4
Broca's area 71
Brooks, Rodney 122, 133, 134
Bruni, E. 235
Burge, Tyler 101, 109
Buzsáki, G. 59

Cao, R. 84, 95n4
Carello, Claudia 31n3
Carey, Susan 101
Carota, F. 234
Carruthers, Peter 260, 262, 270n7
causal abstraction 213–215
Chaitin, Gregory J. 138n7
Chalmers, David 107, 156, 157
Chang, L. 91, 92
character identity mechanisms (ChIMs) 61, 64, 65, *65*, 66; episodic memory 76; paired gnathostome appendage 67

Chirimuuta, M. 158
Chollet, François 139n11
Chomsky, N. 237n1, 254; I-language 222
circuit 200–201; formation phase 207–208
Clark, Andy 137, 176, 177, 185n2
Clark-Polner, E. 292
classical computability theory 12
classical computational theories 83
classical computational theory of mind (CCTM) 149–150
coders 126–128, 136
cognition 19, 122, 136; biological systems 170; character identity mechanisms (ChIMs) 74–76, 78; homologies 59–78; maps 127–130, 139n11; neuroscience 7; psychology 288; science 19; semantics 222–223; signal-processing aim of 122–124; situated approaches to 23
cognitive ontology 10, 59–78; beyond neuroimaging 303–304; bottom-up approaches 60; conservatives 293–294, 296–297; incremental approaches 59–60; radicals 60, 293–295, 297–300; reformists 293–294, 296, 300–303; in terms of cognitive ChIMs 74–77; three stances on 293–294; top-down approaches 60
cognitive representations 98–115; compositionality debates 106–108; efficiency-based explanations 99; Fodor's Picture Principle to 102; formats, computational account of 112–114; formats of 101, 114; implementation of 105; public *versus* 104–106; transformation-based explanations 99; visual demonstratives 108–112
Coltheart, Max 45, 223, 232
computation: digital TM-style of 149–150; equivalence 125; mechanistic accounts 157–159; neural 148, 150–155, 159; old-style accounts of 145–148, 152; old-style theories of 150–155, 157; philosophical theories of 148; physical 145–147; pragmatic accounts 158; profiles 112; semantic accounts 158, 159; theories 83, 145, 148
computational cognitive neuroscientists (CCNs) 33n10
computational-representational vehicle 112, 113, 115
computational theory of mind (CTM) 120, 144, 149; classical computational theory of mind (CCTM) 149–150
conceptualization controller 265
ConDialInt model 265–269, *266*
connectionism 2, 107, 144, 149
Connell, L. 251
conservatives 293–294, 296–297, 305
consistency fallacy 45–46, 55
Constant, A. 176
Continuity Principle 110–112, 116n12
convolutional neural networks 235
Core Affect 300
corollary discharge (CD) 258–260, 262, 269n5, 270n7
covering-law model 44
covert speech 255–259
Cox, David D. 131
Crapse, T. B. 259
Craver, Carl 40, 47, 56n1
Cummins, Robert 38, 147, 148, 150, 155

Dacey, M. 172, 183
data: contamination 191–192; structure 128
Davidson, Donald 281
deep learning 12, 194–195
deep neural networks (DNNs) 190–215
delay period activity 277, 278
Dell, G. S. 255
Description Logics 233, 238n6
developmental homology principle (DHP) 63–64
Dewey, John 19, 31n2
Dewhurst, J. 158
DiCarlo, James J. 131
dictionary learning 212–213
Diffusion Tensor Imaging 268, 307n6
DiFrisco, J. 64, 66, 74, 79n3

Dijkstra, E. W. 128; algorithm 129, 130, 135, 138n9, 139n10
Directions into Velocities of Articulators model (DIVA) 255, 258
distributional AI models 235, 238n6
distributional semantic models (DSMs) 233, 234, 252
dogmas 145, 148–150; architectural 145, 147, 151, 152; computation without 157–159; logical 145, 146, 151; neural computation and 150
Dove, G. 238n5
dual code theory 231, 232
Dymarska, A. 251

ecological psychology 9, 19–20, 23, 30–31, 32n7; action capabilities 21; affordances 21; direct perception 20–21, 31n3; information 20; organism-environment system 21–22; perceptual capabilities 21
ecological rationality 172, 182–183
Edelman, Gerald 24
efference copy 258, 269n5
Egan, F. 84
Ekman, Paul 289–290, 296, 297, 299, 302, 305, 306n2; Theory of Basic Emotions 289, 292, 296, 302, 303
electrocorticography (eCOG) 92
electroencephalography (EEG) 227–228
Elhage, N. 197
Ellsworth, Phoebe 296
emotions 14, 289, 300; basic 289–291, 296, 301; theory 288–306
encoding technique 227, 229
episodic memory 76, 77
epistemic charge 182
event-related potential response (ERP) 259
explicit Theory of Mind 254, 269n4

Facchin, Marco 135
face fusiform area (FFA) 92
face space 87, 88, 95n3
facial expressions 170, 290, 302
Fernandino, L. 229
Fernyhough, Charles 249
fMRI 85, 225–227, 229, 230, 232–233, 303, 306n1
Fodor, J. A. 1–2, 52, 102, 106, 107, 114, 115n5, 115n6, 116n9, 230, 232, 237n2; Picture Principle 115n3; Representational Theory of Mind 83
formalism 146, 156–157
forward model accounts (FMAs) 259, 262–269; ConDialInt model 265–269, *266*; PGG's model 263–265, *264*
Frankish, Keith 260, 270n7
Friston, K. J. 71, 72
functional explanation 38, 55; abstractness response 46–49; analysis 9; Bayesian 44; consistency fallacy 45–46; dilemma 42–46, 55; empirical content, requirement of 43–45; implications for the autonomy of 46; Krakauer's defense of the autonomy of 40–42; mechanistic details 39, 50–51; and mechanistic explanations 42; Minimal Cognitive Grid (MCG) 39, 51–53
functional homology 68
functionalism 19, 31n2
functional/structural ratio 52, 53, *54*, 55
Fusar-Poli, P. 290

Galanter, Eugene 275, 279
Gallego, J. A. 25–26
Gao, L. 212
García, Claudia 67–69, 73, 79n5
Gärdenfors, P. 237n2
Geiger, A. 213
generality 52, 53
Gibson, James J. 19, 21, 22, 31–32n4; ecological psychology 22, 23, 32n7
Ginsburg, S. 172, 178
Gładziejewski, Paweł 133
Glenberg, Arthur M. 76
Godfrey-Smith, Peter 127
GPT-4 191, 211, 216–217n1
grammatical number 197, 198
Grandchamp, R. 263, 265, 267
Green, E.J. 109, 116n11, 116n13
grounded approaches to conceptual processing (GACPs) 246, 247–249, 251, 269

Hamann, S. 292, 298, 306n3
Hardcastle, Valerie 122
Hassin, R. R. 168
Hauk, O. 229
Hayes, Justin 76
Hazineh, D. S. 209
Helmholtz's theory 177
hemodynamic activity 300
heterogeneity 169, 171
hierarchical representations 134, 135–136
hippocampal place cells 127
Hitch, Graham 276, 279
Hobbes 160n2
Hodgkin, Alan 40, 41–42, 44, 47
Hodgkin-Huxley (H&H) model 41, 44–45, 47, 48, 53, 56n1; functional/structural ratio 53, *54*; generality 53, *54*; performance match 53, *54*
Hoffman, P. 235
Hohwy, J. 175, 176
homology 10; *versus* analogous traits 62, *62*; behavioral 68; biological concept of 61; in biology 61–66; functional 68; homologous brain regions 70; ontogenetic/developmental conception of 63, 66; phylogenetic conception of 63, 66, 79n3; serial 63
Hopfield, J.J. 151, 156
Horikawa, T. 302
hub-and-spoke models 248
Hubel, D.H. 153
human: computers 146; -interpretable properties 200; syntax 74, 75
Huxley, Andrew 40, 41–42, 44, 47
hypothalamus 73
hypothesis 25; amodal 230, 231, 234; interdependence of symbols 231; manifold 25; neural manifold 25, 26, *26*, 28, 29; non-sensorimotor 231; null 191; sensorimotor 224, 228–229, 234

indiscriminate ablations 196
infantile amnesia 76
inference 166–177; a-rational association and rational 168–173, 183; Bayesian 177; conception of 184, 185; conscious reckoning 167–168; ecological rationality of 172; flexibility 170; language-like representations 173–174; mistaken 171; in (neuro) cognitive systems 164–185; representations 176; species of 167; terminological extension 177–179; unconscious 177
inferotemporal (IT) cortex 86–89
information: mutual 124–125; processing 19; senomic 186n3; state 181; theory 11, 138n7
inner speech 246–269, 269n7; cognitive functions of 250–255; ConDialInt model 265–269, *266*; forward model accounts (FMAs) 259, 262–269; LRM's model and 256–259; mnemonic effects 251; multiple contents of 259–269; nature of 249–250; PGG's model 263–265, *264*; phenomenological definition (PD) 249–250; predictive motor control and 256–259; processing definition 250; sensory 260; speech comprehension accounts (SCAs) 259–262
intention 275–284; revamping 281–284; and working memory 279–281
interchange intervention 214, 215, 217n7
interventionist methods 190–215; ablation to causal probing 195–199; induction heads 204–205, *206*; modular addition 205–208, *207*; for understanding LLMs 195–215; world models 208–212
intrinsic networks 295, 298
IT cortex 93
iterative nullspace projection 197–200, *199*

Jablonka, E. 172, 178
Jack, B. N. 259
Jack, R. E. 302
Jackendoff, R. 237n2
James, William 19, 31n2, 59
job description challenge 135, 136
Jones, M. 235
Jonides, John 278, 283
Jorba, M. 262, 265

Journal of Neurophysiology 41
Joyce, James 304

Kemmerer, David 257
Kennedy, John 49–50
Khajeh-Alijani, Azadeh 139n10
Kiela, D. 235
Koch, Christof 152, 160n1
Koide-Majima, N. 302
Kolmogorov, Andrei N. 138n7
Kragel, P. A. 292
Krakauer, John 39–43, 46, 54
Kuhnian paradigms 101
Kuhnke, P. 232

LaBar, K. S. 292
language: models 200; of thought 246–269
Language of Thought (LOT) hypothesis 13, 14, 106–108, 115n5, 115n6, 116n7, 116n8, 116n10, 232, 238n4, 246–249, 251, 253, 269
large language models (LLMs) 114, 116n8, 190–191; interpretation methods 191–195; interventionist methods for 195–215; mental syllabary 258; model 256–259
Lazaridou, A. 235
left inferior frontal gyrus (LIFG) 258
Leibniz 160n2
Levelt, Willem 255, 256, 260, 261, 265, 267, 269n7, 270n7
Levy, A. 47, 48
Li, B. Z. 211, 212
Li, K. 208, 209
Li, Ming 138n7
Lieto, A. 52
limited associative learning (LAL) 172–173
Lindquist, K. A. 292, 298
linear operation 216n3
linguistic experience 238n5
Loaiza, J. R. 293
logical dogma 145, 146, 151
Louwerse, M. M. 231
Love, A. C. 74

Machery, E. 299
machine learning algorithm 229
magnetoencephalography (MEG) 227–228
Maley, C. J. 160n10
Mandelbaum, E. 168–171, 173, 174
manifold theory 25
Marconi, D. 237n2
Marr, D. 41, 51, 53
Marshall, R. C. 261
McCaffrey, J. B. 299
McCulloch, W.S. 150
mechanistic details 39–40, 42, 44, 48, 50–51
mechanistic explanations 39, 42, 56, 192–194, 201
mechanistic interpretability 200, 201, 211, 213–215
memorization phase 207
mental functions 6
mental logic 170, 171
mental model 120
mental (neurocognitive) processes 14
mental processes 168
mental representations 10, 11, 14–15, 84, 94
mental syllabary 258
Mesulam, M. M. 227
Meyer, A. S. 255
Michaels, Claire 31n3
Miller, George A. 275, 279, 280, 284
mind 1; classical autonomist picture of 2, 8; connectionist autonomist picture of 2–3; integrationist picture of 5–9; multilevel mechanisms 5–6; normativist autonomist picture of 3–4; philosophy of 1–15; reductionist picture of 4–5; sciences 1–15; situationist autonomist picture of 3; *see also individual entries*;
mind-brain 144–159
Minimal Cognitive Grid (MCG) 39, 51–53, 56; analyzing Krakauer's examples using 53–54; functional/structural ratio 52; generality 52; performance match 52; structural accuracy 52
Mitchell, T. M. 227
modal gating behavior 48
Mollo, D. C. 173
Money, N. P. 165
Montague, R. 237n1
Muller, R. U. 139n10
multilayer perceptrons (MLPs) 144
multimodal distributional semantics 235

multiple object tracking (MOT) 102, 108
multiscale modeling 6
multivariate pattern analysis (MVPA) 226, 284n1
mutual information 124–125

Nanay, Bence 250
Nanda, N. 205–206, 209, 210
naturalistic theories 105
natural language 246–269; basic property 254; syntax 254
network information theory 138n8
neural basis 275–284
neural computation 144–145, 148, 149, 157, 159
neural correlates 289–292, **290**, 300
neural Darwinism 24
neural manifold hypothesis 25, 26, *26*, 28, 29
neural networks 192, 194; black box of 190, 193, 200; convolutional 235; distributed representation models 197; intervention on 195–196; neurons in 196
neural population 24–25, 91; activity 30; dynamics 24, 26
neural representations 10, 11, 15, 89–94, 176
neurocognitive models 255–259, 268
(neuro)cognitive systems 164–185; inferential processes 164; inferential transition in 182; minimalist account of inference in 179–184; philosophy, inference in 166–177
neuroecological cognitive architecture 18–31
NeuroEcological Nexus Theory (NExT) 23–24, 31, 32n7; manifold hypothesis 25–27; neural population dynamics 24–25; organism-environment system 24; organism (neural, body)-environment activity 28–31; pass-through-able 29–31, *29*; synergies 27–28
neuroimaging methodologies 227
neuroimaging techniques 226–228, 232, 236
neurons 152–153, 196
neuroscience 4, 8, 94, 196; affective 307n7; arena of 288–306; confirmation and explanation in 38–56; turn 222–237
neurosemantics 223, 227; cognitive semantics to 222–237; distributional models in 235; growing tendency in 236
neuroskeptics 289, 293, 306
non-sensorimotor hypothesis 231
nucleus laminaris 70, 71
null hypothesis 191

object: recognition 131; tracking 110
Oppenheim, G. M. 255
organism-environment systems 21–22, 24, 30, 32n9
Othello-GPT 208–209, *210*

Paivio, A. 231–232
Panksepp, Jaak 304, 307n7
Parvizi, J. 92
Perceptual Object Representation (POR) 108, 111, 116n11
PET 225, 228, 230, 303
Petersen, S. E. 224
PGG's model 263–265, *264*, 267
phenomenological definition (PD) 249–250
philosophical psychology 281–284
philosophical theory 145
physical computation 145–147
Piccinini, Gualtiero 40, 48, 137, 156–158, 160n11, 176, 235, 238n4, 285n4
Picture Principle 102, 103, 115n5
Pitts, W.H. 150
Plans and the Structure of Behavior (Miller, Galanter, and Pribram) 275
Plato 31n3
Poeppel, D. 259
Poggio, Tomaso 131
Poldrack, R. A. 288
polysemanticity 196
poverty of stimulus (PoS) 20–21
prefrontal cortex (PFC) 72–73
Pribram, Karl H. 275, 279
Price, C. J. 71, 72
principal component analysis (PCA) 86
processing definition 250

production implementer 263, 264
psychological theory 284
public representations 112; analogies to 106; formats of 112, 114; as frames 101–112
Pulvermüller, F. 231
pure word deafness (PWD) 261, 262
Putnam, H. 148
Pylyshyn, Z. W. 106, 107, 114, 115n6, 116n9, 155

Quilty-Dunn, J. 102, 109, 115n3, 116n11, 116n13, 168–171, 173, 174

radicals 293–295, 297–300, 305
Ramsey, William M. 135
rationality, ecological model of 171
Ravfogel, S. 197, 198
Reber, A. S. 165, 178
receptors 135
redeployment 91
reductionism 4–5, 7
reformists 300–303, 305, 306n4
Reid, Allison 127, 129, 139n10
Reid-Staddon model 134
representations 83–85, 94, 98, 174; classical computational theories 83; cognitive 98–115; computational theories 83; counterfactual 198–199; formats 100, 115n1; hierarchical 134, 135; hungry problems 137; and inference 176; mental 84, 94; neural 89–94, 176; node 127; philosophical theories of 195; public 112; sender-receiver model of *126*; structural 176; vehicles of 83–95
Rescorla, M. 149
Restricted Access Sequence Processing Language (RASP) 216n6
revamping intention 281–284
reverse inference 226
Ritchie, B. J. 183
Roelofs, A. 255
Rolls, E.T. 152
Rooij, Iris van 138n5
Roskies, A. L. 85
rotation composition *207*
RSA technique 227, 229
Rubner, Andrew 110, 114

Saarimäki, H. 292, 301
scale separation 6–7
Schultz, Wolfram 41, 50
Schwarz, Georg 155
Schweizer, P. 158, 159
scientific discovery 100–103; frames, public representations as 101–103; inquiry and frames of 100–101
scientific psychology 19, 31n1
Searle, J. R. 148
self-attention 216n5–6
semantics: computational models of 232–236; dementia 230; distributional 233; foundational debates in 237n3; hubs 248, 257; hybrid theories of 231–232; knowledge 233; memory 234; of mental representations 237n2; model 229; multimodal distributional 235; of natural language 237n2; values 237n2
senomic information 186n3
sensorimotor: attributes 229; hypothesis 224, 228–229, 234
sensory 260; experiences 235; motor integration 72; recruitment 278, 283
serial homology 63
Shagrir, O. 158
Shallice, T. 228
Shannon, Claude E. 124, 125; information theory 138n7
Shapiro, L. 174
Shea, Nicholas 135, 139n11
Siegel, Susanna 166, 168, 179–182; Response Hypothesis 179–181, 184
signals 126–128
Sims, M. 177
simulation theory 228
situated approaches 18–19, 22, 23, 30
Skinner, B. F. 19
Smolensky, P. 114
Solomonoff, Ray J. 138n7
Sommer, M. A. 259
space complexity classes 126
sparse autoencoders (SAEs) 212–213
spatiotemporal properties 110
spatiotemporal scale 28, 29
spectrum 18

speech: comprehension representations 269n7; inner 246–269, 269n7; sound map 258
speech comprehension accounts (SCAs) 259–262
Sphex 170, 185n1
spike-triggered average (STA) 86–88, 90, 95n3
splitting strategy 155
Staddon, John 127, 129, 139n10
structural representations (SR) 11, 121, 176; coders 126–127; cognition, signal-processing aim of 122–124, *123*; cognitive maps 127–130; as complexity management 120–138; hierarchical processing 131–133, *132*; liberality challenge 134–136; signals 126–127; transformations, complexity of 124–126; usefulness 133–134
structural sameness 69, 73
synergies 27, 28
syntactic-lemma stage 257–258
syntax 174; birdsong 74, 75; human 74, 75; natural language 254
systematicity 248

targeted ablations 196
Templeton, A. 212
temporal difference (TD) learning 41; functional/structural ratio 53, *54*; generality 53, *54*; performance match 53, *54*
temporal distribution 225, *225*
thalamus 73
Tian, X. 259
time complexity 126
Titchener, Edward 31n1
Tomkins, Silvan 296–297
topological spaces 25
Toribio, Josefa 137
Touroutoglou, A. 298, 299
transformer models 201, *202*, 204, 205, *206*, 208, 233–234
Transformer Programs 216n6
Tremoulet, P. 110
Tsao, Doris 86–93, 95n2, 95n4
tuning curve 153, 154
Turing, A. M. 160n5
turing machines (TMs) 144, 146–147, 155, 160n5

Ultrabithorax 64, 65
unconscious inference 177
uncontrolled manifold (UCM) 27, 28
unified inferential hierarchy 179
univariate analysis technique 226
unlimited associative learning (UAL) 172

Vanderwolf, C. H. 59
Varela, F. J. 122
variable transformers, *see* coders
verbal transformations (VTs) 267–268
Vernazzani, Alfredo 110, 114, 173
Vicente, A. 262, 265
visual working memory 283
Vitányi, Paul 138n7
Von Neumann, J. 160n2
Vytal, K. 298, 306n3

Wagner, G. P. 74
Warrington, E. K. 228
Watson, John 19
wave equation 103
Wiesel, T.N. 153
Williams, D. 179
Wilson, Bret 115n4
Wittgenstein, L. 222
working memory 260, 261, 275–284; activity 276; brief primer on 276–278; cognitive neuroscience of 284; delay period activity 277, 278; dynamics of 280, 282; function of 284n2; intention and 279–281; models of 276; sensory recruitment models of 276, 278; tasks 277; visual 283
Wundt, Wilhelm 31n1

Zeigler, B. P. 52

Kolibri 360 GmbH
Ferdinand-Jühlke-Straße 7
99095 Erfurt, Deutschland
produktsicherheit@kolibri360.de